Electronic Inventions and Discoveries

Electronic Inventions and Discoveries

*Electronics from its earliest beginnings
to the present day*

4th revised and expanded edition

G W A Dummer

MBE, CEng, FIEE, FIEEE, US Medal of Freedom
(former Supt. Applied Physics, Royal Radar Establishment, UK)

Institute of Physics Publishing
Bristol and Philadelphia

First edition 1977 published under the title *Electronic Inventions 1745–1976*
Second edition 1978 (*Electronic Inventions and Discoveries*)
Third revised edition 1983

British Library Cataloguing-in-Publication Data

A catalogue record for this book is available from the British Library

ISBN 0 7503 0376 X (hbk)
ISBN 0 7503 0493 6 (pbk)

Library of Congress Cataloging-in-Publication Data are available

Published by Institute of Physics Publishing, wholly owned by The Institute of Physics, London
Institute of Physics Publishing, Dirac House, Temple Back, Bristol BS1 6BE, UK
US Editorial Office: Institute of Physics Publishing, The Public Ledger Building, Suite 1035, 150 South Independence Mall West, Philadelphia, PA 19106, USA

Printed in the UK by J W Arrowsmith Ltd, Bristol.

Contents

Histories on a Page

Preface

As in previous editions, it is not intended that this book should be a learned treatise on a particular aspect of historical electronics, but rather a wide summary of first dates in electronic developments over a very wide field, both for interest and for ready reference.

Because no one person can be an authority in all fields of electronics, the data given are extracted from a wide variety of published sources, i.e. books, patents, technical journals, proceedings of societies, etc, to whom full acknowledgment is made. This present work covers inventions from Europe, USA and Japan. Obviously, a survey such as this cannot be completely accurate because of, in many cases, the passage of time and, in others, conflicting claims, but gives the opinions of those knowledgeable in their fields. There are a few cases where, because of incomplete data, a date is given in the 'History on a Page' but not in the text. The author is well aware that conflicting dates are inevitable and welcomes factual data to assist future editions. In addition to the summaries of well known inventions, some little known discoveries are included which may, one day, be important.

In this edition, an attempt has been made to trace the development of electronics from its earliest beginnings up to the present day. As far as the author knows, it is the only book in the world to describe concisely the majority of main developments in electronics. The book describes, in nine chapters, developments in electronic components, tubes, transistors, integrated circuits, audio and sound reproduction, radio, communication, avionics, radar, sonar, television, computers, robotics, mechatronics and information technology, in addition to industrial, automobile, medical, educational, office, banking, consumer and security electronics.

How does one define an electronic invention? One can consider the first idea or concept, the construction and operation of the first laboratory lash-up, the first prototype, the 'first in service' or the patent date. In this book the author has decided to use the 'first recorded use' as far as it is possible and the selection has been made on the basis of simple language and explanation.

Throughout the book, the author has used the American term 'tube' in place of the English term 'valve'.

The process of invention has changed from the individual inventor to that of the large research laboratories which have the advantage of funds and cross-fertilization of ideas. Certainly the Bell Laboratories in the USA made the greatest contributions to semiconductor technology, not only by inventing a working transistor, but by producing materials (Si, Ge) of a purity previously unknown. This work, basic to microelectronics, has created entirely new industries. The complexity of modern electronics has brought together chemists, physicists, mathematicians, engineers, and others as the fields of development widen. Research, development, and production are now more closely integrated.

Looking back at the history of electronics, there seem to be two periods of creativity shown by the chart on page 5. They are between 1800–1900 (100 years) and 1950–1980 (30 years). In the author's opinion there are three fundamental inventions on which others depend. They are: first, Faraday's discovery of electromagnetic induction from which the dynamo was developed to generate electricity (imagine a world without electricity today!); second, Lee de Forest's thermionic tube, opening up the fields of communications and computers; and third, the Bell Laboratories transistor, because the modern 'chip', in fact, consists of multiple transistors.

In the production of electronics, two inventions stand out as enabling devices to be mass produced at reasonable cost: the printed circuit with dip soldering and the planar photo-masking techniques for microelectronics 'chip' production.

In preparing this book, one major impression has emerged, instanced by chapters 1 to 9—how deep the penetration of electronics has become into every part of modern life—whilst the 550 inventions described in this book, together with over 1100 additional references, form a background to electronics progress which, with ever increasing tempo, is now changing the world in which we live.

G W A Dummer
Malvern Wells
UK

Acknowledgments

In this book the author has attempted to summarize the development of electronics in chapters 1 to 10 whilst, in chapter 11, his task has been that of a compiler rather than an author.

Chapter 11 has only been made possible because of the cooperation of so many authors. Many books and technical journals have provided extracts which are relevent and the author is indebted to all those detailed in the 'source' following each abstract. Where 'source' is quoted, the words and opinions are exactly those of the authors of the extracts. The page number given in each case is that of the extract and not that of the title page.

Full acknowledgment is made to all the authors quoted. Thanks are due to the many authors and publishers for their permission to quote from their publications and also to the Patent Office and many libraries for their help. Full credit is given to the Institution of Electrical Engineers, London, and the Institute of Electrical and Electronics Engineers, New York, for permission to use material from their published journals. Extracts from *Science at War* are used with the permission of the Controller of Her Majesty's Stationery Office. Acknowledgment is made to *The Book of Inventions and Discoveries* (Associate Editor, Valerie-Anne Giscard d'Estaing, published by Compagne DOUZE, Paris) for permission to reproduce extracts. Acknowledgment is also made to IPC Magazines, London, for permission to publish extracts from *New Scientist*. Also to Van Nostrand/Rheinhold, New York, for the use of data from one of their published books.

The author would like to record his appreciation of the help given on this and previous editions by the Science Museum, London, in particular Dr B P Bowers, W K E Geddes and Dr Denys Vaughn, and also the following for their advice and assistance on the development of electronics in the various fields: S W Amos, W Bardsley, G Bayley, W Bowes, P J Baxendall, C den Brinker, E Chowietz, T A Everist, C Hilsum, N Jones, H G Manfield, A L McCracken, T P McLean, J L Powell, E H Putley, D Sargent, K Thrower, D H Tomlin, N Walter, P L Waters and Professor Dr Jun-ichi Nishizawa, Tohoku University, Japan. The author would like to record the special help given to him by Eryl Davies, acting as a consultant on the contents of the book; to John R Guest and John K Oakley for help on the chapters on radio and computers; to Professor Russell Burns for help on the chapter on radar; to Mark Williams for data on satellites; to Dr P R Morris for help on semiconductor data; to Charles P Sandbank for help on the chapter on television; and to Robert Winton for proofreading the final draft chapters.

It is hoped that the data patiently collected for this book will be found useful, both as a review of electronics development from its earliest beginning to the present day and as a source of reference on electronic inventions.

Geoffrey W A Dummer

Chapter 1

The Beginning of Electronics

For hundreds of years, two phenomena have been known to exist: static electricity and magnetism. These remained unexplained until the early 1700s when many practical experiments commenced on both electrostatics and magnetism. By the early 1800s, work by Galvani, Oersted and Faraday on galvanism, electromagnetism and electromagnetic induction opened up a new field of experimental work which ultimately paved the way to present-day electronics.

1.1 Electrostatics

Static electricity had been known for many centuries as some substances, when rubbed together, produced static charges which could generate sparks and, in other cases, could attract small pieces of paper and other materials. The Greeks knew that friction on amber material by fur gave rise to these attractive forces and the Greek word for amber was 'electron', although the word 'electron' was not really used until after 1897 when J J Thomson discovered the electron as we know it today.

Early in the 18th century, static electricity was being studied by many experimenters. In 1729, Stephen Gray distinguished between conductors and insulators. In 1730, Charles Fry discovered that electricity induced by rubbing could be of two kinds—positive and negative. The earliest method of measurement of static electricity was the gold-leaf electroscope, invented in 1787 by Bennet. This consisted of two strips of gold leaf which moved apart when a charge was applied to it. When a rod of ebonite was rubbed with a piece of fur, the ebonite would have a negative charge and the fur a positive charge. Glass rubbed with silk exhibited a similar phenomenon. Many ingenious methods of generating static electricity were developed. Faraday, in his early experiments, showed the distribution of static charges in hollow conductors. Many attempts were made to collect the charges continuously. The Kelvin replenisher was developed as a rotary device to build up the charges, but the most important device of this time was the Wimshurst machine, built later in 1882 (see figure 1.1). The problem of storing the energy was solved by the first capacitor—the Leyden jar—invented in 1745. Having produced static charges and calculated the potential voltages available (these could be quite high—Wimshurst machines were used for working x-ray tubes), measurement was now becoming important and electrometers of various types based on the earlier gold-leaf electroscope were developed, resulting in the first electrostatic voltmeters. Electrostatics could now be generated and stored for short periods, but could not be further used and attention was focused on the other phenomena—magnetism.

1.2 Magnetism

Magnetism has also been known for centuries. It was exhibited in lodestone, found in the vicinity of Magnetia in Asia Minor and termed 'magnetite'. It has the property of attracting fragments of iron and when a bar of the material was suspended by its centre from a thread of silk, it aligned itself north and south. It was found that when stroked along a piece of steel, the steel also became magnetised and a knitting needle magnetised. In this way it became a magnet and aligned itself north and south when suspended, becoming the basis of the first compass.

Figure 1.1. A Wimshurst machine (The Science Museum/ Science & Society Picture Library).

About 1780, Galvani of Italy began experiments on animal electricity and when performing experiments on nervous excitability in frogs, he saw that violent muscle contractions could be observed if the lumbar nerves of the frogs were touched with metal instruments carrying electrical charges.

The problem of storage was still unsolved. The Wimshurst machine could generate but not store electricity and the Leyden Jar was limited in its storage capacity, but in 1800 Volta invented the electric battery. A 'Volta's pile' consisted of copper and zinc discs separated by a moistened cloth electrolyte. The pile was later improved to consist of paper discs, tin one side, manganese dioxide on the other, stacked to produce 0.75 volt. This was soon followed by the first accumulator or rechargeable battery in 1803 by Ritter in Germany. The time was now ripe for the integration of electricity and magnetism and, in 1820, Oersted in Denmark, reported the discovery of electromagnetism and led him to develop the Galvanometer, allowing accurate measurements of currents and voltages to be made, and from this our present range of ammeters and voltmeters was developed.

In 1831, the most important discovery was made by Faraday of electromagnetic induction. He wound an iron ring with two coils, one connected to a battery, the other to a galvanometer. On connecting and reconnecting the battery, a reading was obtained on the galvanometer, although there was no direct connection. The first application of this discovery was the static transformer and dynamo. By causing a coil of wire (an armature) to rotate in a magnetic field so as to cut the lines of magnetic force, an 'induced' current was produced in the coil. The current changed in direction as the coil turned. through two right angles and an alternating current was produced. Direct current could be produced by using a commutator to reverse one half of the alternating current. The generation of electric power now became possible. An electric motor is similar in construction, but the current is passed through the armature, the force generated causing it to rotate. The early 1800s was a time of great progress in invention. Infra-red and ultra-violet radiation were discovered and, in 1808, Dalton put forward his atomic theory that all chemical elements were composed of minute particles of matter called atoms. Thermoelectricity, electrolysis and the photovoltaic effect were all discovered before 1840. Work on low-pressure discharge tubes, glow discharges, new types of battery and the early microphone took place in the next 20 years. It 1873 James Clerk Maxwell was the first to consider magnetic and electric fields together and formulated his equations from which he predicted electromagnetic radiation on purely theoretical grounds. He predicted wave propagation with a finite velocity, which he showed to be the velocity of light. Heinrich Hertz succeeded in producing electromagnetic waves experimentally in 1877 and confirming Maxwell's predictions. He also added a loop of wire and increased the distance over which sparks could be transmitted, becoming the first radio communication.

It would be true to say that the majority of basic physical phenomena were discovered in the 75 years between 1800 and 1875, culminating in the practical applications of the telephone, phonograph, microphones and loudspeakers. Towards the end of the century, wireless telegraphy, magnetic recording and the cathode-ray oscillograph were all developed.

In 1911 Rutherford proposed the general model of the atom consisting of a nucleus of protons and neutrons, about which electrons rotated in orbits. In 1913, Bohr proposed that various stable orbits corresponded to various permissible energy levels

The early 1900s also saw the beginnings of many present-day electronic technologies. The three-electrode valve opened the way to radio broadcasting and Campbell-Swinton put forward his theory of television.

The advent of the 1914–18 war changed the pace of development and 'electronics' now covered a wider field of applications. New radio tubes and new circuits were developed for communications and after the war, radio astronomy, xerography, early radar, and computer techniques, were all ready to be further developed during the 1939–45 war. Under the pressure of this second war, radar and computer work led to a great increase in electronics research, and both governments and private industry set up large laboratories. From these came MASERS, LASERS, solar batteries and, in particular, in the 1950s, methods of perfecting ultra-pure materials, such as germanium and silicon. The stage was now set for the next major advance in electronics—the transistor, invented by the Bell Laboratories in 1948, enabling all electronics equipment to be miniaturised. The planar process invented in 1959 enabled many transistors to be manufactured simultaneously and the integrated circuit (known as the 'chip') was born.

Chapter 2

The Expansion of Electronics

The industrial revolution of the Victorian age created large industries—steel, ship building, textiles, railways, heavy machinery, etc. Today, these industries are in decline and the changeover to light industries based on later technologies, e.g. electronics, has presented re-employment and re-training problems. The exploitation of electronic inventions now creates wealth for those nations which take up the challenge, such as the USA, Europe, Japan and, more recently, other Far Eastern countries. New electronic industries are being built up employing large numbers of people, as instanced by the world's semiconductor industry, now accounting for 4% of the world economy and already larger than any other manufacturing industry.

Whilst many of the basic electrical and electronic inventions were made in the last two centuries (see figure 2.1), the early 1960s saw the greatest expansion of electronics technology, and the integrated circuit, together with the computer, laid the foundation for the present-day world-wide expansion of electronic applications, of which typical examples are: world-wide communications, satellites, television, interactive and virtual TV, tape recorders, microwave ovens, calculators, microprocessors, data-processing systems, automotive electronics, educational electronics, industrial electronics, electronic cameras, medical electronics, robotics and many others.

Two examples of the present electronics age are the check-out counter till and aeroplane bookings. By using electronics techniques and merely passing a bar-coded object over a detection device, a detailed bill can be produced by an electronic computer in a matter of seconds. The number of seats in the 170 airlines throughout the world, flying to different destinations, must run into millions, yet any particular seat in any particular aeroplane from any destination to any other destination, can be booked from any airline booking office, due to electronics communications and computers.

Microprocessors and minicomputers have applications in accounting, banking, chemical process control, machine tool control, filing and record keeping (in hospitals, police and business firms), data analysis, instrumentation, automatic testing, automatic justifying or printing, and countless others.

In addition, the field of telecommunications and radar, both commercial and military, used in aircraft, tanks, ships, missiles, satellites, etc, is very large. There are also the component manufacturers who supply these applications, making integrated circuits, hybrids and many specialised devices.

The list of applications is ever-widening as new applications—such as the multimedia and Internet—are being exploited. Certainly, the field of electronics will continue to expand into both home and business life, with interactive and television presentations, combined with data-processing systems, becoming more and more widely adopted.

A brief summary of the growth of electronics over the years is given in figure 2.1, while figure 2.2 shows the possible effect of the 'brain drain' after the 1939/45 war.

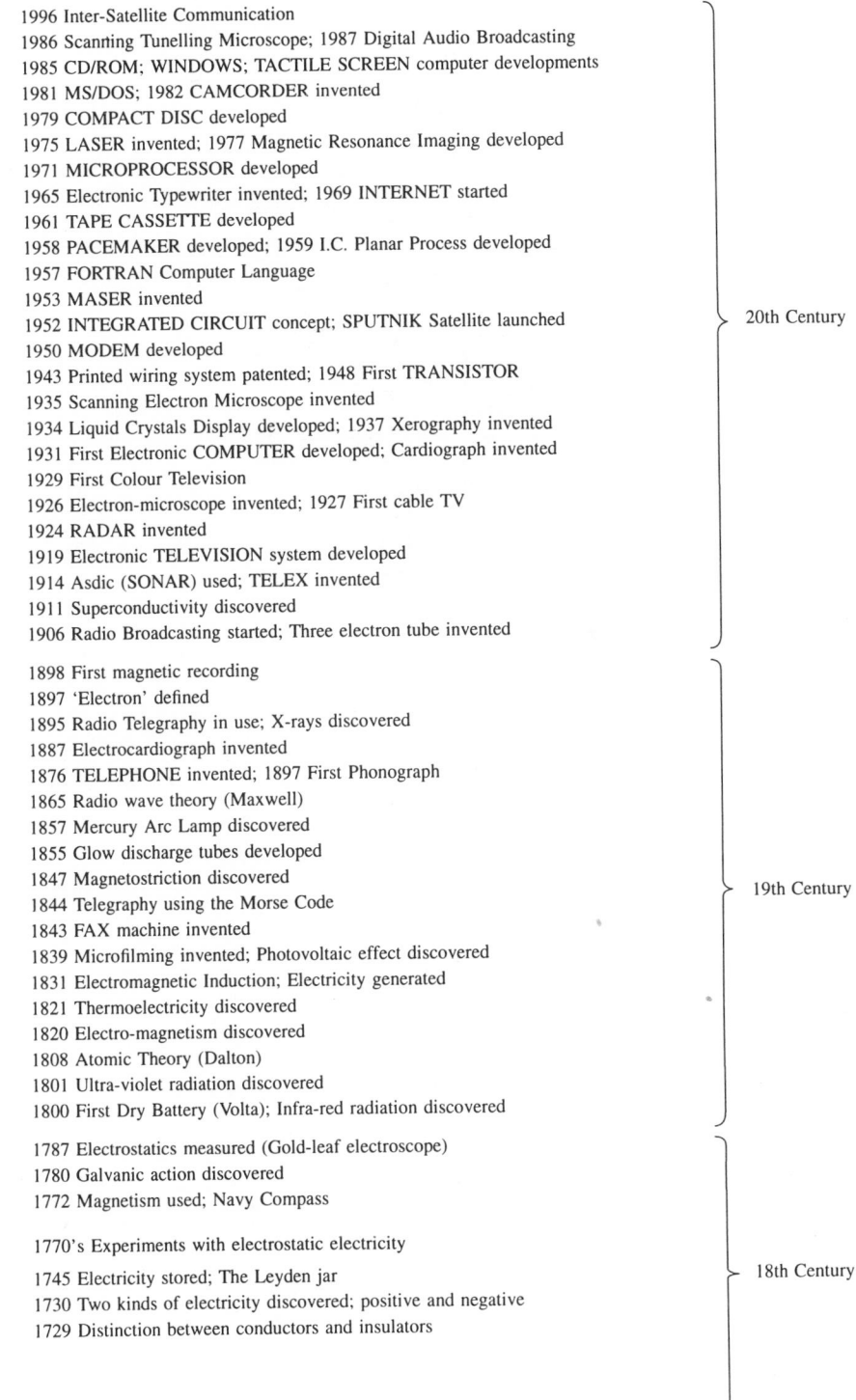

1996 Inter-Satellite Communication
1986 Scanning Tunelling Microscope; 1987 Digital Audio Broadcasting
1985 CD/ROM; WINDOWS; TACTILE SCREEN computer developments
1981 MS/DOS; 1982 CAMCORDER invented
1979 COMPACT DISC developed
1975 LASER invented; 1977 Magnetic Resonance Imaging developed
1971 MICROPROCESSOR developed
1965 Electronic Typewriter invented; 1969 INTERNET started
1961 TAPE CASSETTE developed
1958 PACEMAKER developed; 1959 I.C. Planar Process developed
1957 FORTRAN Computer Language
1953 MASER invented
1952 INTEGRATED CIRCUIT concept; SPUTNIK Satellite launched
1950 MODEM developed
1943 Printed wiring system patented; 1948 First TRANSISTOR
1935 Scanning Electron Microscope invented
1934 Liquid Crystals Display developed; 1937 Xerography invented
1931 First Electronic COMPUTER developed; Cardiograph invented
1929 First Colour Television
1926 Electron-microscope invented; 1927 First cable TV
1924 RADAR invented
1919 Electronic TELEVISION system developed
1914 Asdic (SONAR) used; TELEX invented
1911 Superconductivity discovered
1906 Radio Broadcasting started; Three electron tube invented

20th Century

1898 First magnetic recording
1897 'Electron' defined
1895 Radio Telegraphy in use; X-rays discovered
1887 Electrocardiograph invented
1876 TELEPHONE invented; 1897 First Phonograph
1865 Radio wave theory (Maxwell)
1857 Mercury Arc Lamp discovered
1855 Glow discharge tubes developed
1847 Magnetostriction discovered
1844 Telegraphy using the Morse Code
1843 FAX machine invented
1839 Microfilming invented; Photovoltaic effect discovered
1831 Electromagnetic Induction; Electricity generated
1821 Thermoelectricity discovered
1820 Electro-magnetism discovered
1808 Atomic Theory (Dalton)
1801 Ultra-violet radiation discovered
1800 First Dry Battery (Volta); Infra-red radiation discovered

19th Century

1787 Electrostatics measured (Gold-leaf electroscope)
1780 Galvanic action discovered
1772 Magnetism used; Navy Compass

1770's Experiments with electrostatic electricity

1745 Electricity stored; The Leyden jar
1730 Two kinds of electricity discovered; positive and negative
1729 Distinction between conductors and insulators

18th Century

Figure 2.1. The growth of electronics.

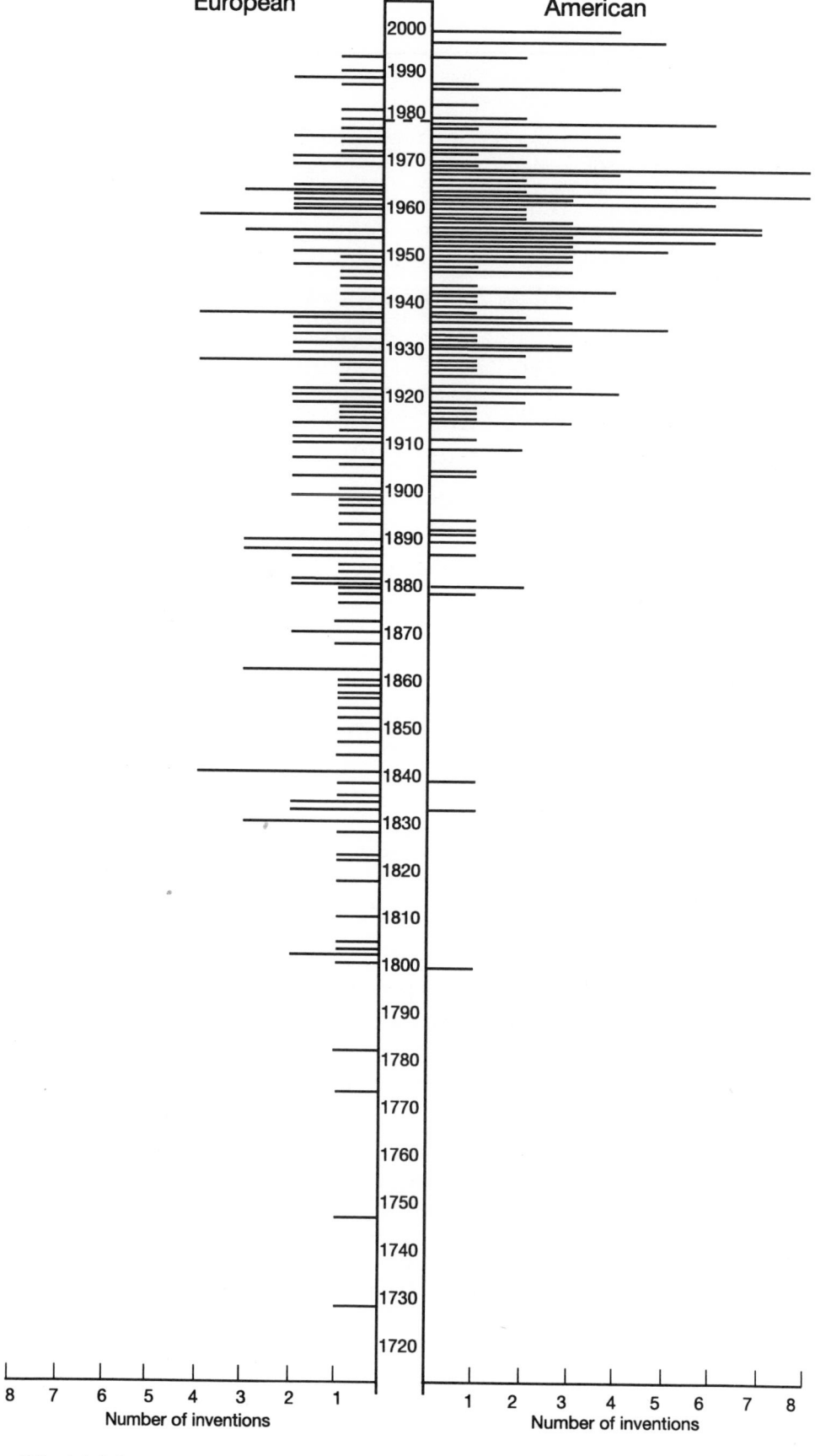

Figure 2.2. A brief summary of the growth of electronics over the years (showing the possible effect of the 'brain drain' from Europe to the USA in the 1960s).

Chapter 3

The Development of Components, Tubes, Transistors and Integrated Circuits

All electronic equipment is composed of components—resistors, capacitors, tubes, transistors, integrated circuits, etc. The development of such components is the story of electronics itself as, with each new invention in electronic techniques, components had to be developed and manufactured in quantity. The earliest components were those developed for the electrical industry—the Leyden Jar, the resistor, the transformer, the relay, the dry-battery, etc, all introduced between 1745 and 1900, originating mainly in Europe.

Radio telegraphy using spark transmitters towards the end of the l9th century was followed by continuous wave generators, such as the Alexanderson alternator, but when the thermionic tube was invented in 1906 continuous oscillation and amplification of radio frequencies became possible. Special tubes, transmitters and receivers were designed and built with the designer of the equipment constructing all the necessary component parts.

The 1914–18 war gave a marked impetus to the development of radio communications, and with the advent of the three-electrode tube in quantity, components such as resistors and capacitors began to assume the form roughly as we knew them up to the 1960s.

The BBC commenced programme broadcasting on 14 November 1922 and from that time, up to about 1930, many component manufacturers began to specialize in individual components, from which the home constructor used to make radio receivers. Most tubes were made initially by electric lamp manufacturers as the techniques of glass blowing and vacuum processes were similar. A typical bright-emitter three-electrode tube of the period is shown in figure 3.1. Bright emitter tubes, in rows, lit many an enthusiastic amateur's living room and, when these were followed by dull emitters, some of the early magic seemed to disappear. Amateur constructors may remember the pungent smell of ebonite drilled at too high a speed although, with the introduction of the screen-grid tube in 1924, a metal chassis rapidly replaced ebonite panels. The stages in construction of radio sets from the breadboard to the screened chassis are shown in figure 3.2. It might be considered that this period (the early 1920s) saw the birth of the components industry. Resistors were produced in large quantities and used as grid leaks, anode loads, etc, and consisted of carbon compositions of many kinds compressed into tubular containers and fitted with end caps. Paper-dielectric capacitors were mainly tubular types enclosed in plain bakelized cardboard tubes, with bitumen or similar material sealing the ends. Bakelite enclosed stacked-mica capacitors, fitted with screw terminals and with the bottom of the case sealed with bitumen, were also in common use. Rectangular metal-cased and plastic-cased types were also used. Electrolytic capacitors were mainly wet types in tubular metal cases. Cracked-carbon film-type resistors were introduced from Germany in about 1928 and by 1934 were being manufactured in quantity in the United Kingdom.

Figure 3.3 shows a front and rear view of the tuner and detector-amplifier circuits of a four-tube receiver built in 1923. Point to point wiring was used between the components. Square section wires were sometimes used with sharp right-angle bends to make all the wiring horizontal or vertical, reminiscent of the wiring patterns of some modern multi-layer printed wiring boards. This soon gave way to round wires and the home constructor grew remarkably adept at wiring up simple radio receivers.

From 1930 onwards the home construction of sets diminished and many component manufacturers and radio-set makers worked together. Techniques for component manufacture in quantities improved, and many millions of radio sets were in use throughout the world in 1939. As the standard to which components were

Figure 3.1. Early bright-emitter three-electrode tube (courtesy Mullard Radio Valve Co. Ltd).

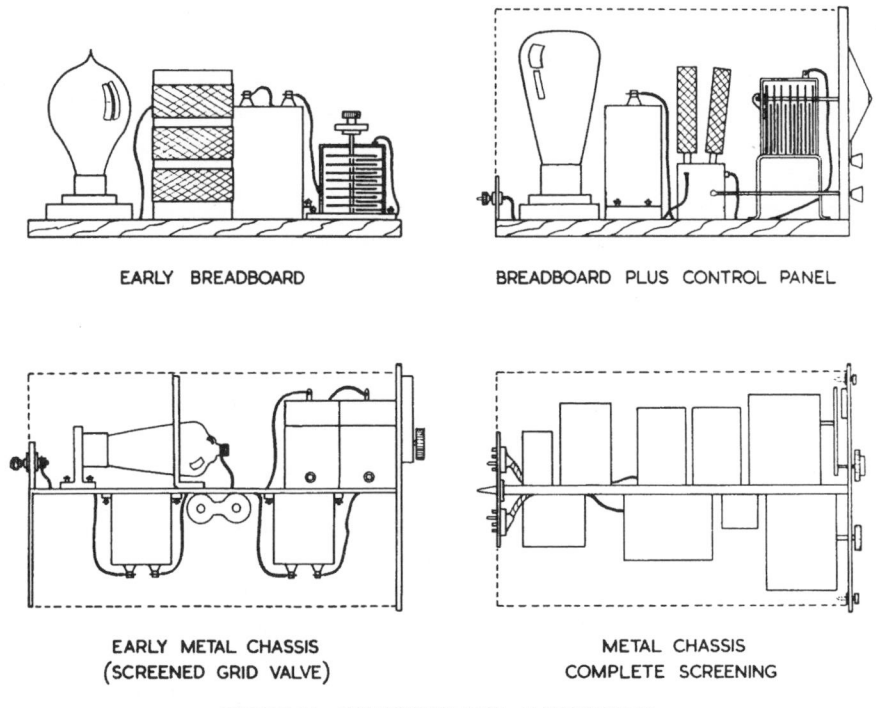

EARLY BREADBOARD   BREADBOARD PLUS CONTROL PANEL

EARLY METAL CHASSIS   METAL CHASSIS
(SCREENED GRID VALVE)   COMPLETE SCREENING

EARLY COMPONENT LAYOUTS

Figure 3.2. Early contributions—breadboard to metal chassis.

Figure 3.3. Four-tube receiver built in 1923 (courtesy Science Museum, London & Burndept Ltd).

made were those of domestic radio, pan-climatic protection was unnecessary and the self-compensating action of the radio tube made wide tolerances and poor stability generally acceptable. With the exception of certain electrical engineering applications, telephone companies, a few sections of the instrument industry and the military, no very high standard was required of the component manufacturer.

The advent of the 1939–45 war had a tremendous effect on components because now operation of equipment in all climates of the world was essential. The spread of the war from Europe to the Far East meant that equipment had to be designed to withstand tropical climates, whereas the war in Russia required equipment to operate in arctic conditions, and the North African desert war exposed equipment to excessive heat and rapid temperature cycling. The war at sea and combined operations made resistance to salt and sea spray necessary. Vibration, rough handling and shock impact affected the mounting of component parts. Particularly damaging was the effect of tropical conditions due to the high humidity which rusted metals, lowered insulation resistance of plastics, grew fungus and swelled some moving parts, making them useless.

Directly due to war requirements, many changes and developments were necessary to produce equipment to meet these arduous conditions. These changes are briefly listed below:

Standardization—increased production of fewer types
Miniaturization—needed for mobile sets, for aircraft, submarines, manpack radios, etc
Reliability—component failures too costly and disastrous
Maintainability—quick replacement needed, often by unskilled personnel
Transport hazards—transport shocks and rough handling

Mechanical shocks—impact of shells, parachute landing, etc
Vibration—in aircraft, tanks and ships affecting components
Storage—long periods before use, all equipment, particularly missiles
Low temperature—use in arctic conditions
High temperature—use in desert conditions
Humidity—use in tropical conditions
High altitude—high-flying aircraft (arc-over, etc)
Combined environments—humidity/high temperature/vibration, etc
High powers—to increase range of radio and radar
EM radiation resistance—to withstand nuclear environment

These stringent environmental and operational requirements, due directly to the 1939–45 war, resulted in major improvements to component design and manufacture. Waxed tubular paper-dielectric capacitors were replaced by metal-cased tubular types with rubber end seals. Metalled paper-dielectric capacitors were developed. Improved control of the temperature coefficient of ceramic-dielectric capacitors was introduced by manufacturers and new high-permittivity ceramic mixes developed. Many manufacturing improvements were introduced into resistor production and testing. Work on sealed variable resistors was sponsored and many types were made available. Transformers operating at higher temperatures, oil-filled and sealed in metal cans, were developed and also resin 'potted' transformers. Sealed relays and indicating meters were also designed and produced to withstand the tropical conditions.

Following the war the commercial success of component companies was concerned with the mass production of components for television receivers and radio sets. Military requirements dropped to a smaller proportion of the total components output but the lessons learned were valuable in improving the standard and reliability of commercial components.

After the war, around 1946, the printed circuit was beginning to be used in conjunction with sub-miniature tubes and many experimental circuits were made using both additive and subtractive printing techniques.

The crystal detector, invented in 1906, forerunner of the modern transistor, was used for many years until the invention of the three-electrode tube. The physics of crystal detectors was not understood until some fundamental research was done by physicists in the 1940s, when it was discovered that certain semiconductors, such as germanium and silicon, contained mobile electrons and 'holes', so that a so-called p–n layer in the crystal would pass current in one direction only as in the normal crystal detector, and many germanium diodes were used in the 1950s as rectifiers.

In 1948, Bell Laboratories' scientists found that amplification could be obtained by means of a third contact to the normal p–n contacts in germanium. This was termed a point-contact 'transistor' or 'transferred resistor' and the contact points were later replaced by an alloyed construction junction transistor. Bell Labs then produced the diffused junction type of transistor, also in germanium, which replaced the point contact transistor. Germanium had one drawback in that temperatures higher than 75°C caused excessive current leakage and silicon, which had a much higher operating temperature, began to be used and is the standard material today.

Field effect transistors were introduced in which the metal oxide formed part of the transistor action and were known as MOSTs (Metal Oxide Silicon Transistors), having a simpler construction and being considerably smaller.

The transistor came into very wide use because of its obvious advantages of low-voltage operation (6–12 volts instead of 250–300), its low current consumption and its extremely small size. In addition, its reliability was very high compared with existing tubes, which had a limited life due to evaporation of the cathode. They were first introduced into deaf aids and then into portable radio receivers; then into computers and were widely used in all electronics until being replaced by the integrated circuit or 'chip', which is in effect a large number of transistors, all made in one operation, described later.

In the early 1950s, glass-dielectric capacitors, metal-film resistors, castellated metallized paper capacitors and subminiature relays were beginning to be used more widely.

About this time subminiature components for use in transistor circuits were being developed. For the first time, as high-tension voltages of 150–300 V were not required, components such as capacitors could be designed to withstand only 6–12 V and (with the low currents at which transistors operated) resistors could be designed for very small power dissipation. This size reduction progressed to the point where handling difficulties and soldering problems arose. However, about this time, it was realized that the actual working volume of components such as resistors and capacitors was only a very small proportion of the total. In the case of a ceramic-dielectric capacitor fitted into a ceramic case only 1/225th of the actual working volume was effective whilst in a plastic-moulded carbon film resistor only 1/280th of the total volume was effective. Attempts to fabricate simple film components led to the early thick-film and thin-film circuits.

Other developments affecting components in the 1950s were automatic assembly techniques and 'potting' or encapsulation techniques. Machines for the automatic insertion of tubular components into printed wiring boards were developed. Axial lead tubular resistors and capacitors were loaded into special feed containers and in the machine their leads were bent over, inserted through holes in the printed wiring board and dip-soldered. Unfortunately the capacity of these machines was so high, e.g. up to 10 000 boards a day, that production quantities were rarely sufficient to make full use of them.

This period saw the exploitation of the junction transistor and its use with subminiature components on a printed-wiring board, with the connections made by dip-soldering. Extensions to the range of transistors and diodes, both in frequency response and power output, were being rapidly made and transistorized equipment was being used for practically all electronic requirements by the mid-1960s. Magnetic core storage using very small toroidal components was also used for computer information storage and retrieval about this time.

The next and most important development in components was the integrated circuit. The idea of integrating miniature components into one solid block was first put forward by the author in a paper read at a Components Conference in Washington, USA, on 6 May 1952 in which he stated:

'With the advent of the transistor and the work in semiconductors generally, it seems now possible to envisage electronics equipment in a solid block with no connecting wires. The block may consist of layers of insulating, conducting, rectifying and amplifying materials the electrical functions being connected directly by cutting out areas of the various layers'.

This proposal followed several years work on miniaturization of components done by the author's Division at the Royal Radar Establishment, Malvern, UK. It was not possible with the 'mesa' techniques of the time to fabricate production methods, but models were made and demonstrated for possible assemblies, termed by the author 'solid circuits'. Figure 3.4 (taken in 1957) showed for the first time the possible size of the integrated circuit. In spite of attempts by J Kilby and others, it was not until 1959 that the invention of the planar process with aluminium metallization by Fairchild's Noyce and Hoeni enabled mass production of circuits, starting the electronics revolution.

Initially, digital circuits were developed for use in computers, this being the maximum market, but linear circuits were also being developed for general purpose amplifiers, etc.

The development of the integrated circuit components now known as 'chips' was extremely rapid. In 1962 bipolar transistors were replaced by MOSFETs (Metal Oxide Semiconductor Field Effect Transistors). Although MOSFETs were slower than bipolar transistors, they were smaller, cheaper and used less power. In 1963, Fairchild introduced a resistor-transistor-logic (RTL) chip, known as a flipflop, which contained isolation channels and buried layers which were forerunners of later chip developments, leading to the first operational amplifier in 1964. An op-amp could not only add and subtract incoming signals, but could also average, integrate and otherwise manipulate them, enabling them to be used for control, measurement and computer systems.

In 1970, the first Random Access Memory (RAM) was produced by Fairchild (the 256-bit RAM), followed in 1972 by the 1024-bit RAM by the new company INTEL. Also produced in the same year by INTEL, was the first microprocessor (the 8008). An Intel Pentium microprocessor of 1992 is illustrated in figure 3.5. Progress in chip design was rapid and, in 1975, the first 4096-bit RAM was produced; and, in 1976, the 16 384-bit RAM was produced by INTEL. Also, in 1976, the first one-board computer was made by INTEL. All these chips were about 0.25 cm wide and 0.5 cm long.

About this time, chips were being built into personal computers and, in 1977, the APPLE 2 was introduced. In 1981, IBM entered the personal computer market with the 'PC' and, in 1984, IBM developed a one million bit RAM. In 1987, the PC MACKINTOSH and the IBM PERSONAL SYSTEM 2 were introduced. Comprehensive, inexpensive software systems such as Microsoft MS-DOS became available at this time. Supercomputers, introduced in 1976, were improved each year. In 1978, the CRAY Y-MP was capable of performing 2000 million operations per second.

Present methods of chip assembly include plastic and ceramic sealing and the chips are mounted on printed circuit boards or ceramic bases. The chips vary in shape, e.g. DIL (Dual-in-Line) Flip-Chip, Leadless chip

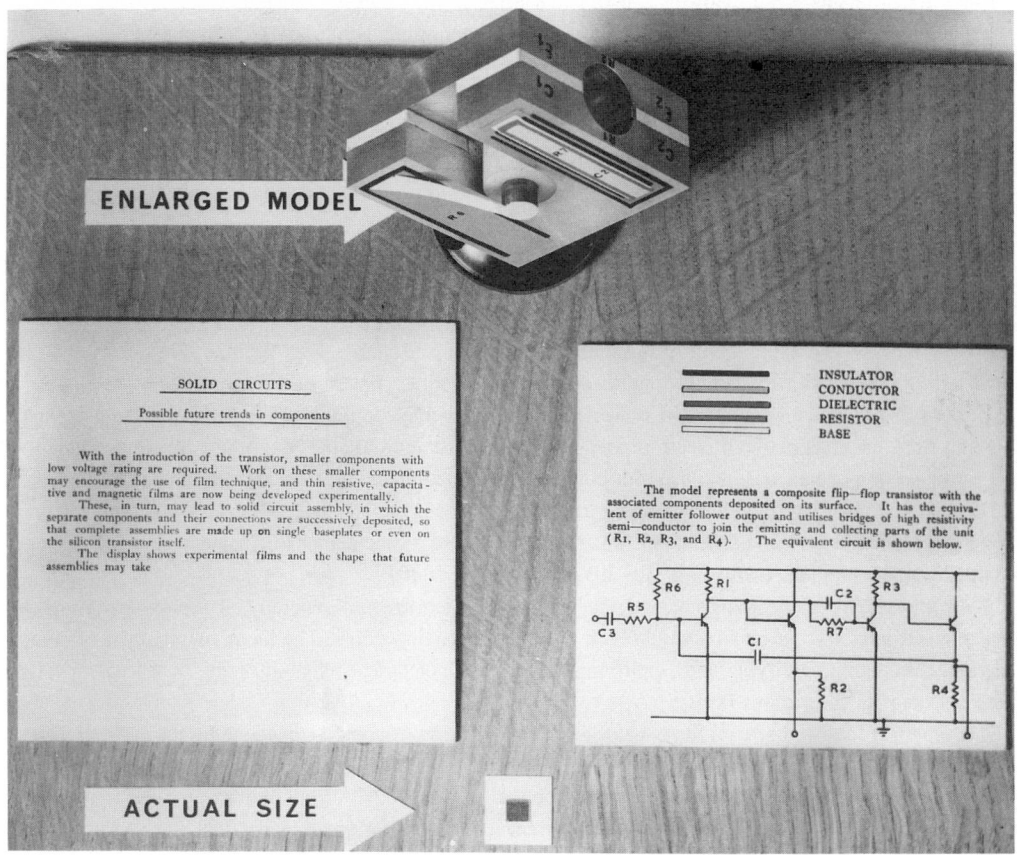

Figure 3.4. Integrated circuit model shown in 1957.

Figure 3.5. An Intel Pentium processer of 1992 (The Science Museum/ Science & Society Picture Library).

carriers, MCMs (Multi-chip Modules), all usually surface mounted. TAB (Tape Automated Bonding) using metal bumps (usually gold) for connection between chip and base is now widely used. These closely packed assemblies use VLSI (Very Large Scale Integration) and also VHSIC (Very High Speed Integrated Circuits), whilst electron beam technologies are used to obtain the very high definition and accuracy required for these assemblies. Close packing brings problems of overheating of individual LSI chips and miniature fans are sometimes necessary.

A cross-section illustrating the state-of-the-art technology in the 1960s is shown in figure 3.6 whilst a bipolar integrated circuit of 1991, with four layers of interconnections, is illustrated in figure 3.7.

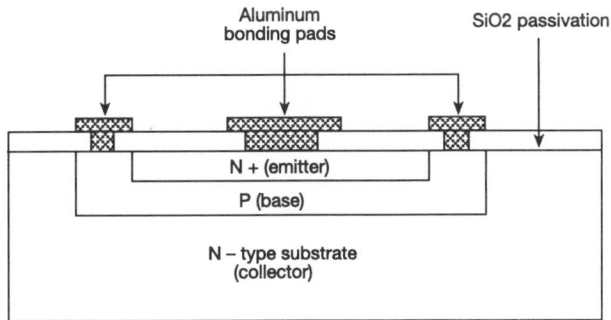

Figure 3.6. Cross-section of a bipolar integrated circuit of the 1960s.

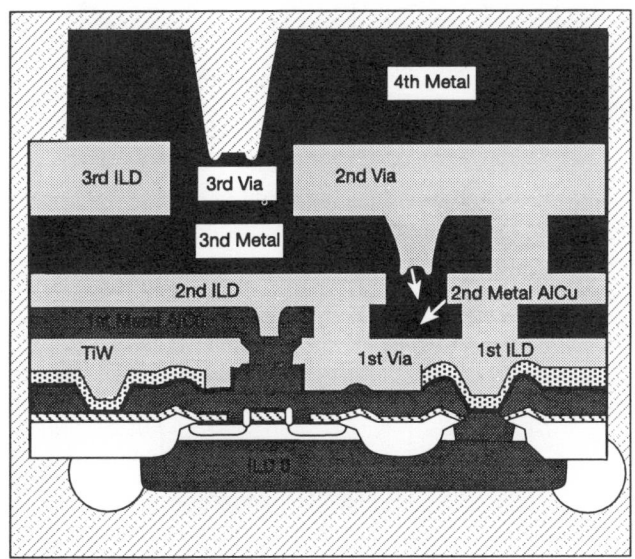

Figure 3.7. Cross-section of an integrated circuit of 1991.

Over the years CMOS technology has become the primary technology for fabricating ICs, whilst bipolar technology, although still used, has become a smaller percentage of present day IC manufacture.

VHSICs, due to their small geometries, can be used at much higher frequencies, and IC operating frequencies have increased from approximately 20 MHz clock rate in 1975 to approximately 200 MHz in 1995. Digital ASICs (Application Specific Integrated Circuits) and VHSICs, now behave like RF circuits and need proper transmission line termination, cross-talk protection and grounding.

The operating voltage is being reduced due to both the physical limitations of materials used to manufacture ICs and the high levels of integration being achieved. The smaller geometries required for state-of-the-art ICs (i.e. storage 64 Mb, 256 Mb, and 1 Gb DRAM) require lower power supply voltages, due to the dielectric and metallization thickness and the transistor breakdown characteristics. The portable personal computer and cellular radios are driving the trend towards reduced operating voltage, from 5 V to either 3.0 or 3.3 V now, and eventually to 1.5 V. This change (5.0 to 3.3. V) will double the battery life by reducing power dissipation. However, as the voltage is reduced, the operating speed (frequency) is reduced, the devices become more sensitive to static discharge handling, and there are smaller noise margins.

Adding to the package style transformation is the rapid increase in the number of external package leads (or pins). There has been a dramatic growth in pin count from 68 in 1980 to 512 in 1990. IC packages with pin counts of 900 to 1000 will be in production by the year 2000. This is shown in the pin count illustrated in figure 3.8. Larger package sizes (measured by the number of leads) are required to accommodate large die sizes, especially of application specific integrated circuits (ASICs) which are input/output (I/O) intensive (versus memories), and which make less efficient use of silicon material. The size reduction of DRAM cells over the years is shown in figure 3.9.

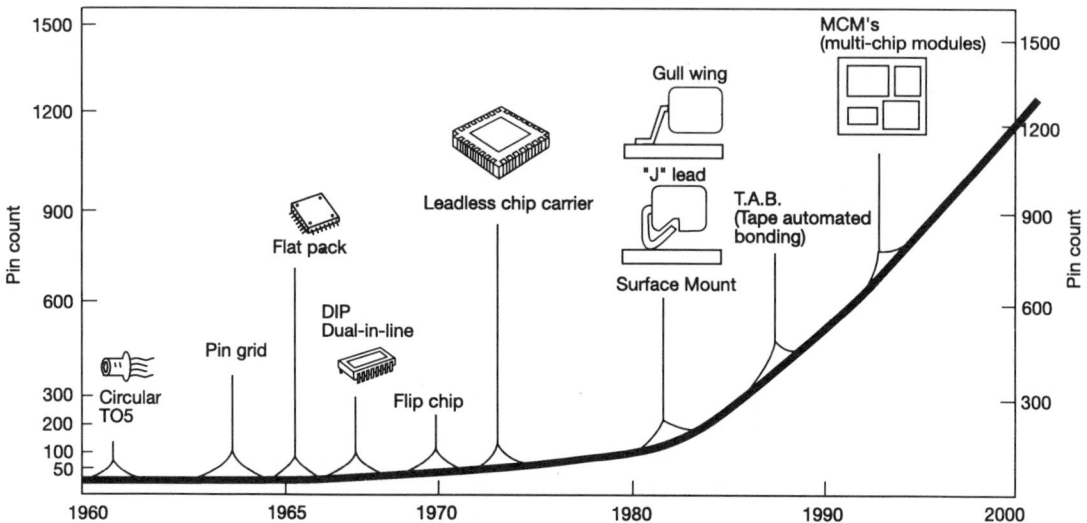

Figure 3.8. Pin count increase.

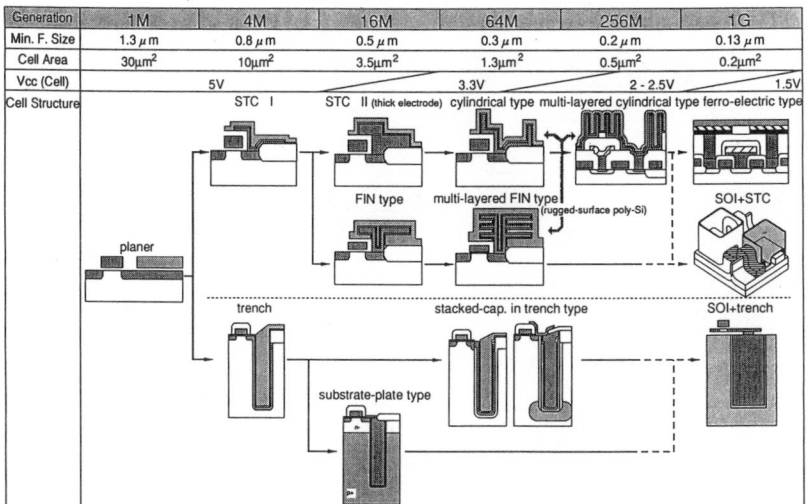

Figure 3.9. Trends of DRAM cells (reproduced by permission from Eiji Takeda, Hitachi Ltd).

The progress of IC technology has been relentless, with rapid advances in processing and packaging. The number of transistors fitted into a single chip has increased up to several millions.

Over the period from about 1900 to the present day, various electronics technologies have been developed, used to the full, and some have declined. Tubes were at their peak in the 1950s but, with the advent of the transistor, began to decline. Similarly, potted circuits and automatic assembly technologies had peak periods in the 1960s. We are now at the stage where TV, Internet, ST, Virtual Reality, mini- and microcomputers and microelectronic devices of all kinds are advancing rapidly.

The history of components shows how the change from passive components (resistors, capacitors, etc,) to active components (transistors, integrated circuits) has affected electronics development. The present explosion of integrated circuits in the form of VLSI and VHSIC has been the most important development in the history of electronics.

The growth of the semiconductor industry has been phenomenal. It is estimated that, in 1995, world semiconductor sales were of the order of 150 billion dollars, thus becoming one of the world's largest manufacturing industries.

The future of electronics, therefore, lies in the increasing use of microelectronic devices. The applications of microprocessors and minicomputers using VLSI are becoming wider and wider and this will affect both home and business life for everyone in the 1990s and beyond.

A summary of developments in components is given in 'History on a Page', charts 1 and 2: Date Chart 1, 'Passive Components' and Date Chart 2, 'Tubes, Transistors, Diodes and Integrated Circuits'.

History on a page
1. Passive Components

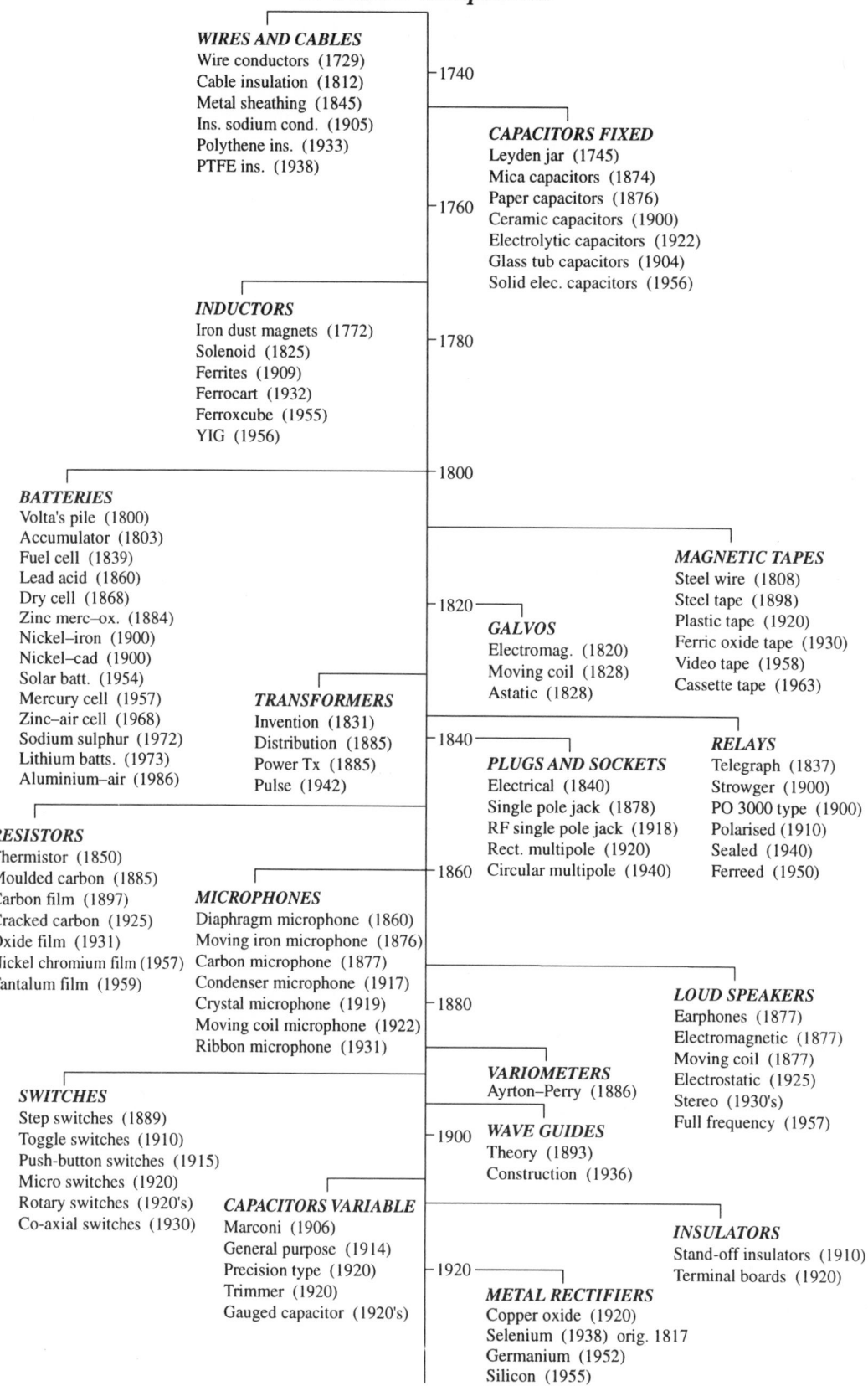

WIRES AND CABLES
Wire conductors (1729)
Cable insulation (1812)
Metal sheathing (1845)
Ins. sodium cond. (1905)
Polythene ins. (1933)
PTFE ins. (1938)

— 1740

CAPACITORS FIXED
Leyden jar (1745)
Mica capacitors (1874)
Paper capacitors (1876)
Ceramic capacitors (1900)
Electrolytic capacitors (1922)
Glass tub capacitors (1904)
Solid elec. capacitors (1956)

— 1760

INDUCTORS
Iron dust magnets (1772)
Solenoid (1825)
Ferrites (1909)
Ferrocart (1932)
Ferroxcube (1955)
YIG (1956)

— 1780

— 1800

BATTERIES
Volta's pile (1800)
Accumulator (1803)
Fuel cell (1839)
Lead acid (1860)
Dry cell (1868)
Zinc merc–ox. (1884)
Nickel–iron (1900)
Nickel–cad (1900)
Solar batt. (1954)
Mercury cell (1957)
Zinc–air cell (1968)
Sodium sulphur (1972)
Lithium batts. (1973)
Aluminium–air (1986)

MAGNETIC TAPES
Steel wire (1808)
Steel tape (1898)
Plastic tape (1920)
Ferric oxide tape (1930)
Video tape (1958)
Cassette tape (1963)

— 1820

GALVOS
Electromag. (1820)
Moving coil (1828)
Astatic (1828)

TRANSFORMERS
Invention (1831)
Distribution (1885)
Power Tx (1885)
Pulse (1942)

— 1840

RELAYS
Telegraph (1837)
Strowger (1900)
PO 3000 type (1900)
Polarised (1910)
Sealed (1940)
Ferreed (1950)

PLUGS AND SOCKETS
Electrical (1840)
Single pole jack (1878)
RF single pole jack (1918)
Rect. multipole (1920)
Circular multipole (1940)

RESISTORS
Thermistor (1850)
Moulded carbon (1885)
Carbon film (1897)
Cracked carbon (1925)
Oxide film (1931)
Nickel chromium film (1957)
Tantalum film (1959)

— 1860

MICROPHONES
Diaphragm microphone (1860)
Moving iron microphone (1876)
Carbon microphone (1877)
Condenser microphone (1917)
Crystal microphone (1919)
Moving coil microphone (1922)
Ribbon microphone (1931)

LOUD SPEAKERS
Earphones (1877)
Electromagnetic (1877)
Moving coil (1877)
Electrostatic (1925)
Stereo (1930's)
Full frequency (1957)

— 1880

VARIOMETERS
Ayrton–Perry (1886)

SWITCHES
Step switches (1889)
Toggle switches (1910)
Push-button switches (1915)
Micro switches (1920)
Rotary switches (1920's)
Co-axial switches (1930)

— 1900

WAVE GUIDES
Theory (1893)
Construction (1936)

CAPACITORS VARIABLE
Marconi (1906)
General purpose (1914)
Precision type (1920)
Trimmer (1920)
Gauged capacitor (1920's)

INSULATORS
Stand-off insulators (1910)
Terminal boards (1920)

— 1920

METAL RECTIFIERS
Copper oxide (1920)
Selenium (1938) orig. 1817
Germanium (1952)
Silicon (1955)

History on a page
2. Tubes, Transistors, Diodes and Integrated Circuits

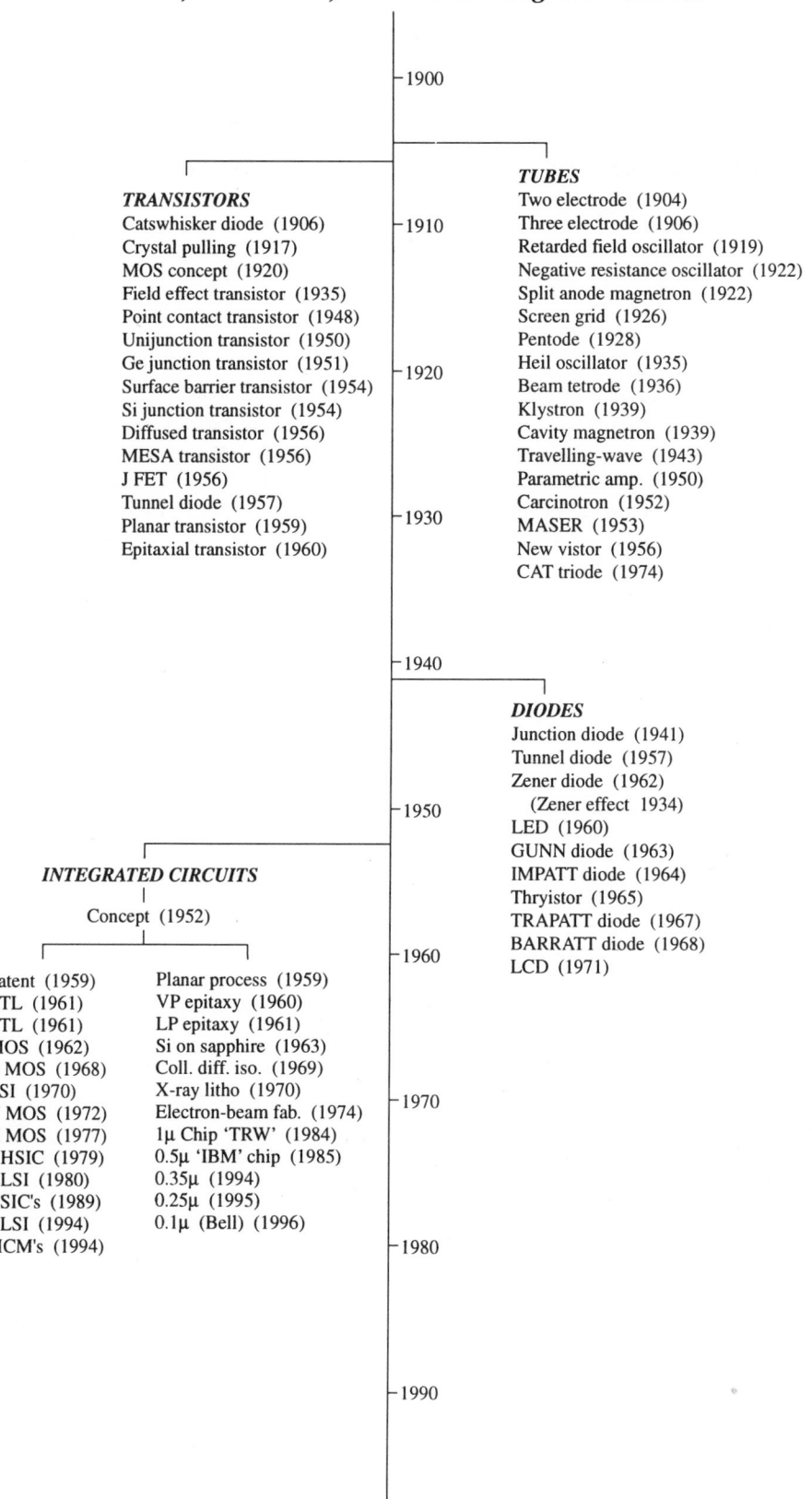

TRANSISTORS
Catswhisker diode (1906)
Crystal pulling (1917)
MOS concept (1920)
Field effect transistor (1935)
Point contact transistor (1948)
Unijunction transistor (1950)
Ge junction transistor (1951)
Surface barrier transistor (1954)
Si junction transistor (1954)
Diffused transistor (1956)
MESA transistor (1956)
J FET (1956)
Tunnel diode (1957)
Planar transistor (1959)
Epitaxial transistor (1960)

TUBES
Two electrode (1904)
Three electrode (1906)
Retarded field oscillator (1919)
Negative resistance oscillator (1922)
Split anode magnetron (1922)
Screen grid (1926)
Pentode (1928)
Heil oscillator (1935)
Beam tetrode (1936)
Klystron (1939)
Cavity magnetron (1939)
Travelling-wave (1943)
Parametric amp. (1950)
Carcinotron (1952)
MASER (1953)
New vistor (1956)
CAT triode (1974)

DIODES
Junction diode (1941)
Tunnel diode (1957)
Zener diode (1962)
  (Zener effect 1934)
LED (1960)
GUNN diode (1963)
IMPATT diode (1964)
Thryistor (1965)
TRAPATT diode (1967)
BARRATT diode (1968)
LCD (1971)

INTEGRATED CIRCUITS
Concept (1952)

Patent (1959)
RTL (1961)
TTL (1961)
MOS (1962)
C MOS (1968)
LSI (1970)
V MOS (1972)
H MOS (1977)
VHSIC (1979)
VLSI (1980)
ASIC's (1989)
ULSI (1994)
MCM's (1994)

Planar process (1959)
VP epitaxy (1960)
LP epitaxy (1961)
Si on sapphire (1963)
Coll. diff. iso. (1969)
X-ray litho (1970)
Electron-beam fab. (1974)
1μ Chip 'TRW' (1984)
0.5μ 'IBM' chip (1985)
0.35μ (1994)
0.25μ (1995)
0.1μ (Bell) (1996)

1900
1910
1920
1930
1940
1950
1960
1970
1980
1990

Chapter 4

A Concise History of Audio and Sound Reproduction

A basic electrical sound system needs a transducer (a microphone) to convert the sound into an electrical waveform, a means for transmission and a means for reproduction of the sound.

The earliest microphones consisted of a stretched flat membrane (Reis used a sausage skin in 1860) actuating a loose metal-to-metal contact which converted the sound vibrations into electric currents. Later microphones replaced the single loose metal contacts by carbon ones (Edison, 1877) and by carbon granules (Hunnings, 1878). The first intelligible human speech was transmitted over wires by Bell, however, in 1876. The microphone he developed consisted of a thin iron diaphragm which vibrated in front of a magnet with a coil wound on it, thus inducing an electric current which varied in sympathy with the voice sounds. This current was sent over a pair of wires to a similar apparatus at the other end. An early Bell telephone used in 1878 is shown in figure 4.1.

Sound could now be transmitted and received by wire and, in 1877, Edison invented a method for recording it. He wrapped a sheet of tinfoil round a cylinder, set a stylus or needle attached to a diaphragm in contact with it, turned a crank to rotate the cylinder and shouted into it 'Mary had a little lamb'. On cranking the cylinder again, his voice was reproduced. Ten years later, Berliner introduced the first flat-disc record and also a method for making many shellac copies.

Sound could now be transmitted, recorded and reproduced. About this time (1877), Siemens in Germany invented the first moving-coil cone loudspeaker, but it was before its time and had to wait for other developments before it could be exploited. The electrostatic loudspeaker, introduced around 1925, failed to gain wide commercial acceptance.

In 1924, Chester W Rice and Edward Kellog, both of General Electric, registered a patent for a voice coil speaker as well as constructing an amplifier capable of providing power of 1 watt for their device. The speaker, known as the Radiola Model 104 had a built-in amplifier and came on the market in 1925.

Ribbon speakers were invented in 1925. The standard cone was replaced by a very fine aluminium ribbon which was concertinered and exposed to a magnetic field. They were brought out in the USA in the 1920's. Boxed-in speakers were introduced in about 1958.

Following Bell's invention, the first public telephone exchange was opened in New Haven, Connecticut, in January 1878 to twenty-one subscribers. These first telephones were leased in pairs, the two telephones being permanently connected together.

In 1889, Strowger invented the first automatic switching system to allow one telephone to work with any other telephone and, in 1919, the improved crossbar selector system was developed in Sweden. In 1952, the first experimental electronic exchange was made by Bell Telephone Laboratories and, by 1958, fully automatic electronic exchanges were in use.

The spread of telephones throughout the world was rapid and there are now over 400 million telephones in service.

In 1898 another method of recording sound was invented by Poulsen in Denmark—magnetic recording. He found that by feeding the current from a microphone through an electromagnet and drawing a piano wire rapidly past it, the wire was magnetized to varying degrees, thus recording the sound. This could be played again and again and could also be wiped off and re-recorded. Poulsen's invention of the magnetized piano wire was revived when tube amplifiers became available and spools of thinner wire were used; but thin wire was inclined to twist and was replaced by a flat metal tape developed by Blattner (the 'Blattnerphone'). However, when the tape broke it was difficult to join, and I G Farben and AEG in Germany produced paper and plastic tapes coated

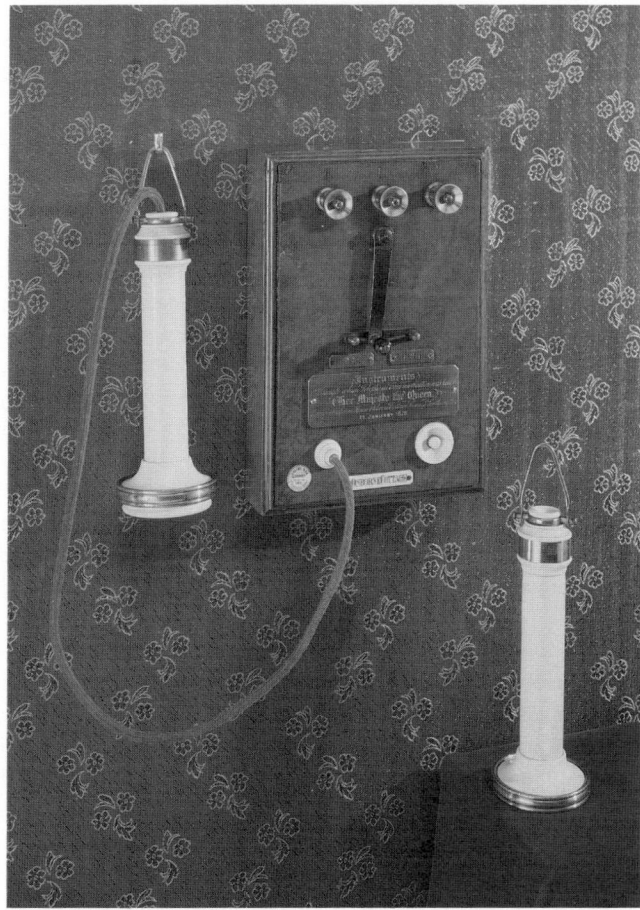

Figure 4.1. An early Bell telephone and terminal panel, as used at Osborne Cottage, 14 January 1878 (The Science Museum/ Science & Society Picture Library).

with iron oxide in the form of a lacquer. Germany, about 1940, introduced the 'Magnetophon' tape recorder, with improved tape and a much improved quality of sound reproduction—the latter largely due to the use of radio-frequency bias as in present-day tape recorders. A plastic-based ferric-oxide tape was used, running at 56 cm/s (22 in./s). Later open-reel tape recorders usually operate at 15 in./s and sub-multiples (7.5, 3.75, etc.,). Cassettes, which are now much more popular than open-reel tapes because of their low cost and convenience (no tape threading), normally operate at 1.875 in./s. Recently, cassettes of half-size have been introduced, but mostly for dictation purposes rather than music. Video recorder tapes can record up to 18 octaves of frequency for colour television, as against 9 octaves for normal sound recording.

Berliner's shellac gramophone discs were later replaced by plastic discs running at 78 rpm, and, about 1948, the 78 rpm disc playing speed was reduced to 45 and 33 rpm, and opened the era of the long-playing record.

The first sound recording to accompany films ('the talkies') was produced in 1926 by Warner Bros. A disc-recording machine called Vitaphone was synchronized to the film and the first successful talking picture was 'The Jazz Singer' in October 1927, with Al Jolson. Fox Movietone News improved on this system by using a sound-on-film system in which the synchronism was automatic, using a variable-intensity light source to record the sound on the film negative. Similar systems are now used in modern cinemas.

In the early 1930s Arthur Keller of Bell Labs in the USA, and Alan Blumlein of EMI in Britain, developed stereophonic disc recording systems quite independently. These gave directional and much improved special effects, but the business depression of those years prevented commercial exploitation before the war. Surprisingly, the first practical demonstration of stereophonic sound reproduction was carried out by a Frenchman, Clement Ader, in 1881. He installed pairs of microphones in the Paris Opera House, connected by telephone lines to pairs of earphones in the Exhibition of Electricity held that year in Paris.

A three-channel system was used by the Bell Laboratories on 27 April 1933, when the Philadelphia Orchestra was reproduced in Washington with a most impressive degree of realism. This may be regarded as the beginning of stereophonic hi-fi.

Quadraphonic or surround sound has been introduced, which aims to give the listener the illusion of being present inside an auditorium more vividly than can be achieved using ordinary two-channel stereo.

A typical hi-fi loudspeaker system consists of two or three loudspeaker units of different cone size, the smallest ones being known as 'tweeters' and the largest ones as 'woofers', with a crossover filter network to direct the relevant frequency band into each unit.

It is necessary to focus attention on the importance of the polar, i.e. directional radiation characteristic of loudspeakers and the relationship between the loudspeaker and the acoustics of the listening room. With improvements in other parts of the audio chain, these aspects are becoming of much greater significance.

Electrical and electronic musical instruments have been evolved over a period of many decades. The major part of this effort has probably been devoted to the electronic organ, but other instruments, such as electronic pianos, have also been produced. A development in this field of creating sound electronically is the synthesizer. This gives the user an unprecedentedly great ease and versatility in the creation and control of audio waveforms with regard to harmonic content, envelope shape, repetition rate and random noise characteristics. Musical and other sound sequences can be built up and controlled from a small keyboard. The BBC Radiophonic Workshop uses synthesizers and other techniques, and a well-known example of radiophonic music created by them is the theme music from 'Dr. Who'.

Electronic organs consist of a series of electrical tone generators controlled by keyboard(s) and stops fed to amplifiers and loudspeakers. An early and commercially successful organ was designed by Laurens Hammond and marketed in 1935. This employed a large number of synchronous-motor-driven alternators to generate the tones. Since then very many different designs have been evolved, mostly of purely electronic nature. The best of these simulate remarkably well the subtle sounds of a good traditional pipe organ, but the largest sales are of smaller instruments aimed mainly at the popular music field. As well as producing a wide range of organ-stop tonalities, often of cinema-organ character, these organs usually also provide various percussive effects, and a 'rhythm box' is frequently added to give repeated drum and cymbal sounds at various selectable tempos. Techniques based on the use of microprocessors are being increasingly employed to relieve the player of the need for a high degree of performing skill.

In Fax machines, teleprinters or computers, the output is in binary form—0s and 1s. As this cannot be transmitted over a telephone line, it has to be converted into suitable audio tones. This is done by a MODEM (Modulator and Demodulator) with a similar MODEM at the receiving end. With a modern computer a modem can be used to connect up to the INTERNET.

For transmitting speech less bandwidth is required than for music. For good speech quality about 16 K bits/s from an A to D converter is required. VOCODERS (or voice coders) are used to reduce this to 2.4 or 3.6 K bits/s by making assumptions about the speech waveform, speech, synthesis. This can be converted back by a D to A converter which is aware of the assumptions made. This produces the Dalek-like speech which is just recognizable. This in turn leads to speech-analysis devices and speaking machines, i.e. synthetic speech. Speech synthesis is based on the theory of visible speech, formulated in 1948 by the Americans R K Potter, G A Kopp and H C Green, who showed how phenomes (vocal sounds) correspond to graphic traces. In 1950, the first machine to recognise ten numbers was built by Bell Labs in the USA. Speech synthesis, which possesses far fewer theoretical problems than speech recognition, is used in many domains, such as industry, cars and games.

The development of video recording techniques for television, together with fast digital techniques in other fields, has had a large influence on sound recording. An early application was the Army Wireless Set No. 10 multichannel transmitter and receiver which used digital techniques and was designed in 1942 and introduced in 1944. A further application of digital techniques to audio was not for recording, but was the BBC's 'Pulse Code Modulation' (PCM) system for sending high-quality stereo and TV sound signals from one BBC centre to another. In the 1960s the instantaneous amplitude of the audio wave was sampled at frequent intervals, and the sample amplitude expressed as a binary number of 12 or more bits, i.e. 0s and 1s. The apparatus for doing this is called an analogue-to-digital (A/D) converter. At the other end, the stream of 0s and 1s is reconstituted into an audio waveform by a D/A converter. A large bandwidth is required to transmit this very rapid succession of 0s and 1s, but the distortion of the system is far lower than that of normal landlines. For example, programmes can be sent from London to Edinburgh and back with no perceptible degradation in quality. These techniques are now being used for sound recording, the digital bits being recorded on a video tape recorder or on video discs. There are several disc systems, the latest one using a laser spot focused on a reflective track, and modulated by a pattern of indentations on a disc. Digital Compact Discs, invented in 1978 by the Philips Co, enabled a 5 inch diameter disc to carry one hour of stereo sound, with much lower distortion and background noise level than could be achieved by analogue methods. In addition, the sound reproduced is unaffected by dust, scratches and fingerprints on the record. The CD system is shown in figure 4.2.

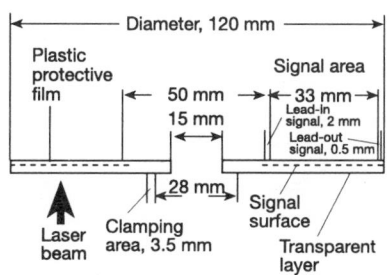

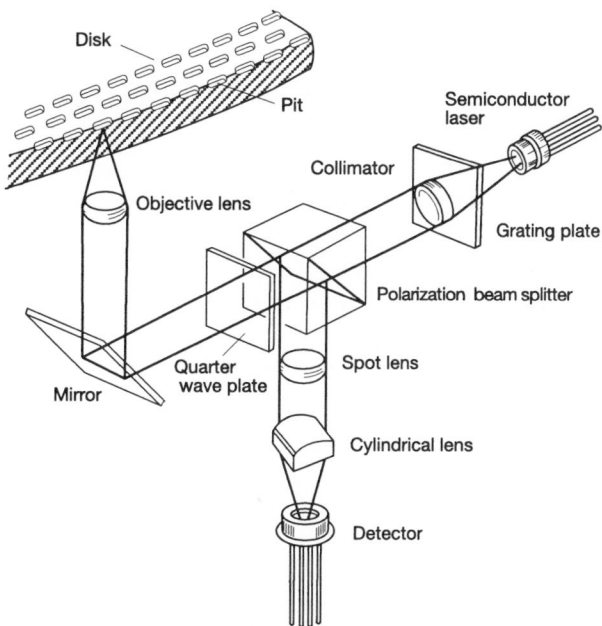

Figure 4.2. The CD (compact disc) system (courtesy IEEE Spectrum, March 1984, and Senri Miyooka, Sony Corporation).

Conventional digital audio, as used for instance on compact discs, requires a large bandwidth for transmission and enormous amounts of storage space. Developments in high-speed digital signal processing chips have made it practical to reduce these requirements by employing sophisticated data compression techniques which reduce redundant and irrelevant information in the source signal.

Over the years the quality of sound reproduction has improved and the standard of excellence in audio reproduction is now the digital compact disc marketed in 1983. In addition to home use, a modern car with four speakers fed from a CD provides a stereo reproduction of excellent quality.

A summary of audio development is given in Date Chart 3.

History on a page
3. Audio and Sound Reproduction

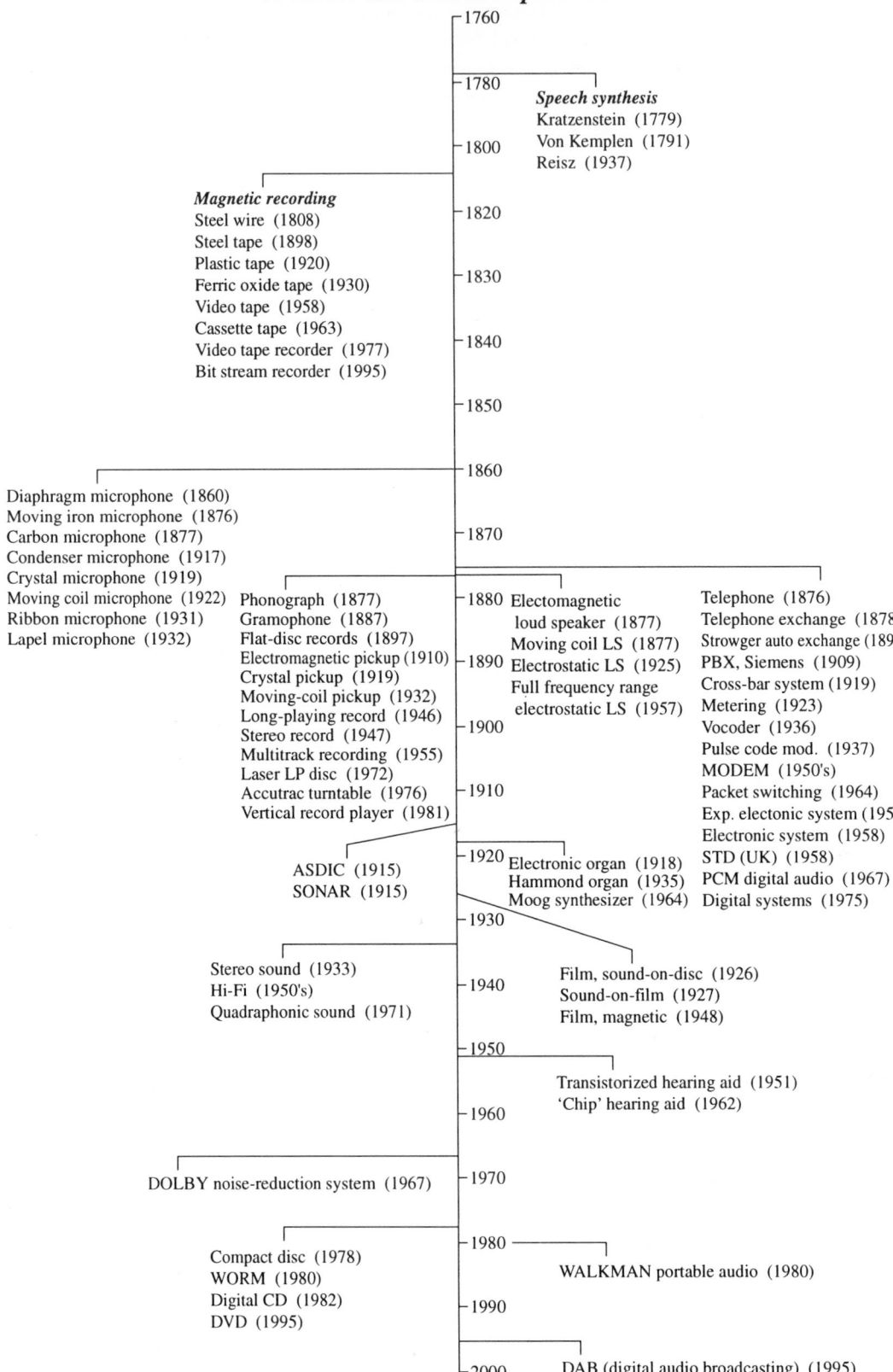

Speech synthesis
Kratzenstein (1779)
Von Kemplen (1791)
Reisz (1937)

Magnetic recording
Steel wire (1808)
Steel tape (1898)
Plastic tape (1920)
Ferric oxide tape (1930)
Video tape (1958)
Cassette tape (1963)
Video tape recorder (1977)
Bit stream recorder (1995)

Diaphragm microphone (1860)
Moving iron microphone (1876)
Carbon microphone (1877)
Condenser microphone (1917)
Crystal microphone (1919)
Moving coil microphone (1922)
Ribbon microphone (1931)
Lapel microphone (1932)

Phonograph (1877)
Gramophone (1887)
Flat-disc records (1897)
Electromagnetic pickup (1910)
Crystal pickup (1919)
Moving-coil pickup (1932)
Long-playing record (1946)
Stereo record (1947)
Multitrack recording (1955)
Laser LP disc (1972)
Accutrac turntable (1976)
Vertical record player (1981)

Electomagnetic
loud speaker (1877)
Moving coil LS (1877)
Electrostatic LS (1925)
Full frequency range
electrostatic LS (1957)

Telephone (1876)
Telephone exchange (1878)
Strowger auto exchange (1892)
PBX, Siemens (1909)
Cross-bar system (1919)
Metering (1923)
Vocoder (1936)
Pulse code mod. (1937)
MODEM (1950's)
Packet switching (1964)
Exp. electonic system (1952)
Electronic system (1958)
STD (UK) (1958)
PCM digital audio (1967)
Digital systems (1975)

ASDIC (1915)
SONAR (1915)

Electronic organ (1918)
Hammond organ (1935)
Moog synthesizer (1964)

Stereo sound (1933)
Hi-Fi (1950's)
Quadraphonic sound (1971)

Film, sound-on-disc (1926)
Sound-on-film (1927)
Film, magnetic (1948)

Transistorized hearing aid (1951)
'Chip' hearing aid (1962)

DOLBY noise-reduction system (1967)

Compact disc (1978)
WORM (1980)
Digital CD (1982)
DVD (1995)

WALKMAN portable audio (1980)

DAB (digital audio broadcasting) (1995)

1760
1780
1800
1820
1830
1840
1850
1860
1870
1880
1890
1900
1910
1920
1930
1940
1950
1960
1970
1980
1990
2000

Chapter 5

A Concise History of Radio, Communications and Avionics

At the end of the nineteenth century, the electron had been discovered and a great deal of experimental work was being done on electromagnetism. Sparks produced by discharges from a Leyden jar through a coil of wire fitted with a spark gap were found to be reproduced in nearby metal objects. This was the first instance of the transfer of man-made electrical energy over a distance without interconnecting wires.

Many theories were put forward for this phenomena, but James Clerk Maxwell was the first to consider magnetic and electric fields together to be responsible and formulated his equations from which he predicted electromagnetic radiation on purely theoretical grounds. In 1873 he predicted wave propagation with a finite velocity, which he showed to be the velocity of light. Heinrich Hertz succeeded in producing electromagnetic waves experimentally in 1877 and confirming Maxwell's predictions. He also added a loop of wire and increased the distance over which sparks could be transmitted.

A spark detector was produced by a French physicist, Branley, who noticed that the resistance of a glass tube filled with metal filings, normally high, became low when an electric discharge occurred in the vicinity, which could be measured. It was now possible to transmit telegraphy by coding the sparks and using Branley's coherer to receive the sparks over a distance. It is interesting that Branley was the first to use the word 'radio' in connection with his experiments.

Guglielmo Marconi was interested in Hertz's experiments and began experimenting in 1895 with a spark induction coil and Branley coherer, and succeeded in sending telegraph messages over a short distance. He travelled to Britain in 1896 to exploit his discoveries with W H Preece, Engineering Chief of the Government Telegraph Service, and Marconi demonstrated his apparatus to the Post Office officials over a distance of 100 metres.

Similar work was being done in Russia and A S Popov gave a demonstration on 12th March 1896, but this was not published until some thirty years later.

Marconi patented his invention in June 1896 and the first Radio Telegraph Wireless Company was formed in 1897 to acquire Marconi's patents and set up the manufacture of spark transmitters and receivers. Other manufacturers such as Siemens and Halske were, by the late 1890s, also making radio equipment. By this time, aerials had become long wires raised as high as possible.

Marconi continued to improve his equipment and succeeded in linking England with France by radio in 1899, and later over longer distances. He also installed equipment on ships for communication between ship and shore.

To explain why radio communication was possible around the curvature of the earth, Heaviside in England and Kenelly in America proposed an ionized layer in the upper atmosphere which reflected the waves. This was in 1901 and was known as the Heaviside/Kenelly layer.

Spark transmitters were almost universal for radio telegraphy, but they were improved when rotary discharger sets were introduced, fed from alternators to give more transmitter power and to radiate continuous waves (CW).

In 1903, Steinmett built a 1 Kw 10 KHz alternator, which was used by Fessenden in experiments with wireless telephony and in 1904, Alexanderson designed an alternator operating at a frequency of 100 KHz (see figure 5.1).

Figure 5.1. The Alexanderson alternator (The Science Museum/ Science & Society Picture Library).

Atmospherics were a great enemy. Larger and larger powers were introduced, such as at Poldhu in Cornwall, operating in 1911 and at the transatlantic station at Clifden in Ireland. In 1912, the sinking of the TITANIC focused attention on radio-telegraphy as, although over 1500 lives were lost, some 700 were saved by ships summoned by wireless.

The Great War of 1914–18 produced a major change in technology with the introduction of the three-electrode tube in quantity, which ultimately resulted in the ending of the spark transmitter era. By 1919, high-power transmitter tubes were being made and the Marconi Company spanned the Atlantic by telephony in daylight. The superheterodyne receiver (see figure 5.2) was in use and provided virtually constant gain and particularly constant selectivity over a wide tuning range. Attempts to make oscillators at much higher frequencies were made, such as the retarded field oscillator by Berkhausen and Kurtz in Germany in 1919 and the negative resistance oscillator by Gill and Morell in the United Kingdom in 1922. Crystal control of frequency was introduced and the screened grid tube and the pentode improved the performance of radio receivers, which were mostly made by home constructors in the 1920s.

Frequency standards were set up, first the quartz crystal itself, then the atomic clock, using the spectral line of ammonia and finally the caesium and rubidium beam magnetic resonance methods intially developed in 1939.

The advent of the 1939–45 war introduced high-sensitivity communication receivers for military use, such as the R1155 for the Royal Air Force. The first production communication receiver in wide use was the National HRO of the late 1930s. Post-war developments benefited from the research which had been done on components and high-quality FM receivers and hi-fi amplifiers became available. Radio astronomy, discovered in 1933 aroused considerable interest and Jodrell Bank was in operation in 1951.

In the 1950s, frequencies in the band 400 to 500 MHz were found to be suitable for Private Mobile Radio (PMR). Mobile handsets and vehicle installations are intended to communicate with a radio Base Station, but

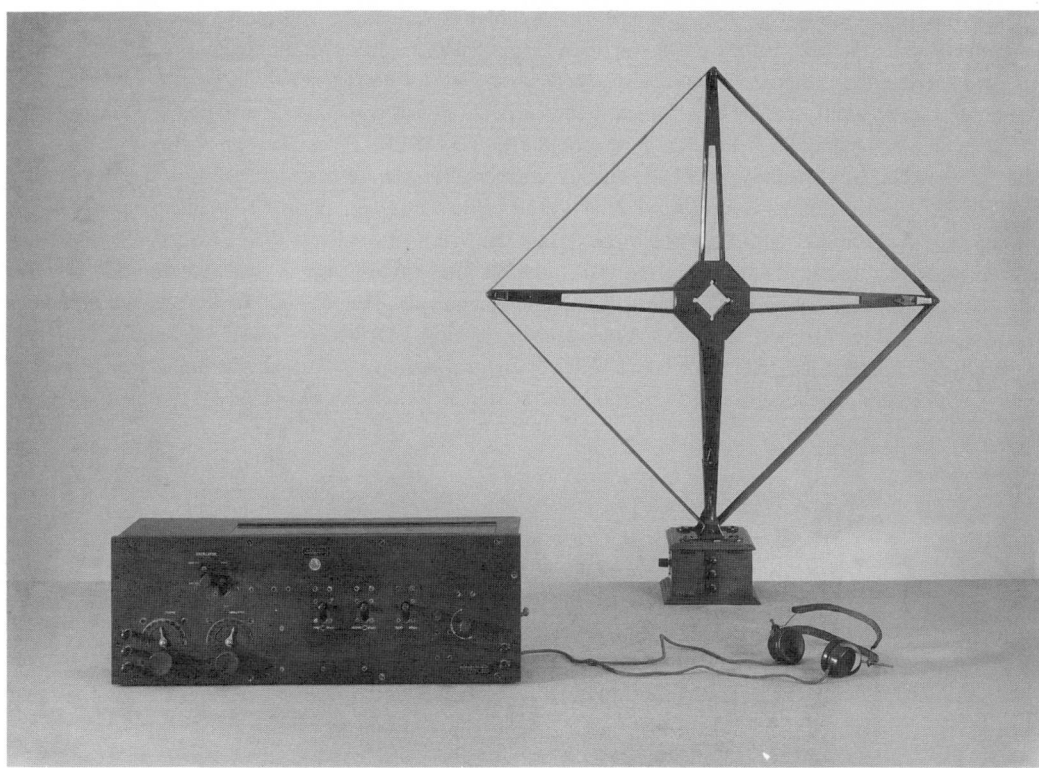

Figure 5.2. A superheterodyne receiver of the 1930s (The Science Museum/ Science & Society Picture Library).

may also 'talk through' the Base Station to other mobiles. The radio Base is usually connected to an office or Headquarters by land-line.

Cellular radio was developed to provide access for mobiles into and out of the public network by direct dialling at the largest possible number of 'mobile' sites. This was achieved by providing a large number of transmit and receive Base Station sites, each covering a small area or 'cell', (usually drawn as a hexagon and a large group of cells as a honeycomb). The cell sites were interconnected and connected to the Public Switched Telephone Network by mobile switching centres.

In the United Kingdom, the cellular Tactical Access Communication System (TACS) commenced in 1985 with two rival licenced organisations, which adopted the names Vodafone and Cellnet.

A pan-European, 900 MHz cellular system which has been under development since 1982 and is now known as the 'Global System for Mobile Communications' (GSM) (see figure 5.3) should solve many of the technical problems experienced with current cellular systems and offer other advantages.

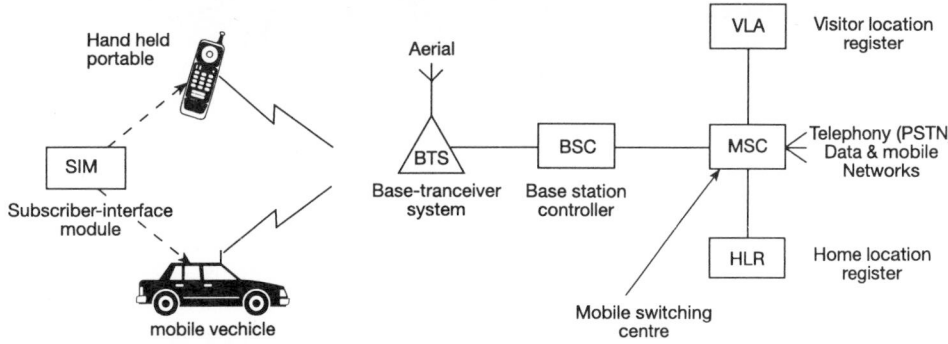

Figure 5.3. Arrangement of the European GSM cellular radio system.

The invention of the transistor revolutionized radio techniques, with its low voltage and current requirements, leading to cheap and reliable battery power supplies for portable sets. In 1954, the Regency transistor set was introduced in America. Printed circuits were widely used with ferrite rod aerials in portable receivers. In 1957, the Russian Sputnik satellite was launched and carried a 1-watt transmitter, becoming the forerunner of modern satellite communication. For low-noise amplification, MASERS, parametric amplifiers and cryogenic techniques were used, particularly in satellite signals receivers. Microwave systems were now developed from the earlier oscillators and the travelling-wave tube was incorporated in high-performance microwave sets. The USA launched EXPLORER I in 1958 which discovered the Van Allen radiation belt. Also in 1958, the first communication satellite was launched by the USA—the SCORE satellite, which transmitted taped messages for 13 days. In 1960, ECHO-1 was launched as a passive communication satellite, which relayed both voice and TV signals and the first active repeater communication satellite COURIER-1B was also launched in 1960, followed by TELSTAR-1 in 1962. Since then a series of communication satellites have been launched by many countries, enabling worldwide coverage of radio and TV signals to be achieved. Intelstat I, known as 'Early Bird', is shown in figure 5.4.

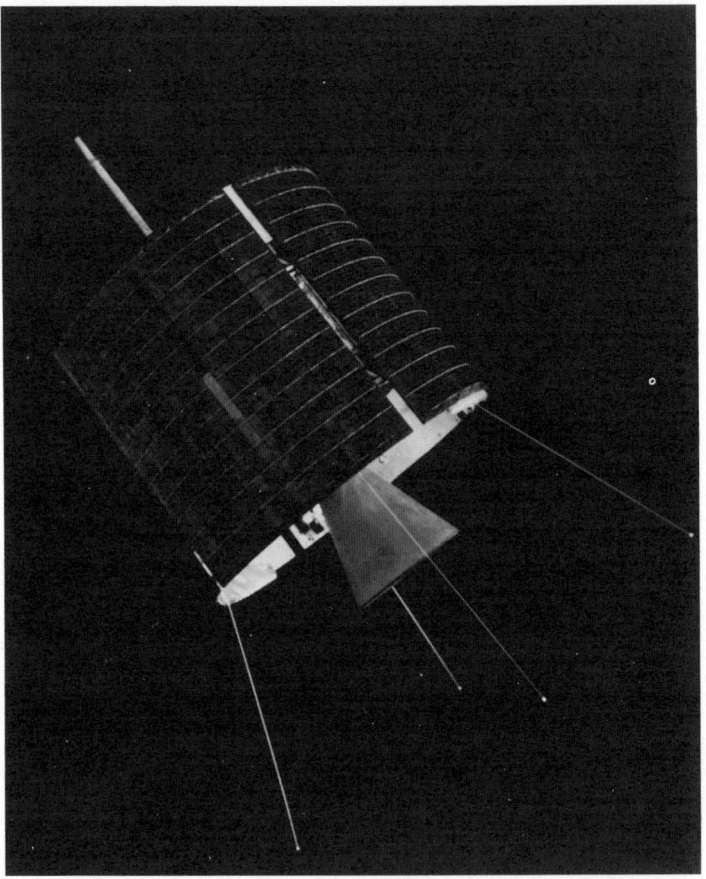

Figure 5.4. Intelstat I, known as 'Early Bird' (courtesy Mark Williamson, Space Technology Consultant, and Hughes Aircraft).

Inmarsat is an international organisation which used several geostationary satellites to provide nearly worldwide coverage for private and business users. It offers an extension to the public telephone, data, Telex and Fax services to and from ships at sea and more recently land mobiles.

Integrated circuits were now being used in radio receivers, commencing in the early 1960s, with circuit designs, which often used transistors in place of many resistors and capacitors. Improvements in circuit techniques and in packing density and complexity resulted in modern radio receivers of high performance and great reliability. LSI (Large Scale Integration), VLSI (Very Large Scale Integration), microprocessor devices, etc, are now used in radio systems.

Spread-spectrum communication techniques are increasingly being used, particularly for satellite communications. A spread-spectrum system is one in which the transmitted signal is spread over a wide frequency band. Various spreading techniques are used, e.g. digital direct code sequences (DSN), frequency hopping (which can also be 'time-hopping' or 'time-frequency hopping'), and chirp or frequency sweeping. Because the system is based on coding, privacy may be obtained by selective de-spreading. There is also improved resistance to interference.

High-speed digital signal processors (DSPs), are now being designed to translate analogue signals into digital data for microprocessor based control systems such as high-frequency video, radar signal processors, etc. The fastest method is known as parallel or flash encoding, where stray capacitances are reduced to a minimum by 1 micron lithography and other techniques.

The use of digital data systems in which machines communicate directly with other machines over radio paths is increasing rapidly, often because of the improved bandwidth/information rate that can be achieved (compared with human beings) and the trade-off flexibility that can be achieved between bandwidth and data rate.

Digital Audio Broadcasting (DAB) is the result of a standard devised and developed by a group of European broadcasters and consumer electronics industries and their research institutes to provide a reliable, multi-service digital sound broadcasting system for reception by mobile, portable and fixed receivers, using a simple rod aerial.

There are now strict and very detailed international OCIR (Comite Consultatif Internationale Radio) rules governing the use of radio transmission from a few KHz. to tens of GHz. Developments in radio communication are summarised in Date Chart 4: 'History on a Page'.

5.1 Avionics

The first use of avionics in aircraft was probably D/F (direction finding) initially developed by Marconi as far back as 1905. He used a ring of inverted 'L' aerials and, by selecting the aerial receiving the strongest signal, the direction of the transmitting station could be found. When short wave radio became available in the 1920s, a directional beam system using the morse letters 'A' and 'N' was transmitted from aerials right and left of the runway approach, enabling the aircraft to fly down the beam by keeping the tone steady, thus becoming the first ILS (Instrument Landing System) (see figure 5.5.)

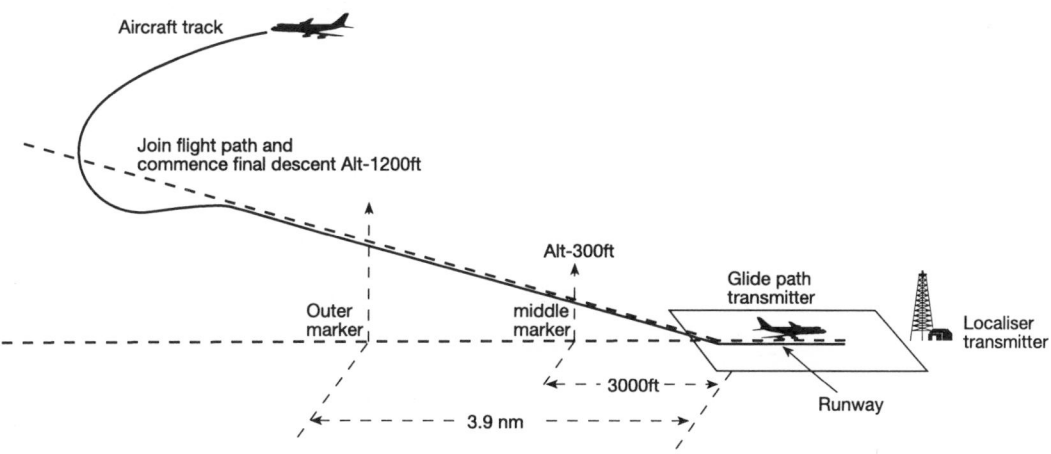

Figure 5.5. ILS system layout and approach path.

During the 1939–45 war, interlocking morse signals using a series of dots and dashes was used as a beam navigation system by Germany, known as 'Knickebein' whilst, in the United Kingdom, radar systems were developed. A hypobolic grid navigation system known as GEE gave position coverage up to 350 miles. The USA developed a similar system under the name 'LORAN'. This was followed by a United Kingdom precision navigation system known as 'OBOE' which used a dots and dashes system. A high definition radar picturing the ground below capable of being fitted in on aircraft, known as H2S was developed by the UK for accurate bombing raids.

After the war, intercontinental flying increased considerably and more aircraft were needed. As more aircraft were built, the need for air traffic control became paramount. The chief method of control was by radar from a control centre. In addition to the main radar system a secondary surveillance radar (SSR) was also used to identify each aircraft. Using the information from the radar screen and in direct radio communication with the aircraft's navigator, the controller could have continuous control of all aircraft movements.

On small airports, VHF Ohni-range (VOR) is used as a navigational aid in conjunction with Distance Measuring Equipment (DME) carried in the aircraft. Today's ILS systems operate on a frequency of 108.0 to 112 MHz, whilst the Microwave Landing System (MLS) operates on a frequency band of 5031.0 MHz to 5090.0 MHz.

The increase in aircraft production has been phenomenal and the variety of aircraft has expanded to cover all forms of flying. The use of electronics has become essential as the modern aeroplane is now controlled and flown by the use of computers (see figure 5.6). Several microprocessors control communications, multimode radars, cabin temperature and air pressure, flight data recorders, cockpit displays (in colour), engine monitoring, flap controls, onboard maintenance systems, including fly-by-wire linkages to control systems and geostationary satellite navigation all designed to reduce the workload of the aircrew.

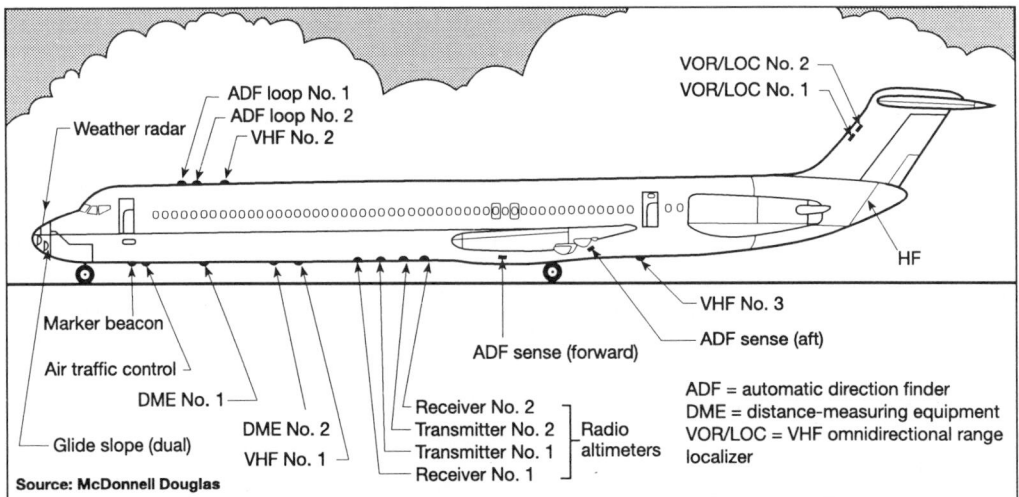

Figure 5.6. Modern civil aircraft electronics.

In addition, warfare electronics in military aircraft have created an entirely new field of sophisticated electronic systems, e.g. surveillance airborne radar, including sideways looking and synthetic aperature radars, terrain following radar, fire control systems, ground mapping, missile approach warning systems, thermal imaging systems, jamming countermeasures against heat-seeking or radar controlled missiles, electronic head-up display in the glareshield, digital moving map cockpit displays in colour and global position navigation systems (GPS). All this data is controlled by microprocessors. An illustration of the offensive and supporting elements of a defensive avionics system in a modern warplane is shown in figure 5.7.

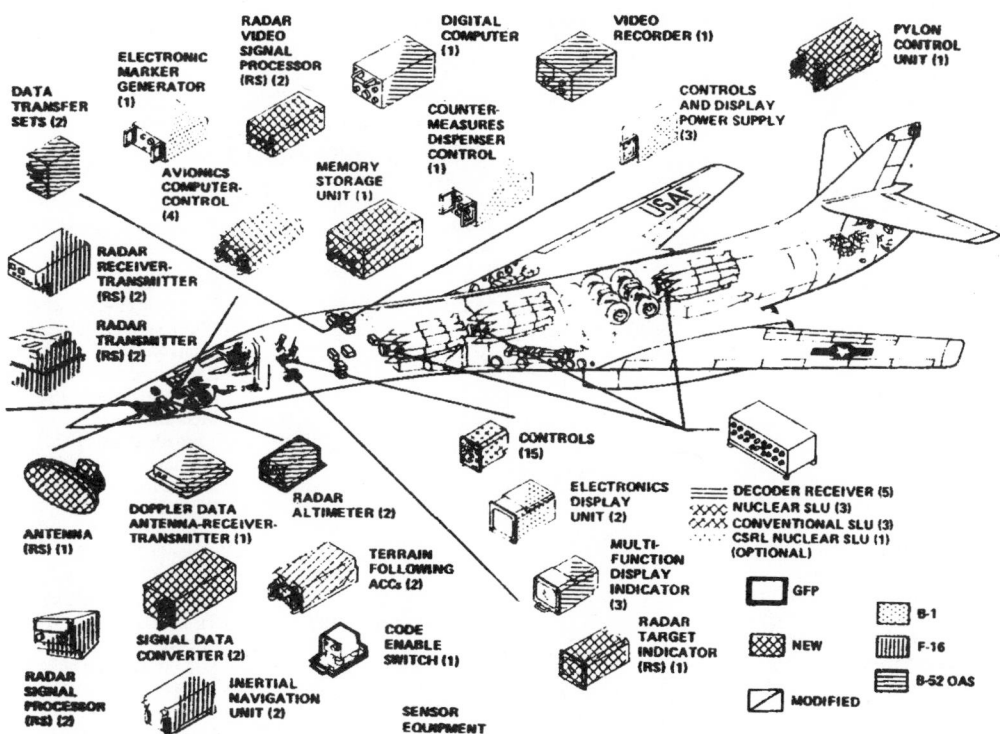

Figure 5.7. Offensive and supporting elements of defensive avionics systems. (Reprinted with permission from *Jane's Avionics 1995/96*, published by Jane's Information Group.)

History on a page
4. Radio and Communications

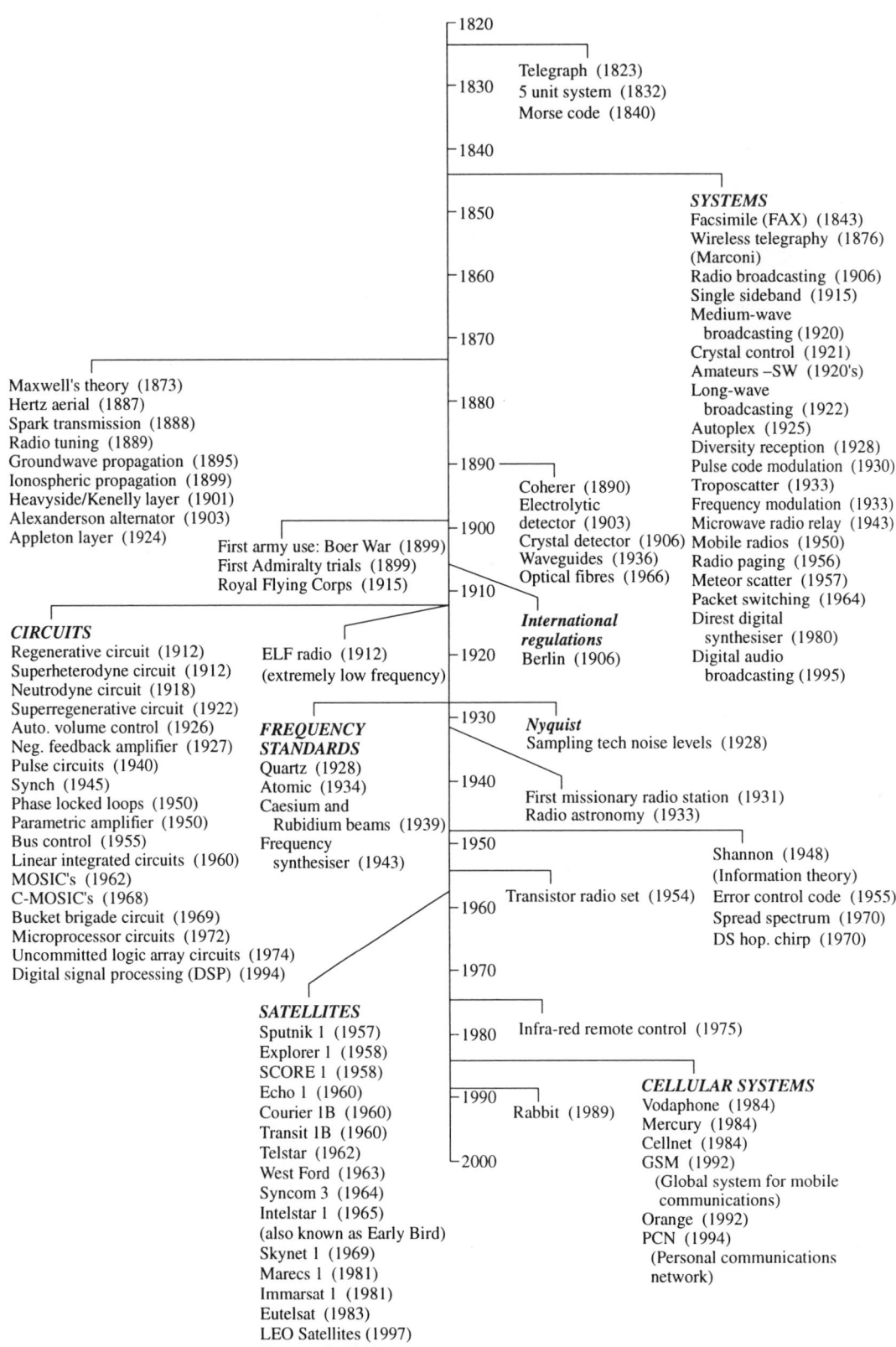

1820

1830 Telegraph (1823)
5 unit system (1832)
Morse code (1840)

1840

1850 **SYSTEMS**
Facsimile (FAX) (1843)
Wireless telegraphy (1876)
(Marconi)

1860 Radio broadcasting (1906)
Single sideband (1915)
Medium-wave
broadcasting (1920)

1870 Crystal control (1921)
Amateurs –SW (1920's)

Maxwell's theory (1873)
Hertz aerial (1887)
Spark transmission (1888)
Radio tuning (1889)
Groundwave propagation (1895)
Ionospheric propagation (1899)
Heavyside/Kenelly layer (1901)
Alexanderson alternator (1903)
Appleton layer (1924)

1880 Long-wave
broadcasting (1922)
Autoplex (1925)
Diversity reception (1928)

1890 Pulse code modulation (1930)
Coherer (1890) Troposcatter (1933)
Electrolytic Frequency modulation (1933)
detector (1903) Microwave radio relay (1943)

1900 Crystal detector (1906) Mobile radios (1950)
First army use: Boer War (1899) Waveguides (1936) Radio paging (1956)
First Admiralty trials (1899) Optical fibres (1966) Meteor scatter (1957)
Royal Flying Corps (1915) Packet switching (1964)

1910 Direst digital
synthesiser (1980)

CIRCUITS *International*
regulations
Regenerative circuit (1912) ELF radio (1912) Berlin (1906)
Superheterodyne circuit (1912) (extremely low frequency)

1920 Digital audio
broadcasting (1995)
Neutrodyne circuit (1918)
Superregenerative circuit (1922)
Auto. volume control (1926)

1930 *Nyquist*
Sampling tech noise levels (1928)
Neg. feedback amplifier (1927) **FREQUENCY**
Pulse circuits (1940) **STANDARDS**
Synch (1945) Quartz (1928)

1940 First missionary radio station (1931)
Radio astronomy (1933)
Phase locked loops (1950) Atomic (1934)
Parametric amplifier (1950) Caesium and
Bus control (1955) Rubidium beams (1939)

1950 Shannon (1948)
(Information theory)
Linear integrated circuits (1960) Frequency
MOSIC's (1962) synthesiser (1943) Transistor radio set (1954) Error control code (1955)
C-MOSIC's (1968)

1960 Spread spectrum (1970)
Bucket brigade circuit (1969) DS hop. chirp (1970)
Microprocessor circuits (1972)
Uncommitted logic array circuits (1974)
Digital signal processing (DSP) (1994)

1970

SATELLITES
Sputnik 1 (1957)

1980 Infra-red remote control (1975)
Explorer 1 (1958)
SCORE 1 (1958)
Echo 1 (1960)

1990 **CELLULAR SYSTEMS**
Courier 1B (1960) Rabbit (1989) Vodaphone (1984)
Transit 1B (1960) Mercury (1984)
Telstar (1962) Cellnet (1984)

2000 GSM (1992)
West Ford (1963) (Global system for mobile
Syncom 3 (1964) communications)
Intelstar 1 (1965) Orange (1992)
(also known as Early Bird) PCN (1994)
Skynet 1 (1969) (Personal communications
Marecs 1 (1981) network)
Immarsat 1 (1981)
Eutelsat (1983)
LEO Satellites (1997)
(Low earth orbit)

Chapter 6

A Concise History of Radar and Sonar

Between the 1920s and the 1930s, it was noticed that signals received from high-frequency radio transmitters were affected by aircraft flying overhead. Appleton and Barnett when measuring the height of the ionosphere in the United Kingdom in 1924, and Briet and Tuve, in 1925, doing similar experiments in the USA, both noticed this effect, although Briet and Tuve used a pulse technique for their measurements. During the following ten years, experimental work on the detection of aircraft and of vessels at sea by radio was done in the USA, France and Germany. In Britain, in 1935, a committee was set up under Sir Henry Tizard to make a serious investigation into the air defence of the United Kingdom. Sir Robert Watson-Watt suggested lines of research and with other scientists developed the first effective radar system in the summer of 1935.

Radar operates by sending out a pulse of energy at radio frequencies, some of which is reflected back by any object in its path. By measuring the time interval between the transmitted and received pulse on a time base, the range can be accurately calculated. Three main wavelengths were used—around 10 m for initial early warning floodlight radars; 1.5 m for more accurate positional radars and, later, around 10 cm for pencil beams of high resolution accuracy.

Because the United Kingdom developed and used the first practical radar system in the world, the following descriptions are primarily of wartime developments in British work. The success of United Kingdom radar was due to the close working liaison between the fighting forces and the scientists. A series of CH (Chain Home) early warning radar stations was constructed, using wavelengths in the 6–15 m band. In 1938, five stations covered the East Coast and, by 1940, the whole of the East and South Coasts were covered against aircraft flying at 15,000 feet out to a range of 120/140 miles. The CH transmitting aerials were mounted on metal towers, 360 feet high, and the receivers on 240 feet wooden towers. A peak transmitter power of 200 kW later 750 kW was used with a pulse repetition frequency of 25 and 12.5 pulses per second. The receiving aerials had a DF system to give approximate bearing. The complete co-operation between the scientists and the RAF, ensured that fighter aircraft were able to be alerted with time to intercept enemy bombers, enabling the 'Battle of Britain' to be won in 1940.

Low-flying enemy aircraft could avoid the 10 m CH aerial coverage so a chain of stations operating on 1.5 m was set up in 1939, known as CHL (Chain Home Low).

Figure 6.1 shows some low-flying enemy mine-laying in the Humber tracked by CHL. The Navy then swept mines on these plots. The 1.5 m beam was directional and was swept round on its turntable one or more times per minute, searching each sector of the sky in turn. The time was now ripe for a rotating time base to be linked with the aerial rotation and the first PPI (Plan Position Indicator) was installed in 1940. The night bomber was posing a problem, and directing the fighter towards a bomber became a possibility with this equipment. This was known as GCI (Ground Controlled Interception) and a controller with a GCI station was able to see the echoes of both fighter and bomber on the PPI in their relative positions and advise the fighter pilot by radio. A long afterglow CRT was used to 'hold' the signals.

In 1936, a start was made on airborne radar and an experimental 1.5 m ASV (Air to Surface Vessel) radar fitted in an Anson aircraft observed echoes from ships at a range of 5 miles on 4th September 1937. By 1939, a fully developed ASV system was installed in Coastal Command.

Also in 1937, the first AI (Air Interception) 1.5 m radar flew in a Battle aircraft. This was the first of a series of AI sets and the AI Mark 4 was used in the early part of the war with good success. With the invention of the cavity magnetron (shown in figure 6.2) initially giving about 400 W C W, by J T Randall and H A H

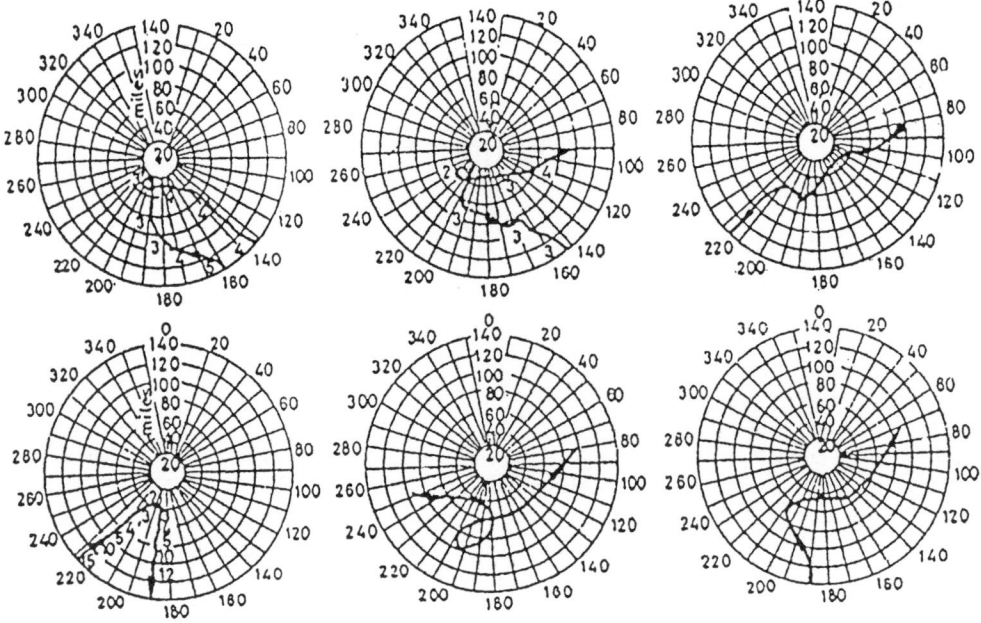

Figure 6.1. CHL tracks of low-flying enemy mine-laying in the Humber.

Boot in 1940, both AI and ASV later became centrimetric and the magnetron was one of the turning points in radar development. Up to the discovery of the cavity magnetron, klystron oscillators were the only source of high-frequency power. Using the same resonator principle, Randall and Boot developed the first magnetron high-power oscillator which was improved to produce hundreds of kilowatts in short pulses at a wavelength of around 10 cm.

Figure 6.2. The original cavity magnetron (The Science Museum/ Science & Society Picture Library).

With a small reflector, a pencil beam was now available for an AI set which could be scanned to present the pilot with a picture showing the position of the enemy aircraft. Two methods of scanning were used: a helical and a spiral scan. As an experiment, the spiral scanner was pointed downwards and a recognisable picture of towns was seen on the PPI tube. Following this, a Halifax bomber was fitted with a perspex blister underneath carrying the radar. It was flown on 27th March 1942 and the equipment was code named H2S. It became the RAF's most successful blind navigation and bombing aid.

It was realized soon after the development of the early radar system that some form of identification was needed to avoid shooting down our own planes. A beacon system was devised (IFF)—'Identification Friend or Foe', which automatically responded to small transmitted signals and returned a stronger coded signal on the same frequency which identified the aircraft as friendly. All were fitted with detonators in the event of a crash.

In 1941, the navigational problem of finding the target town in Germany for night bombers was solved by the invention of GEE (Grid Navigation) by R J Dippy. This was a transmitted hyperbolic grid pattern from three pulsed transmitters, one a master station which could activate two slave station timing pulses (shown in figure 6.3). The aircraft carried a cathode-ray tube display on which the arrival of the pulses was recorded. By referring the pulse positions to a map marked with the hyperbolic curve lines, the navigator could find his position and also the target. GEE could guide hundreds of bombers simultaneously to a target 350 miles away with an accuracy of about 2 miles. The first 1000 bomber raid was carried out using GEE navigation on 30 May 1942. A modified system (GEE/H) used two stations only and the aircraft transmitted to both. The signals were transmitted back with different delay times according to the aircraft's position. A still more accurate system was code named OBOE. This was so accurate that it was used to guide Pathfinders to the target for dropping marker flares, so that following bombers could unload their bombs. It used two transmitters, one enabling the aircraft to fly along a beam centred on the target. If too far one side, the navigator heard a series of dots; if too far, the opposite side, a series of dashes was heard. A continuous note meant the correct course. The second transmitter measured the distance along the track and signalled the bomb aimer exactly when to release the bomb. Bombing raids from a height of 30 000 feet, at a distance of 250 miles, controlled by 'Oboe' Pathfinders were accurate to within 150 yards.

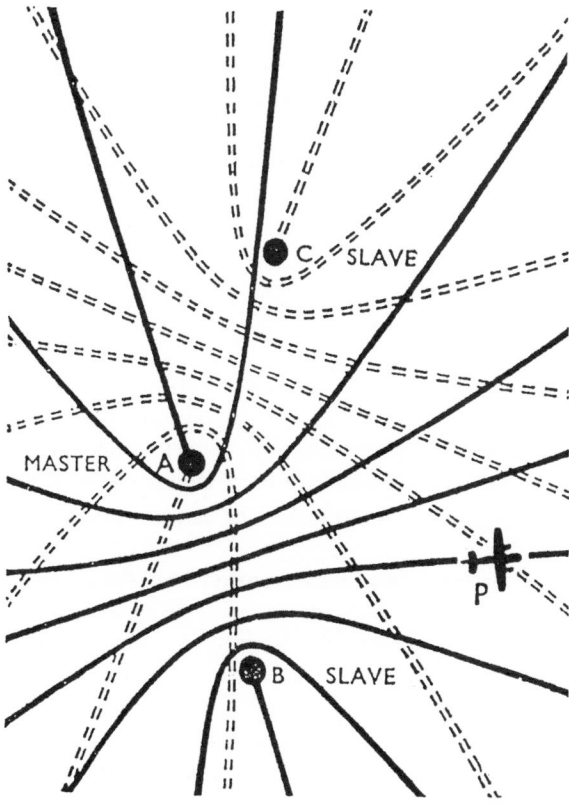

Figure 6.3. Grid pattern of GEE navigational system.

Many attempts were made by Germany to jam or render ineffective the radar-guided bombing raids and, in turn, methods of confusing German radar screens were adopted by the British. One was 'WINDOW' which consisted of thin tin foil strips scattered from a single aircraft giving multiple echoes to create the impression of a large bombing raid. This was first dropped in 1943 and was extremely successful. It was again used in 1944 with the invasion of France in conjunction with a further deception device called 'MOONSHINE', in which an IFF set was modified to imitate the echoes from a large formation of aircraft, instead of a single one. There were also many deceptions towards the end of the war as radar equipment became more sophisticated.

After the war surface-to-air guided missiles were developed such as 'Bloodhound' for the RAF, and 'Thunderbird' for the Army and small air-to-air missiles were designed for the RAF, some carrying infra-red homing heads which homed on to the engine exhausts of jet aircraft. Many sophisticated devices were produced after the war, such as TV guided air-to-ground missiles, cloud- and collision-warning systems, terrain-following radars and even more sophisticated systems.

Since the war, both in the USA, the UK, and France, there have been great advances in radar systems, and digital techniques have been widely introduced. Video integrators, displays and ranging equipment were the first functions to be converted to digital techniques. This digitization is necessary to handle the large amount of information collected by modern search radars. Digital computers are widely used and with an electronically scanned or phased array antenna there are no moving parts. Array antennas have many advantages: rapid scanning, beam shaping, step scanning, so that radar aerials of this type can obtain multiple target capability and function time-sharing.

In the USA, electronically scanned antenna arrays, together with microwave devices using Gallium Arsenide multichip MMIC's at X band, have been developed and are being extended to millimetre wave Ka band (35 GHz).

So far the equipments described were used by the RAF, but parallel developments were taking place in the Army and Navy.

6.1 Army Radar

Following the original optical sightings of targets, the AA guns were aimed and fuses set by information from predictors which were, in effect, mechanical analogue computers. The Army's first use of radar in 1938/39 was the CD (Coastal Defence) followed by the GL1 to feed information to the predictor at night when the target could not be seen. The first radar set GL1 (Gun Laying Mark 1) provided range and bearing only in which two cabins were used, one containing the transmitter and one the receiver; the receiver cabin rotating to obtain the bearing. GL1 provided elevation by adding two more vertically displayed antennas and an extra display. GL2 was an improved system with range, bearing and elevation all built-in and was later used to track the German V2s under the code name BIG BEN although CH was also used.

Following the completion of the tests on the early GL3 equipment, a centimetre height finding equipment (CMH Mark 1) was designed to give increased range of detection on very low flying aircraft beyond the capability of a CHL.

In 1940, the searchlight-control system, SLC (known as ELSIE) using Yagi transmit and receive antennas (some mounted directly on to a 90 cm or 150 cm searchlight), was also used in conjunction with Bofors AA guns. The first 10 cm AA radar GL3 was deployed in 1943, leading to auto-follow, auto-gun-aiming, auto-fuse setting and automatic firing in later models. Close co-operation between the Army and the Radiation Laboratory, MIT, in the USA, resulted in an auto-following anti-aircraft fire-control radar, called SCR 584, which was widely used.

In 1940, a VT (Variable Timing) proximity fuse was proposed by W A S Butement in the United Kingdom, to fit into a shell. A small radio transmitter and receiver inside the shell measured reflections from the target and, when within a certain range, exploded the shell. The shells were manufactured in the USA, and were fitted in the UK, in time to meet the V1 flying-bomb attack. 75% of the German VI's approaching London were ultimately shot down by AA guns using VT fuses.

Radar was also used for field artillery fire correction. With a 3 cm pencil beam, the ground ahead could be scanned to give a rough picture of the terrain. In the same way that the Coastal Artillery sets could locate shell splashes in the sea, the field artillery could observe shell bursts on the ground, enabling fire correction to be carried out. Another type of field army radar set 'Watch Dog' allowed movement to be detected in any zone in which its beam could search. 'Watch Dog' used the Doppler effect to differentiate between fixed echoes and moving echoes.

After the war, target-tracking radars for AA guns were improved in accuracy. For light anti-aircraft fire, a radar was developed called 'Blue Diamond', which used a 'Foster' scanner, which produced a rapid azimuth scan of 60°, 20 times per second. The elevation scan at a lower speed was done by a nodding arrangement and the whole equipment formed part of a Fire Control System called 'Yellow Fever' for the L70 AA guns. Other search and tracking radars were developed and put into service with the Army for the tactical control and fire control of medium and heavy AA guns.

Mortar locators were developed about one year before D-day, the first being 'Green Archer', which measured the range, bearing and elevation angles of two points on the trajectory of the mortar shell in order to extrapolate back to its point of origin, so that it could be attacked. This was followed by 'Cymbeline' which was a lightweight and improved version of 'Green Archer'.

To deal with low-flying attacking aircraft or helicopters, a one-man fire-control and missile system equipment known as RAPIER was developed. Four radar-guided missiles were mounted on a lightweight trailer that could be towed by a Land Rover.

6.2 Naval Radar

Before the advent of radar, the Navy had to rely on good weather and high-power telescopes to detect enemy planes and ships. This gave reasonable co-ordinates, but range was difficult to estimate, errors of 1000 to 2000 yards being common at ranges of 20,000 yards. At the beginning of the war, Type 79 radar operating on 7 metres and Type 81 on 3.5 to 4.0 m were fitted in battleships and cruisers. The GL1 Army radar was added to provide accurate range for fire-control against aircraft. For close ranges, multiple pom-poms fitted with a Type 282 radar with Yagi aerials allowed ranges up to 6000 yards against dive-bomber attacks. For surface fire in capital ships and cruisers, multiple antenna systems were used with a Type 284 radar. It was fitted on the Direction Control tower in HMS King George V and gave ranges on cruisers of 20,000 yards and on destroyers between 12,000 and 14,000 yards.

By the end of 1940, 50 cm radar range-finders were fitted in HMS Suffolk and Radar 281 was responsible for tracking the Bismarck in June 1941.

About this time, a 10 cm set Type 271 had been fitted in Corvettes for anti-U-boat attack and Type 273 in most battleships and cruisers for surface warning.

Beam-switching was adopted to improve the accuracy and for blind-firing the guns at night. This technique could provide tracking in bearing of ships and aircraft with an accuracy of about ±1/2°. Conical scanning was used in a 10 cm high angle set against aircraft. A 3 cm radar set (Type 262) was developed to provide blind-fire with twin Bofors guns. It was designed to pick up a target at 7000 yards and track it in all three co-ordinates before it reached 5000 yards. The first 'out-of-sight' fighting using radar involved the sinking of the Scharnhorst in 1942. Radar echoes could be obtained from shell splashes in the sea, and in a low-angle gunnery set, a special display was provided for the observation of 'fall-of-shot' echoes.

Because the Navy initially developed high-power tubes, such as the silica tubes manufactured at Portsmouth in 1939, responsibility for designing and producing transmitter and all other tubes was allocated to the Navy and the CVD (Committee for Valve Development) co-ordinated the requirements for all three services.

After the war, surface-to-air guided missiles were developed such as SEASLUG using beam-riding guidance. Other marks followed such as SEACAT and SEAWOLF. SEACAT was designed as a close-range missile to deal with low-flying aircraft or missiles and is controlled manually or by TV along a line of sight controlled by radar.

Missiles launched from submarines such as POLARIS have ranges of a thousand or more miles. POLARIS is a two-stage solid propellant missile with an inertial guidance system. It is 'shot' from its launch tube by a small rocket in the submarine; its own first-stage motor fires only as it leaves the water.

6.3 Civil Maritime Systems

Civil maritime radars normally operate at a frequency in the 9 GHz range or in the 3 GHz range with a CRT display giving information on the position of other ships, buoys or coastlines. Harbour surveillance is also carried out by land-sited radars giving a picture of the movement of all ships in or approaching the harbours.

6.4   Civil Aviation

Since the war, development of radar systems had also been applied to civil aviation and used for precision approach (PAR), for airfield control (ACR), for airfield surface movement indication and for air traffic control, both long-range and terminal areas surveillance (TMA). Airborne cloud and collision-warning radars were also developed.

In addition, the IFF system was adopted and known as SSR (secondary surveillance radar). Through the agency of the International Civil Aviation Organization (ICAO), there is an internationally agreed spacing and connotation for the civil system. This enables each aircraft to be identified.

6.5   Navigational Aids

Radio aids to navigation have progressed from early direction finding (D/F) and goniometer systems to satellite navigation systems such as NAVSAT. Following D/F systems, the pre-war Lorenz system in Germany used a dots-and-dashes guidance system similar to that described in OBOE, but using CW signals instead of pulses.

The war-time German Knickebein long-range navigation beams were a long-range version of the Lorenz system, with greater receiver sensitivity.

Inertial navigation depends on a stable platform and high-grade gyroscopes in order to measure acceleration from a known starting point. From an accurate knowledge of acceleration in three planes and also of lapsed time, position can be calculated.

The very long-range navigational system OMEGA is based on VLF CW transmitters radiating signals on a grid system throughout the world. The first positional system was the hyperbolic grid system GEE invented in 1941, followed by the American system LORAN (Long Range Navigation) in 1942. The first commercial system with high accuracy was DECCA introduced in 1946.

The early LORAN A system operated on frequencies close to 2 MHz, and suffered from sky-wave effects. The later LORAN C operated on 100 kHz, and multipath (which arrives late) by tracking only the early cycles of each pulse. The system has recently been enlarged. Satellite navigation (Global Positioning Systems (GPS)) are now in use.

6.6   Meteorological Radar

Meteorological radar echoes were first noticed in February 1941 when a rain shower was tracked to a distance of 7 miles by a 10 cm radar located on the English coast. Since then, rain detection by radar of showers or heavy storms has become a useful tool for the meteorologist.

6.7   Angels

Echoes from birds and insects are obtained at wavelengths of 20 cm or less and in the early years were known as 'Angel Echoes', as no obvious aircraft or solid targets were present.

6.8   Satellite Surveillance Radars

Satellite surveillance radars operate at frequencies between 200 to 5000 MHz; below 200 MHz sky noise occurs and above 5000 MHz atmospheric absorption effects occur.

6.9   Radar Astronomy

Echoes from meteor trails were observed in the early 1940s and a radar echo was obtained from the moon in 1946. In 1961 echoes from Venus were observed at its closest approach to the earth and since then from Mercury, Mars and the Sun.

History on a page
5. Radar

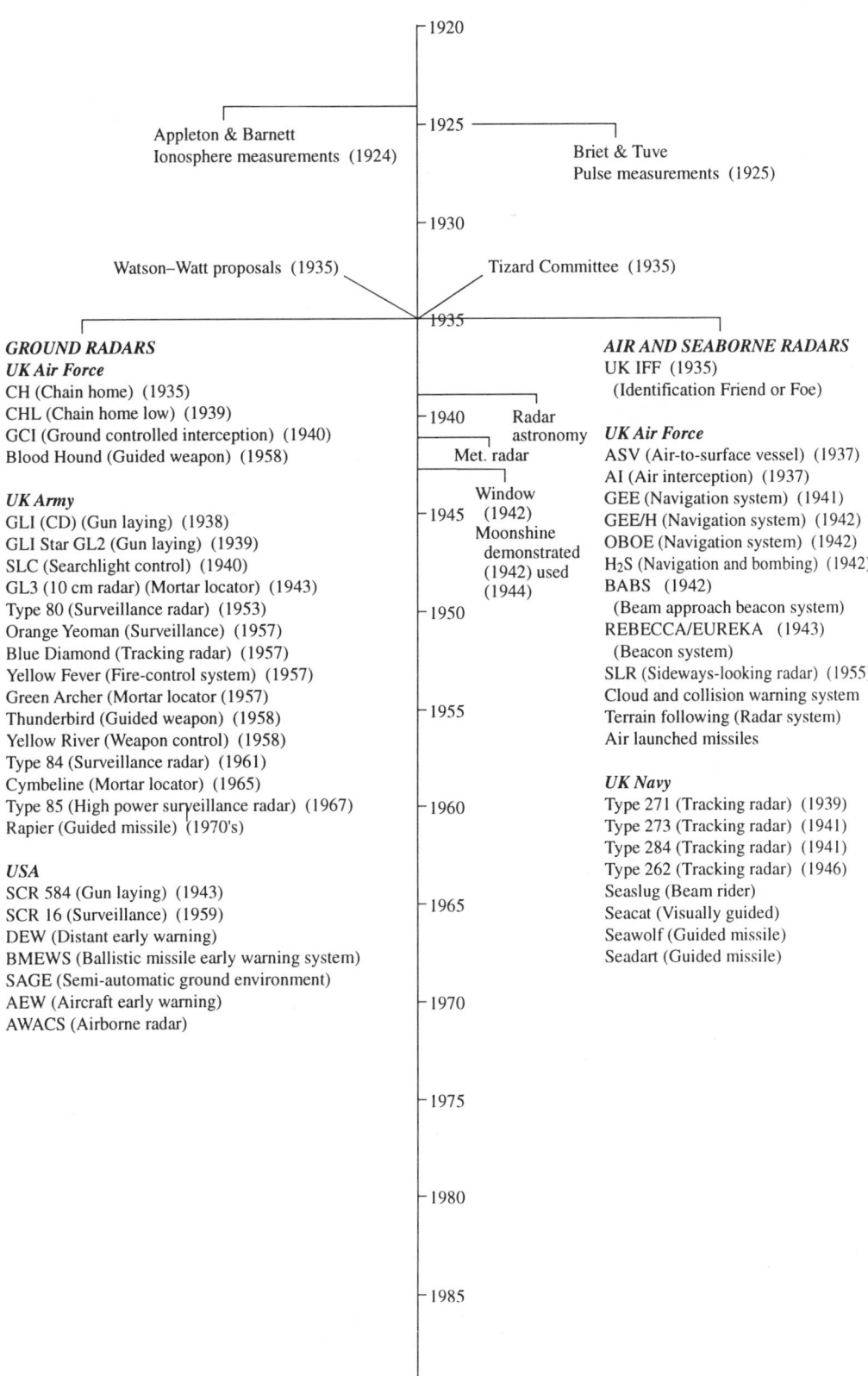

1920

Appleton & Barnett
Ionosphere measurements (1924)

1925

Briet & Tuve
Pulse measurements (1925)

1930

Watson–Watt proposals (1935)

Tizard Committee (1935)

1935

GROUND RADARS
UK Air Force
CH (Chain home) (1935)
CHL (Chain home low) (1939)
GCI (Ground controlled interception) (1940)
Blood Hound (Guided weapon) (1958)

1940 Radar
astronomy
Met. radar

AIR AND SEABORNE RADARS
UK IFF (1935)
  (Identification Friend or Foe)

UK Army
GLI (CD) (Gun laying) (1938)
GLI Star GL2 (Gun laying) (1939)
SLC (Searchlight control) (1940)
GL3 (10 cm radar) (Mortar locator) (1943)
Type 80 (Surveillance radar) (1953)
Orange Yeoman (Surveillance) (1957)
Blue Diamond (Tracking radar) (1957)
Yellow Fever (Fire-control system) (1957)
Green Archer (Mortar locator (1957)
Thunderbird (Guided weapon) (1958)
Yellow River (Weapon control) (1958)
Type 84 (Surveillance radar) (1961)
Cymbeline (Mortar locator) (1965)
Type 85 (High power surveillance radar) (1967)
Rapier (Guided missile) (1970's)

Window
(1942)
1945 Moonshine
demonstrated
(1942) used
(1944)

UK Air Force
ASV (Air-to-surface vessel) (1937)
AI (Air interception) (1937)
GEE (Navigation system) (1941)
GEE/H (Navigation system) (1942)
OBOE (Navigation system) (1942)
H_2S (Navigation and bombing) (1942)
BABS (1942)
  (Beam approach beacon system)
REBECCA/EUREKA (1943)
  (Beacon system)
SLR (Sideways-looking radar) (1955)
Cloud and collision warning system
Terrain following (Radar system)
Air launched missiles

1950

1955

USA
SCR 584 (Gun laying) (1943)
SCR 16 (Surveillance) (1959)
DEW (Distant early warning)
BMEWS (Ballistic missile early warning system)
SAGE (Semi-automatic ground environment)
AEW (Aircraft early warning)
AWACS (Airborne radar)

UK Navy
Type 271 (Tracking radar) (1939)
Type 273 (Tracking radar) (1941)
Type 284 (Tracking radar) (1941)
Type 262 (Tracking radar) (1946)
Seaslug (Beam rider)
Seacat (Visually guided)
Seawolf (Guided missile)
Seadart (Guided missile)

1960

1965

1970

1975

1980

1985

1990

6.10 Radar Imaging

Known now as Surface Penetrating Radar, it is a non-destructive technique using electromagnetic waves to investigate the composition of non-conductive materials under the earth's surface. The technique is now being used to penetrate ground to some depth to discover buried objects. Interferometric techniques are being developed to provide three-dimensional images of target scenes. Aerial survey is used in archaeological studies to discover traces of early settlements. It has been used to detect buried metallic mines, as well as for measuring the thickness of the moon from Apollo space missions. Technical improvements are being made to enable clearer radar images of the internal structure of materials, enabling a wider field of applications to be developed for the future.

Some, but not all, developments in radar and navigational systems are summarised in Date Chart 5 'History on a Page'. The dates given are those when the first radars were demonstrated, not those in operational use.

6.11 Sonar

Prior to the first world war, hydrophones were used for the detection of submarines. They consisted of normal microphones insulated from the water. Because of ship noise, they were often towed behind the ship. SONAR (Sound Navigation and Ranging) was originally known as Asdics (Anti-Submarine Detection) during the 1914/18 war. The word SONAR was invented by F V Hunt of the Harvard Underwater Sound Society in 1942.

In the period between the two world wars, the Royal Navy had an operational asdic system with a quartz transducer housed in a streamlined dome and connected to a chemical range recorder. All systems are based on an electro-acoustic transducer, which converts an electrical signal into an acoustic signal, or converts an electrical signal into an electrical signal. The frequency range for anti-submarine detection by the Royal Navy was 14 to 26 KHz using a 15 inch quartz steel transducer. Initially the transducer was rotated manually, but later scanning transducers were developed. A typical World War II Asdic installation for destroyers is shown in figure 6.4.

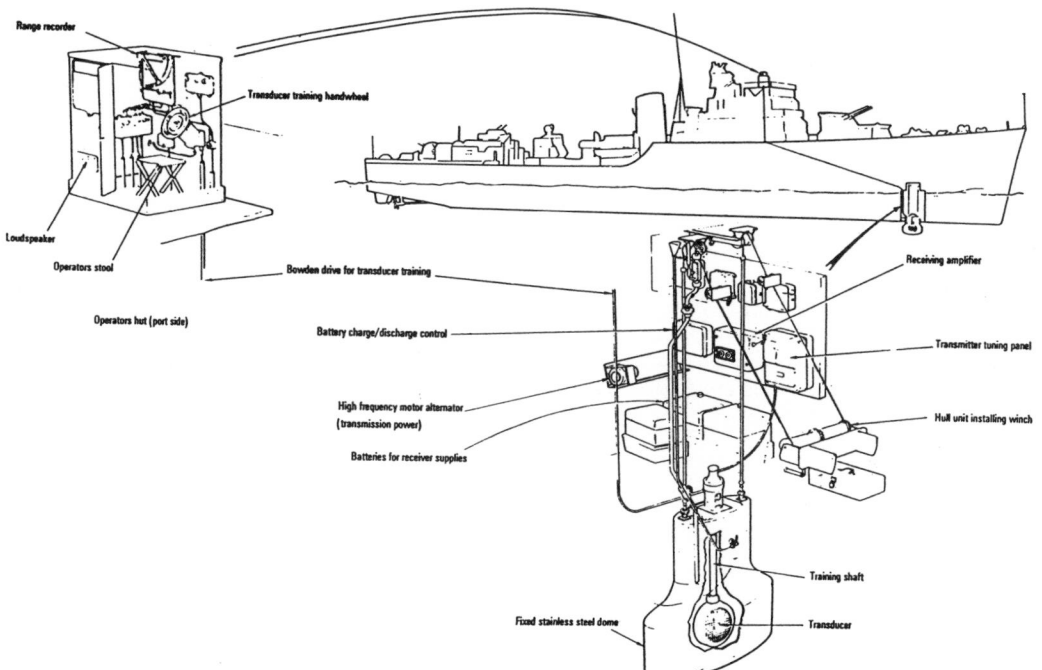

Figure 6.4. Typical World War II Asdic installation.

Three types of Asdics were in use at the beginning of World War II: one for destroyers, one for escort sloops and one for trawlers. Many technical improvements were made as the war progressed. The Type 144 (shown in figure 6.4) was the first attempt at an integrated weapon system with a certain degree of automation, i.e. the transducer pattern was automatically tapered in 5 steps. Automatic recording of the signal was also relayed to the bridge.

In 1919 the average echo range was about 500 yards (457 m); it increased to about 1500 yards (1.2 km) during the second World War and to several miles in 1944. Asdic installations were also used for harbour defence placed on the sea bed and connected by cable to the receiving gear on the shore to give warning of submarine attacks.

Willem Hackman, of the Museum of History and Science, Oxford University and the Science Museum, London, on whose book 'Seek and Strike, SONAR, anti-submarine warfare and the Royal Navy 1914/1954', HMSO (1984), this short summary is based, says 'Of all the techniques tried during the period to take away the submarine's most important weapon, its ability to hide in a mass of water, those based on underwater acoustics were the most effective'.

Since the war Synthetic Aperture radar has been developed by the US Navy to produce pictures of the sea floor with greatly improved detail. The technique creates the illusion of a long aperture by moving the array through the water and then combining the data from several snapshots taken from different positions. Whilst still experimental, it can pick out features measuring 90 cms from a range of up to 400 metres.

Chapter 7

A Concise History of Television

The heart of all modern television receivers is the cathode-ray tube, on which the transmitted moving pictures are built up in colour. It is a far cry from the initial discovery of cathode-rays by Sir William Crookes in 1878 and its development into an oscilloscope by Ferdinand Braun in 1897, to the sophisticated cathode-ray tube in present-day television receivers. The other major development was the 'iconoscope' storage camera tube scanned by an electron beam, originated by Valdimir Zworykin in 1919.

Although A A Campbell-Swinton had proposed a system of electronic television, using cathode-ray tubes for both transmission and reception, it was Philip Farnsworth with Zworykin who first developed a practical system which was demonstrated in 1927. By 1929, television receivers were on sale in the USA, under the name 'Radiovisor' sets, and low-definition television broadcasting was established. The BBC opened the world's first regular high-definition (405 line) television service in 1936 using a system developed by EMI (Electrical & Musical Industries) and, although the service was shut down during the 1939–45 war, it was resumed, in black and white, as soon as the war was over.

The principle on which television operates is that a beam of electrons scans across the screen in horizontal sweeps, whilst at the same time moving down the screen until it reaches the bottom, when one 'frame' or 'field' has been completed. The beam is then deflected back to the top of the screen to start the next frame. The horizontal sweeps are slower than the corresponding fly-back strokes and are synchronized with the incoming picture signals to produce light and dark shades and therefore build up a picture. The faster fly-back strokes are made invisible by blanking out the beam. The picture in contemporary British practice is built up with 625 horizontal lines and in order to save bandwidth, normally several Mhz, but give acceptably low flicker, successive frames are interlaced. The television receiver itself has signal circuits to amplify and detect the incoming signals from the aerial in order to modulate the beam of electrons in the tube. The signal also provides a sound channel and synchronizing pulses for triggering the horizontal and vertical directions is provided by time bases, triggered by the synchronizing pulses, which generate currents in the picture tube yoke coils to deflect the beam of electronics.

The first public television service in colour was transmitted in 1951 in the USA, and it was necessary for it to be compatible with black and white television. The colour system adopted was that specified by the NTSC (National Television System Committee). In Europe, two systems were adopted in 1956; the PAL (Phase Alternation Line) and SECAM (Sequential and Memory). These three systems are now used throughout the world, digital converters being used to convert one system to another.

In colour, the visible spectrum of light starts at violet, passing through blue, green and yellow and orange to red. An approximately equal mixture of all colours gives white light. Any colour can by synthesized by mixing the three primary colours red, green and blue. The varying level of brightness of the picture elements is known as luminance, whilst in colour television when hue and colour saturation information is added, the extra signal required is known as chrominance. The television camera analyses the light from the scene to be transmitted in terms of its red, green and blue constitutes by means of optical filters. The transmitter separates the picture information into two parts, a colour signal and a brightness signal. These are decoded in the television receiver to give a colour display.

Television receivers are one to the greatest technological achievements of the age with their complexity of circuitry, ingenious design, accurate alignment and great reliability. Transistorised television receivers became available in 1959 and the television set has become one of the most important items in the modern home and became more useful with the introduction of TELETEXT systems such as CEEFAX in 1972 and ORACLE in 1973.

The first satellite television pictures were transmitted across the Atlantic on 19 July 1962 by Telstar 1 and, by July 1963, sixteen European countries were exchanging television programmes with the USA. Intelstat 1, the first of the internationally owned satellites, was launched in 1965, and since then a series of television and communications have been in use, relaying television programmes throughout the world. The family of Intelstat satellites is shown in figure 7.1. Recently introduced cable TV offers an increased number of channels, less use of other space and the possibility of interactive communication.

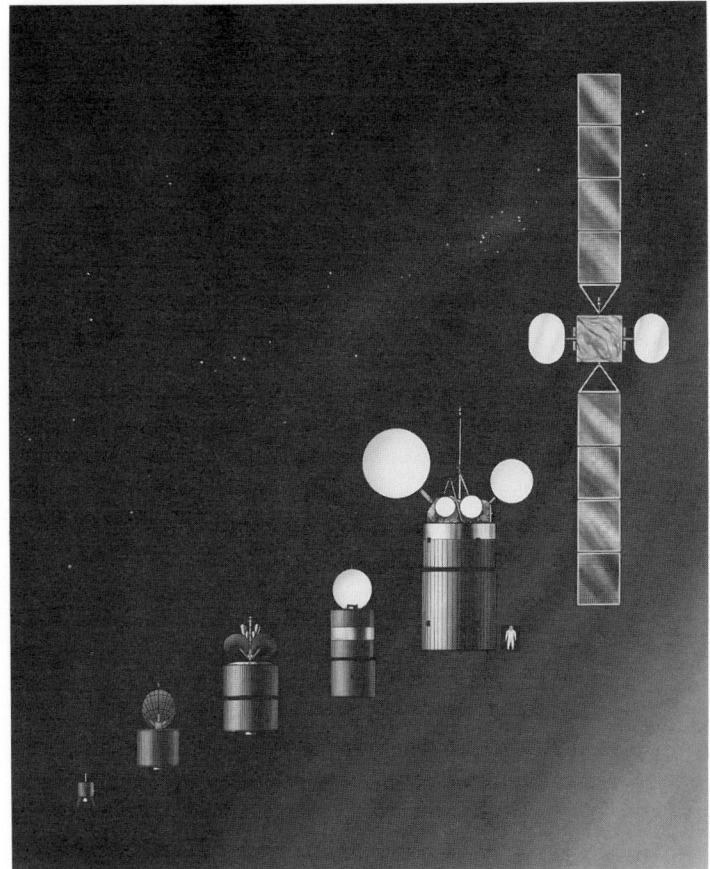

Figure 7.1. The family of Intelstat satellites (courtesy Mark Williamson, Space Technology Consultant, and Hughes Aircraft).

The change from analogue to digital television took place at the end of the 1970s. Digital television derives its origin from the Pulse Code Modulation System (PCM) invented in 1937 by Alex Reeves as a means of encoding signals. With the introduction of semiconductor circuits, the technology of digital storage made possible the transition from analogue to digital television.

The production of special effects for television due to digital technology was now possible, i.e. the original full screen image could be shrunk to occupy a small portion of the screen, moved around and even rotated about a variety of axes. The ability to draw directly on to the 'screen' with an electronic palette, 'cut and paste' parts of images from digital stores, combining drawings with live actors, all gave more artistic freedom to television producers.

In 1984, a 'Cable and Broadcasting Act' was passed to enable cable television to be further developed. Many towns now have cable TV, but it cannot be provided economically in rural areas. In 1977, channels for DBS (Digital Broadcasting Systems) were internationally allocated and are now being used in some systems. However, improvements in receiver sensitivity made it possible to use the channels intended for lower power point-to-point telecommunications services. Many DBS services using the conventional PAL or NTSC standards on the telecommunication frequencies were introduced during the 1980s.

Improvements in technology are always subject to the restriction that, whilst changes in transmission are possible, technical changes which could cause obsolescence of the millions of television sets, can only rarely be introduced. During the 90s, world wide agreement was reached on a means of encoding the digital TV signal using bandwidth compression based on the Discrete Cosine Transform (DCT) algorithm. This led to standards for transmission of digital TV by satellite, cable and terrestrial means. Services using these new digital broadcasting standards are being introduced in the latter half of the 1990s generally with a 'set top box' receiver converting the signals into a form suitable for existing receivers.

The new digital encoding standards made flexible provision for the service provider to choose between conventional definition services, enhanced definition and high definition, all with the option of a new format of 16×9 (compared to the existing 4×3 format) depending on the preference for programmed quantity or picture quality.

Experimental work is being done throughout the world on High Definition Television (HDTV) and substantial archives of HDTV programmes have been built up. It is generally accepted that HDTV should have twice the vertical and twice the horizontal resolution of conventional TV systems using the 16×9 format. Whilst the new digital transmission standards make provision for advanced or higher definition services, domestic displays which do justice to HDTV are still not available at consumer product prices. Thus, for the time being, services are more likely to use the standard or conventional definition, rather than the higher definition options of the digital broadcasting standards.

A summary of television developments is given in Date Chart 6.

History on a page
6. Television

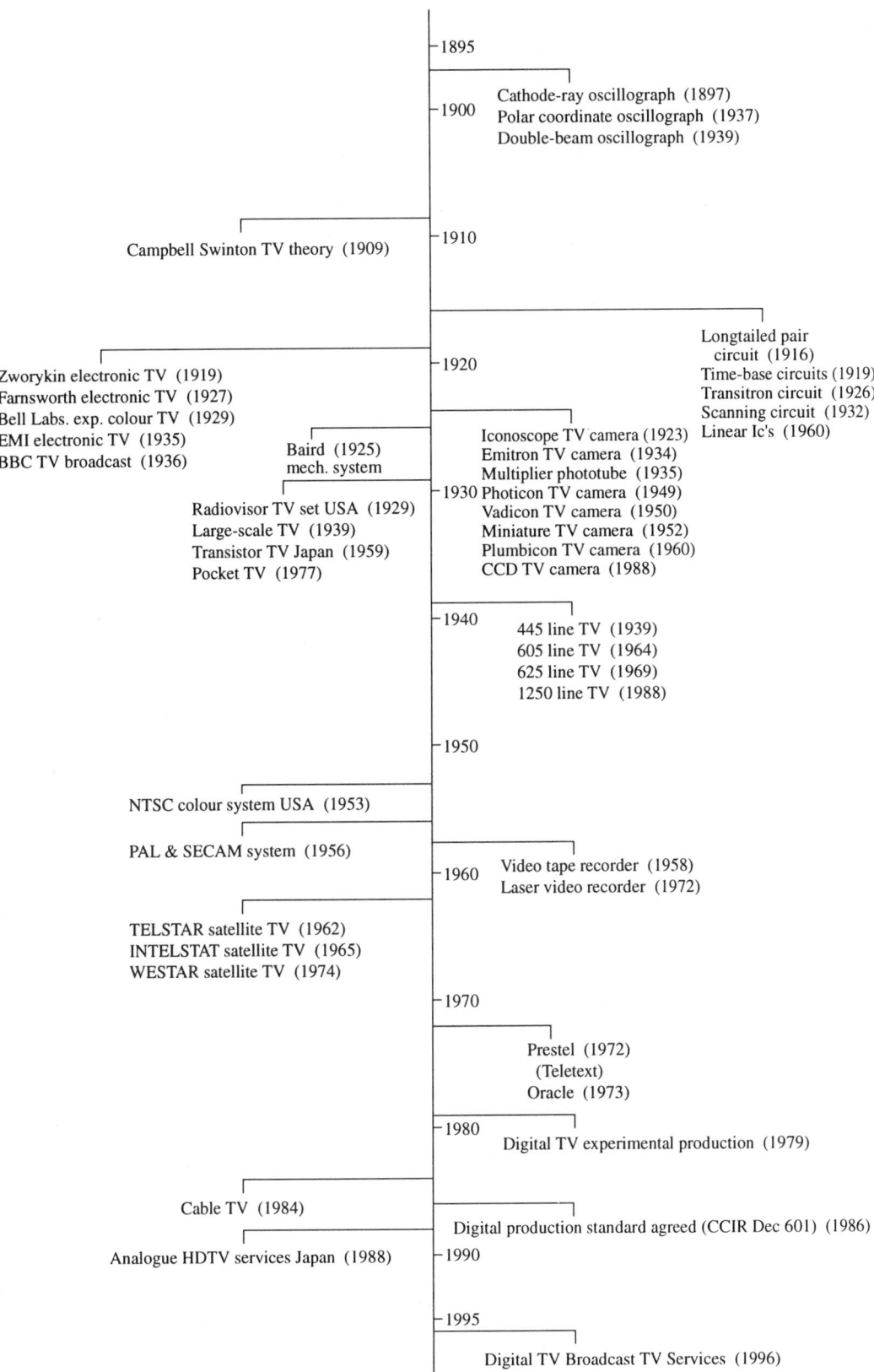

1895

Cathode-ray oscillograph (1897)
Polar coordinate oscillograph (1937)
Double-beam oscillograph (1939)

1900

Campbell Swinton TV theory (1909)

1910

Longtailed pair
circuit (1916)
Time-base circuits (1919)
Transitron circuit (1926)
Scanning circuit (1932)
Linear Ic's (1960)

1920

Zworykin electronic TV (1919)
Farnsworth electronic TV (1927)
Bell Labs. exp. colour TV (1929)
EMI electronic TV (1935)
BBC TV broadcast (1936)

Baird (1925)
mech. system

Iconoscope TV camera (1923)
Emitron TV camera (1934)
Multiplier phototube (1935)
Photicon TV camera (1949)
Vadicon TV camera (1950)
Miniature TV camera (1952)
Plumbicon TV camera (1960)
CCD TV camera (1988)

Radiovisor TV set USA (1929)
Large-scale TV (1939)
Transistor TV Japan (1959)
Pocket TV (1977)

1930

1940

445 line TV (1939)
605 line TV (1964)
625 line TV (1969)
1250 line TV (1988)

1950

NTSC colour system USA (1953)

PAL & SECAM system (1956)

1960

Video tape recorder (1958)
Laser video recorder (1972)

TELSTAR satellite TV (1962)
INTELSTAT satellite TV (1965)
WESTAR satellite TV (1974)

1970

Prestel (1972)
(Teletext)
Oracle (1973)

1980

Digital TV experimental production (1979)

Cable TV (1984)

Digital production standard agreed (CCIR Dec 601) (1986)

Analogue HDTV services Japan (1988)

1990

1995

Digital TV Broadcast TV Services (1996)

Chapter 8

A Concise History of Computers, Robotics, Mechatronics and Information Technology

There is almost a 300 year gap between the invention of the first mechanical computer and the invention of the first electronic computer. In 1642, Blaise Pascal, in France, who was 19 at the time, grew tired of adding long columns of figures in his father's tax office and he designed a mechanical device consisting of a series of numbered wheels with gears for decimal reckoning, capable of adding and subtracting the long columns of figures. Thirty years later, a German, Gothfried Wilhelm Leibniz, invented the Leibniz wheel using similar principles which could not only do subtraction and addition, but also multiplication and division.

Almost a hundred years passed before Sir Charles Babbage conceived the first design for a universal automatic calculator. Again, a mechanical device using counting wheels, coping with 1000 words of 50 digits each, but with one vital difference: he used punched cards similar to those used in a Jacquard loom to control the programme. Punched cards were also used as input and output devices. The machine contained all the functions necessary in a modern computer—an input unit, a store or memory, an arithmetic unit, a control unit and an output unit.

Improvements were made by Pehr Georg Schuetz in Sweden and a machine similar to Babbage's was built in 1854, which was capable of printing out its own tables. Almost forty years elapsed before H Hollerith, in America, developed a machine for tabulating population statistics for the 1890 census. Holes in punched cards were used to denote age, sex, etc., and the size of the cards were made the size of a dollar bill. Another forty years later, Vannevar Bush in the USA, developed an early analogue computer for solving differential equations, and analogue computers were built by several universities (e.g. Manchester University in the UK (1934)).

The stage was now set for the first of a long line of electronic digital computers, although initially electromechanical rather than purely electronic. Howard H Aiken of Harvard University designed in 1937 an electromechanical automatic sequence-controlled calculator which was built by IBM and presented to Harvard 7 years later. A relay-operated computer was built by Stibitz, of Bell Laboratories, about the same time.

The first truly electronic computer was the ENIAC (Electronic Numerator Integrator and Computer) begun in 1942 by the University of Pennsylvania and completed in 1946. It used 18 000 tubes, mostly flip-flops and pentode gates, and was 51 feet long and 8 feet high. The numbers used in this machine had 10 decimal units and the numbers could be added in 200 microseconds and multiplied in 2300 microseconds, this being the fastest calculator developed up to this time. Computer memories progressed from tubes, mercury delay lines, CRT stores, to magnetic cores, tapes, drums and discs. From the middle 1940s, a series of computers were built each using later electronic techniques as they were developed, i.e. tubes to transistors, transistors to integrated circuits, each becoming smaller and smaller, until the present range of microcomputers was reached.

8.1 Computer Systems

A modern computer comprises two basic parts—the hardware and the software. The hardware is the physical manifestation and comprises the computer and all its components and peripherals. This includes the monitor, keyboards, mouse, printer, etc. The software are the 'instructions' to control the functions of the computer. Software is generally stored on magnetic discs, CDs or tapes. Software can be subdivided into 'control' or 'operating system' software which, in some form or other is present in all computers' and application software, which is specific to a particular task, e.g. word-processing. In small modern PCs, the operating system may be 'DOS' (Disk Operating System) or 'Windows'. Software is not inexpensive.

All digital computers operate by adding, subtracting, multiplying and dividing numbers at incredibly high speeds. The architecture or arrangement of a basic computer consists of an input device to feed in the data, a device to hold temporary results until they are required (an accumulator), an arithmetic and logic unit to perform calculations, a memory to hold data and program as required, and an output device to display or print data. A control unit is used to decode instructions from a predetermined schedule known as a 'program' to control the arithmetic and memory processes.

Calculations are normally effected by using binary arithmetic. Two numbers only are used 0 and 1 in binary instead of 0–10 in the decimal system. In the decimal system, a number such as 469 is made up to successive powers of 10. It could equally be written $(9 \times 10^0) + (6 \times 10^1) + (4 \times 10^2) = 9 + 60 + 400 = 469$. Binary numbers are made up of successive powers of 2 and can be compared with their decimal equivalents as follows:

Binary	2^0	2^1	2^2	2^3	2^4	2^5	2^6
Decimal	1	2	4	8	16	32	64

therefore 469 would be 111010101.

The operations carried out by most computers are basically as shown in figure 8.1.

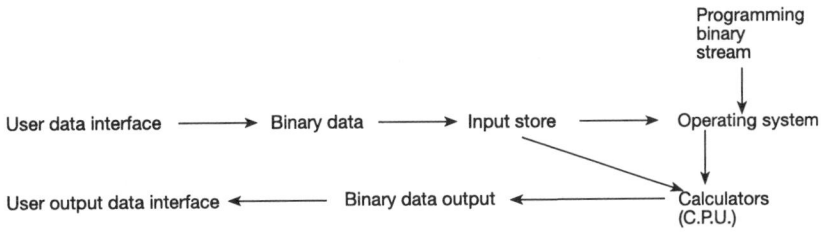

Figure 8.1. Stages in computer operation.

A single binary digit is known as a BIT (Binary Digit). Groups of 8 bits are known as 'Bytes'. The binary equivalent of a word or working length of the accumulator is typically between 1 and 8 bytes depending on the computer.

The use of the binary code greatly simplifies the design of electronic computers, for basic circuits having two states only: 0 and 1 can be used throughout. Linearity and ageing are not problems.

Transistors have two properties which make them ideal for computers. They are able to store charges and release them when required, i.e. a memory, and they can also act as switches for manipulating codes.

Because the computer operates in binary code, a programming language is necessary to convert words or instructions into suitable binary codes. Initial low-level language used simple mnemonics such as SUB for SUBTRACT; with these, every action of the computer has to be described in detail, but with modern high-level languages, a simple instruction can generate complicated actions. To make sure that instructions are presented in a logical order a flow chart is used sometimes to specify the program. Programming languages such as FORTRAN (FORmula TRANslation) for scientific use, COBOL (COmmon Business Oriented Language) for commercial use and a BASIC (Beginners All-purpose Symbolic Instruction Code) for teaching, were all developed to make the work of the programmer easier.

When a computer is switched on, a pulse generator produces a stream of millions of pulses per second known as clock pulses, for timing. Pulses representing BITS of information are operated upon throughout the computer by gates which permit or inhibit the passage of the pulses being distributed through the various stages of calculations. In operation, instructions from a programmer are fed into the computer by means of compact discs or floppy discs in conjunction with a keyboard. These are then stored in the form of binary numbers in rows of locations. The arithmetic unit then multiplies, adds, subtracts, etc., as dictated by the processor, stores the intermediate answers and finally after processing decodes the binary digits and feeds to a printer or VDU (Visual Display Unit) in the form of printed words and numbers.

Great advances have since been made in the storage or memory system of a computer from the earlier systems such as magnetic tapes, cores and drums. Disc storage is used to give high access speed to a number of locations. Compact discs use a laser spot focused on a reflective track modulated by a pattern of indentations on the disc. Digital Compact Discs, invented in 1978 by Philips, are unaffected by dust, scratches and fingerprints and are now widely used because of their high storage capability.

The integrated circuit memory chip consists of many thousands of transistors. It is able to store and extract a given pulse or bit when required. The chip is composed of a number of cells and a chip with 256 individual cells, with each cell containing a byte (a number composed of eight bits), would have eight input lines to tell which cell was being referenced and eight separate lines for the data. ROM (Read Only Memory) and RAM (Random Access Memory) chips have a facility on the chip to deliver their stored programs when requested. In the case of the ROM, this data is fixed, whereas the RAM chip can be instructed to remember patterns (i.e. the ROM dictates, the RAM remembers).

In 1970, the first Random Access Memory (RAM) was produced by Fairchild (the 256 BIT RAM), followed in 1972 with the 1024 BIT RAM by the new Company INTEL. Also produced in the same year by INTEL was the first microprocessor (the 8008). Progress in chip design was rapid and, in 1975, the first 4096 BIT RAM was produced; and, in 1976, the 16 384 BIT RAM was produced by INTEL. Also, in 1976, the first one-board computer was made by INTEL.

About this time, chips were being built into personal computers and, in 1977, the APPLE 2 was introduced. In 1981, IBM entered the personal computer market with the 'PC' and, in 1984, IBM developed a one million BIT RAM. In 1987, the PC MACKINTOSH and the IBM PERSONAL SYSTEM 2 was introduced. Comprehensive, inexpensive operating systems such as Microsoft Disc Operating System (MS-DOS) became available at this time. Supercomputers, introduced in 1976, were improved each year. In 1978, the CRAY Y-MP was capable of performing 2000 million operations per second.

8.2   Types of Computer

At the present moment, there are four general categories of computer which vary widely in cost and performance.

The Supercomputer. In the first and most expensive category are the 'super' computers such as the CRAY, where not only is computation done at very fast speeds, but many computation processes occur in parallel—hence enabling a high throughput of work.

These computers are used for tasks where an extremely large number of mathematical equations need to be solved numerically in a reasonably short time, often involving the simultaneous input of large volumes of data; such a task is weather forecasting.

The Main Frame. In the second category are the so-called 'main frame' computers; examples of these are the IBM 3033, Univac 90/30 and ICL 2900. In these machines computations are performed fast but with little parallelism. These machines are used for large payrolls and other accounting tasks, factory management, financial planning and for scientific and design problems where a large number of equations need to be solved.

The Minicomputer. In the third category are the minicomputers; calculations in these machines are still reasonably fast: however, the basic word length is short, typically 32 bits so that the accuracy of calculation for some purposes has to be improved by using several programming steps. These machines are used in the main for tasks of moderate complexity in the accounting, scientific and computer-aided design fields. They are also used for controlling large chemical plants, steel rolling mills and other complex continuous processes. Examples of these machines are VAX 750, GEC 4080 and HP 1000.

The Microcomputer (usually referred to as the PC). In the fourth category are the microcomputers, the central processing unit of which is usually on a single chip. These computers have a wide range of applications, and today both computers and microprocessors are widely distributed throughout businesses. Microprocessors are the heart of home video games, and are used as the main processing element in many automated machines such as banking cash dispensers, store checkout tills and industrial robots.

Computer Aided Design (CAD). Computer Aided Design is using computers to aid the designer to do his job efficiently; for example, in the electronics industry, computers are used to analyse the parameters of proposed circuits, and translate that design on to a printed circuit board including component layout and interconnecting tracks. Computers are used extensively in the design of integrated circuits and hence used to design themselves.

History on a page
7. Computers

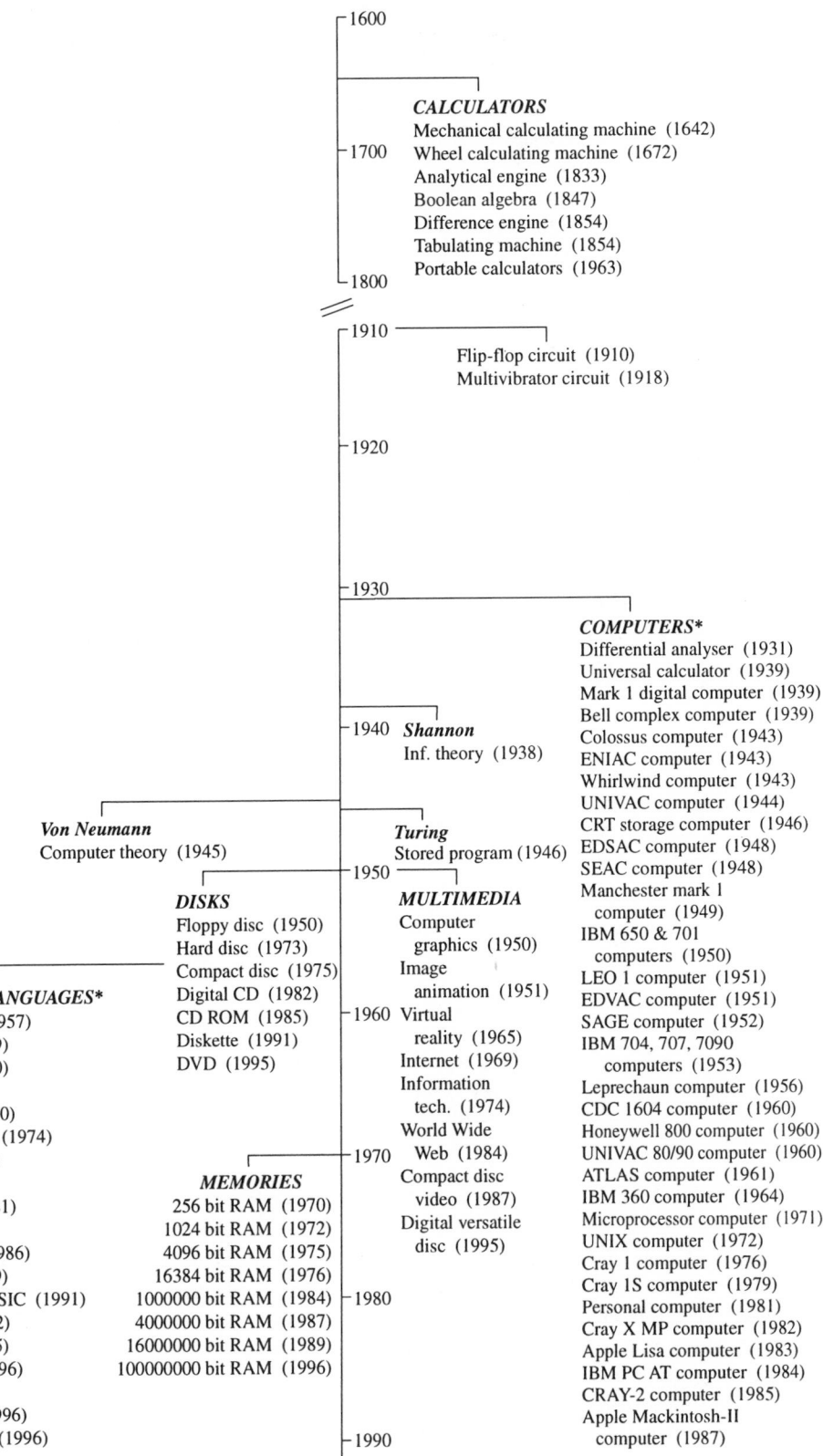

— 1600

CALCULATORS
Mechanical calculating machine (1642)
Wheel calculating machine (1672)
Analytical engine (1833)
Boolean algebra (1847)
Difference engine (1854)
Tabulating machine (1854)
Portable calculators (1963)

— 1700

— 1800

— 1910

Flip-flop circuit (1910)
Multivibrator circuit (1918)

— 1920

— 1930

COMPUTERS*
Differential analyser (1931)
Universal calculator (1939)
Mark 1 digital computer (1939)
Bell complex computer (1939)
Colossus computer (1943)
ENIAC computer (1943)
Whirlwind computer (1943)
UNIVAC computer (1944)
CRT storage computer (1946)
EDSAC computer (1948)
SEAC computer (1948)
Manchester mark 1
   computer (1949)
IBM 650 & 701
   computers (1950)
LEO 1 computer (1951)
EDVAC computer (1951)
SAGE computer (1952)
IBM 704, 707, 7090
   computers (1953)
Leprechaun computer (1956)
CDC 1604 computer (1960)
Honeywell 800 computer (1960)
UNIVAC 80/90 computer (1960)
ATLAS computer (1961)
IBM 360 computer (1964)
Microprocessor computer (1971)
UNIX computer (1972)
Cray 1 computer (1976)
Cray 1S computer (1979)
Personal computer (1981)
Cray X MP computer (1982)
Apple Lisa computer (1983)
IBM PC AT computer (1984)
CRAY-2 computer (1985)
Apple Mackintosh-II
   computer (1987)

— 1940 **Shannon**
         Inf. theory (1938)

Von Neumann
Computer theory (1945)

Turing
Stored program (1946)

— 1950

DISKS
Floppy disc (1950)
Hard disc (1973)
Compact disc (1975)
Digital CD (1982)
CD ROM (1985)
Diskette (1991)
DVD (1995)

MULTIMEDIA
Computer
   graphics (1950)
Image
   animation (1951)
— 1960 Virtual
   reality (1965)
Internet (1969)
Information
   tech. (1974)
World Wide
   Web (1984)
Compact disc
   video (1987)
Digital versatile
   disc (1995)

PROGRAM LANGUAGES*
FORTRAN (1957)
COBOL (1959)
ALGOL (1960)
APL (1962)
PASCAL (1970)
MICROSOFT (1974)
BASIC (1975)
ADA (1979)
MS-DOS (1981)
LISP (1981)
NETTALK (1986)
ANSI-C (1989)
VIRTUAL BASIC (1991)
DYLAN (1992)
ADA 95 (1995)
ANSIC++ (1996)
JAVA (1996)
JAKARTA (1996)
KRAKATOA (1996)

— 1970

MEMORIES
256 bit RAM (1970)
1024 bit RAM (1972)
4096 bit RAM (1975)
16384 bit RAM (1976)
1000000 bit RAM (1984)
4000000 bit RAM (1987)
16000000 bit RAM (1989)
100000000 bit RAM (1996)

— 1980

— 1990

* As there is much controversy on first dates, these dates must be regarded as approximate.

For mechanical design, computers are used as drafting aids and for mathematically representing solid objects allowing such properties as mass, centre of gravity, etc, to be evaluated. They can also be used to analyse stresses in components and assemblies; for example, bridges, aircraft components and structures. Solid modelling techniques also allow real-looking computer pictures to be produced which can be used for marketing activities before a prototype has been produced.

Computer Aided Manufacture (CAM). Computer Aided Manufacture is where computers are used to aid the manufacturing process; for example, the automatic programming of numerically controlled machines, such as lathes, two- or three-dimensional milling and drilling machines. Programmes are also produced for printed circuit manufacture, mask production, drilling, assembly, etc. Computers can also produce manufacturing schedules, assembly instructions and work plans for processes to be performed manually.

The Date Chart 7 ('History on a Page') has been compiled as a useful quick reference on major developments in computers.

8.3   Robotics

Robotics has been the subject of science fiction for many years, e.g. the book and film 'RUR' by Karel Capek, but it is only recently that the microcomputer has enabled more sophisticated devices to be made. Servo systems were an early form of what is now termed robotics, e.g. power-assisted steering in cars, servo controls in aircraft and in industrial processes. The first robot to imitate the grasping motion of the human hand was used in a US nuclear plant in 1960 (known as 'Handyman'). Definitions of robots vary considerably, but a general description is 'a reprogrammable multifunction manipulator designed to move materials, parts, tools or specialised devices through variably programmed motions in space for the performance of a variety of tasks'. The popular conception of an industrial robot is an articulated arm with a total of 5 or 6 degrees of freedom achieved through a series of revolute and/or sliding joints. The position of these joints is controlled by a microcomputer which is capable of being variably programmed by the user of the robotic equipment. A typical arm made by Unimation (Europe) Ltd is shown in figure 8.2.

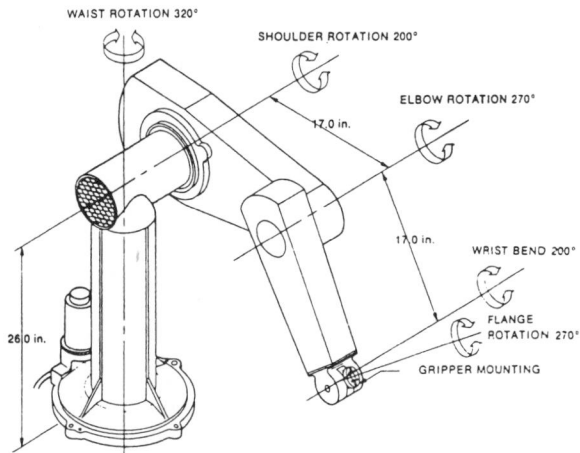

Figure 8.2. A typical robotic arm.

Each robotic application requires its own 'endeffector' at the end of the arm, usually in the form of gripper, spray gun or welding torch; this device is also controlled by the computer from a user-defined programme. The robotics controller also has to be able to control the processes being performed by the robot, i.e. control of welding current or paint flow. The arm power source for the larger robots, intended for the heavier tasks, is frequently hydraulic, with a few notable devices being electrically powered. However, robots intended for lighter tasks, such as part sorting and assembly, are more commonly electrically powered and are rapidly becoming the preferred device.

An important part of the modern industrial robot is the software associated with the control computer. This software consists of a suite of system programmes that enable the robot user to teach the required endeffector paths, operations and the control of other external equipment such as welding plant, numerically controlled machines, etc, as a series of recorded steps in the computer's memory. The computer can then move the endeffector through these recorded paths, initiating endeffector and external equipment operations at the required points of robot articulation. The method of teaching the robot varies considerably with the intended application; for example, teaching the current generation of paint-spraying robots entails moving the robot arm through the required spraying motions at the proper speed, the computer recording these movements to be played back during operation. On the other hand, pick and place robots are taught the positions where operations have to occur, usually with a few intermediate points to ensure collision free operation, leaving the robot to derive the path between these points. The more sophisticated robots have software refinements that allow the arm to move from point to point along pre-defined straight or curved paths, possibly with a change in attitude of the endeffector; the simpler machines use joint interpolation (i.e. driving the joints at constant velocity) to move from point to point, the actual trajectory of the end-effector path depending on the robot's geometry.

The load-handling capacity, overall size and reach of these robots varies considerably. One of the largest robots is the Cincinnati Milacron: this robot can handle loads of 102 kg and position them in space with the required orientation, to within less than 1.25 mm and has a total reach of 2.5 m; this machine has a hydraulic powered source. One of the smaller machines is the Unimation PUMA 250 which can only handle 1.5 kg, and has a reach of just 0.45 m with positional repeatability of better than 0.05 mm, this machine being electrically powered. The operational speed of the industrial robot (the maximum controllable velocity of the endeffector at the end of the arm) varies considerably with the task they are designed to do. For example, a fast pick and place robot can achieve speeds of up to 1.5 metres per second, whereas the welding robot will weld at a maximum of 1 metre per minute; between welds, a faster speed is, of course, required.

Walking robots, such as that used on the moon, have been developed. Attempts were first made to produce an exact simulation of human leg movements, but the main problem was keeping an effective balance. In 1977, a six-legged walker was designed at the Ohio State University which could climb over obstacles and up shallow stairs. A one-legged robot was built in 1983 by Carnegie-Mellon University to study the problem of balance.

A recent development in which a Canadian company will design and build two flight robotic workstations that include display and control functions will enable astronauts to operate the Space Station's robotics (known as the Mobile Servicing System); one robot will be the Space Station Remote Manipulation System and the other the Special Purpose Dextrous Manipulator. They will be used in the assembly and maintenance of the Station.

Many attempts have been made to reproduce a programmed robot which can carry out domestic tasks, but without success. In the main, industrial robots are used for such tasks as loading, unloading, inspection, maintenance, welding, painting and precision manufacturing.

Artificial Intelligence (AI) is thought to be a key factor in the future of robotics. AI is the use of computers in such a way that they perform functions normally associated with human intelligence, such as learning, adapting, self-correction and decision taking.

There is considerable controversy on whether, or how soon, AI will become commercially viable. Present work focuses on game-playing problem solving, the use of natural language in controlling robotics and the 'expert systems' of computer programming. Expert systems may help in the specialised intelligence used by human experts, such as a doctor diagnosing a disease.

The technological revolution brought about by the increasing use of robots, brings the problem of increasing unemployment, when factories become automated. This is a problem for society to solve.

8.4 Mechatronics

Just as the integrated circuit in electronics brings together transistors, passive components, heat dissipation, circuit design reliability, etc, in one overall function, 'mechatronics' applies the same principle to mechanical systems. Mechatronics brings together mechanical engineering, electronic engineering, electrical engineering, control engineering and computer technology, covering a wide range of manufacturing products and processes. It involves sensors, drive and actuation systems, control and measurement systems, behaviour analysis and computer technology, e.g. microprocessors.

Examples of mechatronics can be seen in the design of cars, washing machines, automatic cameras, machine tools, robots and many others. A recent example is the use of mechatronics in surgery—using automated tools for orthopaedic applications; mechatronic techniques for minimal access surgery, and mechatronic methods for tool guidance, registrations and control of processes during surgery. All these techniques have to be incorporated as early as possible in order to design more flexible and cheaper systems for production.

8.5   Information Technology

The definition of Information Technology (IT) is generally assumed to be: the use of technology to provide the capture, storage, retrieval, analysis and communication of information, whether in the form of data, text, image or voice.

With the invention and exploitation of the integrated circuit or 'chip' since the 1960s, the growth of applications using electronics has been phenomenal. The modern electronic computer can process data, graphics and speech at extremely fast rates. The microprocessor is at the heart of what is known as the IT revolution. In the hand it controls washing machines, cookers, televisions, telephones, home computers, cameras, video games, digital watches and many other devices.

Offices and factories now use microprocessors in their everyday life, as do cars, fax machines, aircraft flight control, railway signalling, police computer databases, including the Armed Forces with guided missiles, battle control systems, submarine control and countless others.

The effect of this IT revolution has been to transform labour-intensive work, such as mining, agriculture, iron, steel and cotton industries, hardware manufacturing, etc, into an industry where a few highly-skilled workers manage large factories with mainly automated labour. The manufacturing labour force has largely transferred to the Service Industries, whilst improved methods of rapid communications have enabled home-working to become increasingly possible.

The influence of the Multimedia is part of the IT revolution. The change from analogue to digital television enabled special effects to be developed, such as the original full screen television image which could be shrunk to occupy a small portion of the screen, moved around and rotated about a variety of axes. Combining drawings with live actors, all give artistic freedom to television producers.

Compact discs can record complete encyclopaedias, as well as providing sound and pictures.

The development of IT in its various forms has meant a major change in the working lives of the population, with its social implications and it has truly been named the IT revolution.

Chapter 9

A Concise History of Industrial, Automobile, Medical, Educational, Office, Banking, Consumer and Security Electronics

9.1 Industrial Electronics

The earliest general use of electronics in industry was for control of processes, such as induction heating for the hardening of steels and setting of glues, electroplating, resistance welding for the joining of metals and ultrasonics for cleaning castings, crack testing in metals, drilling glass, etc, Photocells were used for a wide variety of devices, e.g. the automatic opening of doors, smoke-density control, sound on film cine, sorting small objects by colour, facsimile transmission of photographs and many others.

Process control needs three functions—a transducer to convert changes in pressure, flow, temperature, colours, or levels into electrical voltages, a signal process or amplifier and a control mechanism to open and close valves, regulate power, etc, so that the required equilibrium is obtained. A feedback system is usual to regulate the amount of control necessary.

With the introduction of the transistor and later the integrated circuit, process control expanded and, with the development of computers, particularly minicomputers and microcomputers, a new field of control was opened up. Large masses of data could be accommodated by the computer and assessments could be made of the effects of possible changes in production before actual decisions were made. The computers can be used to control production processes with great accuracy and there is little limit to the type of industry which can now use electronics in one form or another. The first computer built for business use, calculated taxes, payrolls, and even helped with tea blends for J Lyons & Co., a victuals purveyor in England. Lyons modelled the room-size system on Maurice Wilke's EDSAC. LEO (Lyons' Electronic Office) stayed in service for fourteen years, computing employee paychecks three hundred times faster than a human clerk.

Signal processors enable all computers to cope with process variations required in controlling production outputs. Among the industrial applications are the metal industry, the mining, glass, pulp and paper, machine tool, cement, textile, rubber, plastics and chemical industries, in addition to aircraft and space applications, land and marine transport applications. Mini or microcomputers are used in automatic machines and systems on the factory floor to carry out many different processes. For example, industrial robots are used for paint spraying, spot welding, handling dangerous and difficult material such as hot glass and making cores for casting processes, fettling, etc. Computers are also used to control automatic test equipment which tests the electronic components coming into the factory and the completed products, and are able to isolate and report the type and position of a fault. Another important area of use is the transfer of parts between automatic machines in the factory for automatic warehousing.

9.2 Numerically Controlled Machines

The so-called numerically controlled (NC) machines are a special group of industrial machines controlled by electronics, usually a microcomputer, which receives commands in the form of a sequence of numbers. The computer then reads the input sequence of commands, one item at a time, then uses interpolation (the form of which depends upon the application) on the positional commands to precisely control the velocities of each of the axes to achieve the required tool path. Some of the commands in the sequence will, of course, tell the computer to change the cutting speed, the tool type, shaft speed, etc. Examples of machines having this type of control are NC lathes which normally have two axes, NC milling machines which have three, four or possibly five axes and automatic printed circuit drilling and routing machines which are three-axis devices. Numerically controlled machines come in many different forms: lathes, mills, drills, routers, etc.

The control sequence or programme for these machines can be created by a variety of methods: (*a*) the NC machine itself can be used to create the programme sequence by moving the cutter to the required positions, recording these and the type of interpolation to be used between this and the next point; (*b*) the sequence can be programmed off-line by a production planner, using a human readable form of programme which is later translated by the computer into a numerical sequence; (*c*) the sequence can be generated directly from a computer-aided design (CAD) system.

In induction heating, the design of high-power supplies has been influenced by the availability of fast-switching high-power devices, such as insulated gate bi-polar transistors, and metal oxide semiconductor field-effect transistors.

Machine vision systems in which optical sensing is used to control machine output, are increasingly being used. Efforts are being made to apply neural networks to such tasks as pattern recognition and the control of robot welding systems.

Whilst the production of heavy units is still necessary, production techniques are turning to the automatic computer controlled assembly of integrated circuits, for which there is an enormous demand. Machine vision systems are being used in today's production systems. The machine vision industry requires the manufacturer to pinpoint an object, e.g. a component on a printed circuit board, with extreme accuracy (of about 1 pixel). Image processing rates for this are now in billions of operations per second; the resultant assembly speeds are now considerably higher than would be thought possible a few years ago.

9.3 Automobile Electronics

The first electronic device to be fitted in cars was the car radio, first mass-produced by Motorola in 1930, but the greatest impact of electronics was the use of the electronic computer to control the car's functions. Microprocessors, in conjunction with specialized sensors, today control automatic gear shifting, monitor fuel consumption, ignition timing and duration, instrumentation, warning devices, airconditioning, anti-lock braking (ABS), suspension characteristics, air bags and car security systems. These are all examples of the influence of modern electronics.

In the USA, there are up to 12 microcomputers on many of the luxury cars, which control ride, handling, transmission and air conditioning, in addition to engine management and control to meet stringent emission requirements. There are body control computers, as well as electronic control of lighting, door locks and window lift. Cars now include facia liquid crystal displays of fuel level, water temperature, battery voltage, oil pressure, lights, etc.

Engine analysers used in service workshops electronically measure such characteristics as r.p.m., timing, ignition, cylinder leakage, combusion efficiency, dwell angle and many other engine-related parameters.

Using navigational satellites, such as the GPS (Global Positioning System), it is now possible for a car to locate itself, using the dashboard video map, with an accuracy of a few metres. The driver enters his destination into the car's navigational computer; the computer then chooses the fastest route, taking into account current traffic conditions and guides the driver with verbal instructions over the car's sound system and through the car's video display screen. Such systems are now available as standard on some upmarket vehicles in Japan.

The Irradium satellite network to be launched in 1997 will comprise a network of 66 satellites to give uninterrupted coverage of the globe for personal communication purposes. This will obviously revolutionize mobile voice and data communications.

There is a trend to multivoltage systems—the standard 12 volt battery may give way to other voltages to improve efficiency. For instance, it may be advantageous to supply the electronics at 5 volts DC; lighting at 6 volts AC; active suspension at 350 volts DC; and motors and activators at 42 volts DC.

Matching the electrical and electronic supplies to the car's requirements with the highest efficiency will be a requirement in the near future.

Electric cars are also under development in which electric motors are powered by fuel cells which generate electricity by combining hydrogen with oxygen from the air. It is estimated that 5% of California's vehicle fleet will be electric by the year 2000. There are, however, many technical problems to be solved before the electric car replaces the present system.

The trends in automotive electronics are illustrated in figure 9.1, reproduced by permission of Newnes/Butterworth/Heinemann, from the book *Automobile Electronics* by Eric Chowanietz.

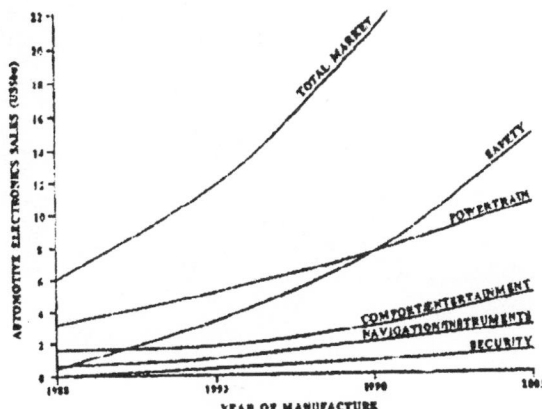

Figure 9.1. Trends in automotive electronics.

9.4 Medical Electronics

Heart sound amplifiers have been used for many years, together with cardiograph displays of heartbeat waveforms. Devices can now sense when a spontaneous beat arises and take appropriate action, e.g. by inhibiting their output stimulus so as not to compete with the natural beat. The electroencephalograph (EEG) displays brain rhythms and computers analyse their waveforms. The electrocardiograph records the heart's beats. Since the introduction of the first cardiac pacemaker in the early 1900s, great strides have been made in their use. Figure 9.2 shows a portable electrocardiograph (ECG) operating in 1947. The ECG detects and records the tiny electrical signals that co-ordinate the heart's beats and which can indicate the heart's disorders. The signals 'ripple' outwards and are picked up by metal sensors stuck to the skin. William Einthoven (1860–1927) developed the ECG during the 1900s. Portable ECGs date from around 1928. One of the first implantable pacemakers used in 1962 is shown in figure 9.3.

Defibrillators, providing natural counter shocks to the heart and electrocardiographs observing blood circulation and functional heart problems, are in use. Intensive-care patient monitoring systems are standard in most hospitals and closed-circuit TV systems are also used. Other electronic aids are laser retinal photocoagulators, radio pills, ultrasonic echo analysis and hearing aids. Fibre optics are used to illustrate areas normally inaccessible. Pressure-sensitive radio pills are used for measuring conditions in the gastrointestinal tract. A major electronic technique is non-invasive diagnosis by computerized scanning, using nuclear, ultrasonic, fluoroscopic and x-ray equipment. MRI (Magnetic Resonance Imaging) is a medical imaging technique that relies on the response of hydrogen atoms to a magnetic field to distinguish between various types of soft tissue. Computerized axial tomography can provide 'slices' of patients anatomy and yield valuable diagnosis, enabling physicians to visualize anatomic structures of live patients. Figure 9.4 shows the first to be made and the experimental model with which the earliest trials of patients were undertaken at Atkinson Horley's Hospital in 1970–71. Techniques developed with this machine established computed tomography (CT scanning) as a key tool for studying the brain.

Telemedicine systems are used for transmitting data from one hospital to another and for telesurgery. Spare parts of the human anatomy are increasingly being used with implants, such as cardiac valves, balloon angioplasty, digital hearing aids, hip joints, pacemakers, and many others. Computerized electrically driven artificial limbs are now possible. Virtual reality is being increasingly used in three areas: virtual humans for training; the fusion of virtual humans with real humans in performing surgery; the virtual telemedicine shared decision environments for training of physicians, nurses and other professionals. The electronic artificial eye

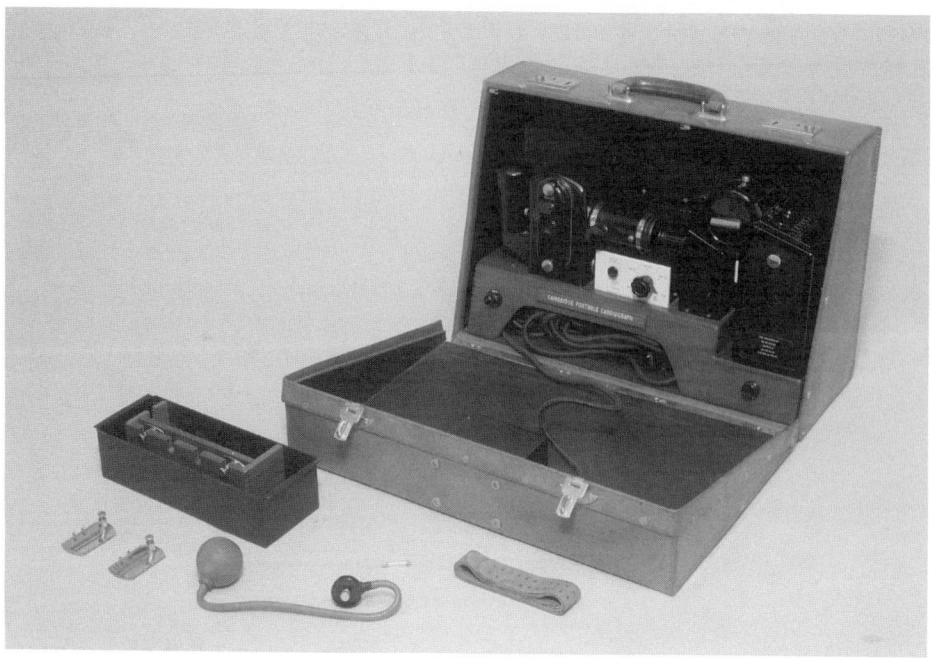

Figure 9.2. A portable electrocardiograph of 1947 (The Science Museum/ Science & Society Picture Library).

Figure 9.3. An implantable pacemaker of 1962 (The Science Museum/ Science & Society Picture Library).

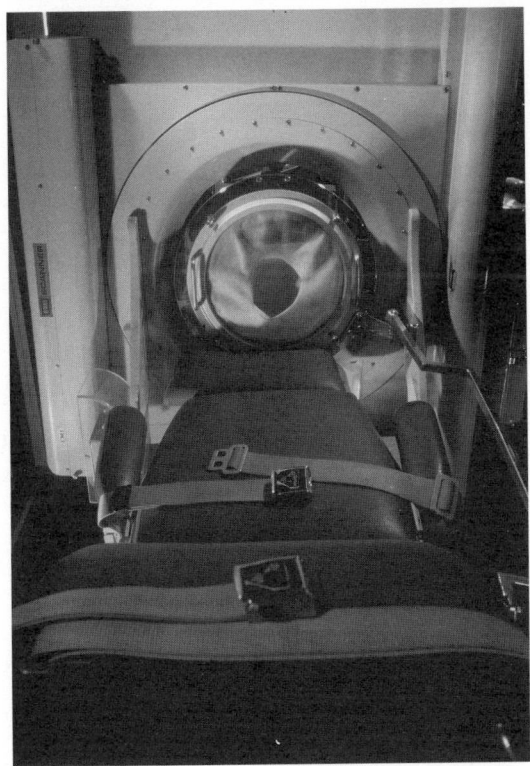

Figure 9.4. The first computerised tomography scanner (The Science Museum/ Science & Society Picture Library).

uses ocular implants for certain types of blindness, leading to all-silicon electrode arrays implanted directly into the brain for the profoundly blind.

Computer analyses using expert systems are finding their way into more medical applications and, in the USA, a software package can include 2200 diseases and 5000 symptoms in its knowledge base. Computerized detail recording of patients' medical information and treatments can now be kept by medical practitioners, together with computer link-up with Government records for accurate record-keeping.

The part played by electronics is inceasing day by day and, eventually, may lead to computer systems replacing parts of most of the human brain.

9.5 Educational Electronics

The electronic calculator may be considered to be the introduction of electronics into education. There was a course at that time on electronics of A-level standard, but very few teachers were capable of teaching it. Television and videos found considerable use, but it was the introduction of computers into schools, initially as an experiment, which revolutionized the teaching of electronics. In the United Kingdom today there is one computer for every 17 pupils in primary schools and one for every 8 in secondaries.

Basic programmes were taught and electronic games were not discouraged. 'Information Technology' as a subject is now taught in schools; word-processing replaced typewriting, and data-handling skills were acquired. Programmed learning of foreign languages is available. 'Electronic Technology' as a separate subject is now taught. Experience is gained by assembling kits of parts or modules to make up a complete oscilloscope, transistor receiver, chips as operational amplifiers, logic gates, etc. More complex kits can demonstrate standing waves, magnetic resonance, electron spin resonance, gas lasers, etc.

It is predicted that every secondary school will have some form of Internet connection in the near future, allowing wider access to the information superhighway, whilst video-conferencing in language lessons would allow link-ups with France, Germany, and other countries.

The ability to draw pictures on Graphic TV enables students to learn directly without the necessity for many textbooks. Mathematics can be presented pictorially as algebraic equations or as large numerical calculations. Computer programmes can be written to cover all stages of learning, either as simulation and modelling programmes, or more general application packages. Synthetic software for educational purposes has been developed and many are available over the Internet.

9.6   Office and Banking Electronics

Computers can now do tasks such as cost accounting, ledger records and retrieval, updating and filing. Data collection and storage on magnetic tape or compact discs with data processing can present analyses or visual displays or printout. Telex systems with data printers, facsimile document transmission and time-sharing computers allowing multiple use, all point the way to the electronic office. In the typing of letters or documents, word processors are now used. The typist starts by typing a draft document on to a VDU (Vidual Display Unit). At the same time, a copy of the document is also written on a floppy disc or CD-ROM. When the final version of the document is returned with all its corrections, modifications, additions, etc, having been made, the typist has only to type in the corrections or additions. The original copy on the disc, now corrected, is used to type automatically, at 175 words per minute, a perfect copy of the document. The word processor makes logical decisions such as automatically deleting, respacing and repositioning words and sentences, inserting hyphens, determining line lengths and controlling format and tabulation.

The modern office can now provide normal and networked computers, scanners and laser printers. Spreadsheets are used for record keeping, etc, whilst graphics and desktop publishing are facilities now available in an electronically assisted office.

In banking, banknote counting and dispensing machines deal with both notes and cash, computers deal with cheque transactions, mortgages, loans, and dividend payments, credit-card systemization, security and customer recognition safeguards, all data being sorted on magnetic tapes or CD-ROMs. Page-reading systems used flying spot scanning of characters from a cathode-ray tube and a group of photocells picked up reflections and output them to a decision network to make the white, black or grey decision (video processing). These are now replaced by modern computer systems using visual image systems, which can scan print at high rates and print out in different type and letter size. Electronic machines are available which can identify genuine banknotes and reject those which are counterfeit. Electronic bill paying services are being developed and the use of smart cards is increasing.

9.7   Consumer Electronics

The microcomputer is being built into more and more customer products. They range from the very simplest built as a single chip to more complex units with several dozen chips.

A simple example of an application which uses a computer on a chip is the electronic door chime. The programme for this device is stored in an RCM which is programmed when the chip is made. Twelve different tunes can be played, the tune depending an which door the chime was initiated from and on the tune selected. Other applications are controllers for central-heating systems which may be programmed with different on/off cycles, home games and teaching aids—a good example is the 'speak and spell' unit which synthesizes speech! More complex micros are used in gaming machines found in arcades and pubs in which sophisticated graphics and sound effects are generated as well as complex game strategies. Micros now find their way into such mundane devices as washing machines, cookers and videotape recorders.

The change to digital audio radio (DAR) in 1996 will affect consumer radio and television sets, whilst high definition TV now being developed will mean more changes for consumers. Compact discs have replaced most other forms of recording data and music and high density compact discs (DVDs) offer wider applications, including movies. For the home user, camcorders are now widely used, including recently developed digital camcorders such as the cybercam.

With the introduction of DVDs replacing CD-ROMs for games, many new improved games will be available, some being interactive. Home automation, using modern communication techniques to close the curtains or switch the cooker on from afar will undoubtedly increase.

9.8 Security Electronics and Surveillance

Intruder alarms now use sound detectors, magnetic switches on doors, light beams, laser beams, pressure mats and hoses, infra-red detection, microwave radar, surveillance scanners, security pass systems and TV coverage. For surveillance, millimetre wave cameras can penetrate walls to see subjects from a distance. Digital video cameras can pick out faces in a crowd using computer-recorded information. Microphones (or BUGS) can be hidden in buildings, even hidden in steel girders or fed into the water to lodge in taps or drains.

Conversations in a room can be overheard from outside by bouncing a laser beam off a glass window picking up the sound waves. Surveillance satellites can watch over large areas of the world. There is little now which can be secret from prying eyes using electronics.

Chapter 10

List of Inventions by Subject

1. Assembly Techniques and Packaging
2. Circuitry
3. Communications
4. Components
5. Computers
6. Industrial and Medical
7. Microelectronics
8. Physics
9. Radar
10. Sound Reproduction
11. Television
12. Transistors and Semiconductors
13. Tubes, Lamps, etc

1. Assembly Techniques & Packaging

1940	Thick film packaged circuits	(Centralab)
1943	Printed wiring	(Eisler)
1945	Potted circuits	(Various)
1947	Automatic circuit making equipment	(Sargrove)
1949	Dip soldering of printed circuits	(Danco & Abramson)
1950	Thermo compression bonding techniques	(Anderson *et al*)
1951	'Tinkertoy' automatic assembly system	(Nat. Bur. Standards)
1953	Wire wrapping of connections	(Mallina *et al*)
1953	Automatic assembly systems: Autofab, Mini-Mech, etc,	(General Mills *et al*)
1956	'Flowsoldering' of printed circuits	(Fry's Metal Foundries)
1958	Micro-module' assembly system	(US Army)
1960	Printed wiring multilayer boards	(Photocircuits)
1962	'Flat-Pack' integrated circuit	(Tao)
1964	DIP or DIL (dual in line package)	(Rogers)
1964	Etch-back plated-through hole technique	(Autonetics)
1964	Beam lead connections	(Lepselter)
1966	'Flip-chip' bonding technique	(Wiessenstern *et al*)
1971	Ceramic chip carrier	(3 M Co.)
1973	Dry etching technique	(Mitsubishi)

2. Circuitry

1826	Ohm's law	(Ohm)
1843	Wheatstone bridge	(Wheatstone)
1848	Boolean algebra	(Boole)
1945	Kirchhoff's laws	(Kirchhoff)
1912	Regenerative circuit	(de Forest *et al*)
1912	Heterodyne & superheterodyne circuits	(Fessenden, Armstrong)
1915	Filter networks	(Campbell, Wagner)
1918	Neutrodyne circuit	(Hazeltine)
1918	Shot effect noise	(Schottky)
1918	Multivibrator circuit	(Abraham & Bloch)
1918	Dynatron circuit	(Hull)
1919	Retarded field microwave oscillator	(Barkhausen & Kurtz)
1919	Flip-flop circuit	(Eccles, Jordan)
1919	Miller time base circuit	(Miller)
1921	Crystal control of frequency	(Cady)
1922	Negative resistance oscillator	(Gill, Morrell)
1922	Super regeneration	(Armstrong)
1923	Squegger circuit	(Appleton, Herd *et al*)
1924	Linear saw-tooth time base circuit	(Anson)
1925	Johnson noise	(Johnson)
1926	Transitron oscillator	(van der Pol)
1926	Automatic volume control circuit	(Whceler)
1927	Negative feedback amplifier	(Black)
1932	Energy conserving scanning circuit	(Blumlein)
1933	Hard valve time base circuit	(Puckle)
1935	Constant RC stand-off circuit	(Blumlein)
1936	Long tailed pair circuit	(Blumlein)
1939	Radio altimeter	(Bell Labs)
1942	Miller integrator circuit	(Blumlein)
1942	Phantastron circuit	(Williams & Moody)
1942	Sanatron circuit	(Williams & Moody)
1943	Magnetic amplifier	(A.S.E.A)
1947	High quality amplifier circuit	(Williamson)
1952	Darlington pair circuit	(Darlington)
1952	Digital voltmeter	(Kay)
1952	Negative feedback tone control circuit	(Baxandall)
1960	Neuristor circuit	(Crane)
1960s	Logic circuits: microelectronics	(Various)
1960s	Linear circuits: microelectronics	(Various)
1967	Rotator circuit network	(Chua)
1968	Mutator circuit network	(Chua)
1969	Bucket-brigade delay circuit	(Sangster and Teer)
1979	Satellite echo cancelling circuit	(Bell Labs)
1980	Fibre-optic laser driven superheterodyne	(Saito *et al*)

3. Communications

1832	6-unit telegraph system	(Schilling and Cooke)
1837	Telegraphy: Morse code	(Morse)
1860	Microphone	(Reis)
1865	Radio wave propagation	(Maxwell)

1866	Transatlantic telegraph cable	(T. C. & M. Co.)
1876	Telephone	(Bell)
1879	Diode detector	(Hughes)
1887	Aerial	(Hertz)
1890	Coherer	(Branley)
1893	Waveguide: theory	(Thomson)
1896	Wireless telegraphy	(Marconi)
1901	Radio: Heaviside/Kenelly layer	(Heaviside/Kenelly)
1906	Radio broadcasting	(Fessenden)
1906	Crystal detector	(Dunwoody)
1907	Crystal detector (Perikon)	(Pickard)
1912	Ionospheric propagation	(Eccles)
1915	Single sideband transmission	(Carson)
1916	Telex	(Markrum Co.)
1918	Alexanderson alternator	(Alexanderson)
1918	Ground wave propagation	(Watson)
1921	Short wave radio (amateur)	(Amateurs)
1921	Crystal control of frequency	(Cady)
1925	Short wave radio (commercial)	(van Boetzelean)
1925	Ionosphere layer	(Appleton)
1926	Yagi aerial	(Yagi)
1928	Diversity reception	(Beverage *et al*)
1928	Frequency standards: quartz clock	(Horton, Morrison)
1929	Microwave communication	(Clavier)
1930s	Meteor scatter (burst) systems	(Shanker *et al*)
1933	Radio astronomy	(Jansky)
1933	Frequency modulation	(Armstrong)
1934	Frequency standards: atomic clock	(Cleeton, Williams)
1936	Waveguides	(Southworth *et al*)
1937	Pulse code modulation	(Reeves)
1939	Frequency standards: caesium beam	(Rabi)
1945	Satellite communication theory	(Clarke)
1948	Information theory	(Shannon)
1950s	MODEM	(MIT, Bell Labs)
1950s	Global positioning system	(Getting)
1954	Transistor radio set	(Regency)
1956	Transatlantic telephone cable	(Various)
1956	Radio paging	(Multitone
1957	SPUTNIK I satellite	(USSR)
1958	EXPLORER I satellite	(USA)
1958	VANGUARD I satellite	(USA)
1958	PIONEER I satellite	(USA)
1958	SCORE satellite	(USA)
1959	LUNIK I satellite	(USSR)
1959	DISCOVERER I satellite	(USA)
1960	TIROS I satellite	(USA)
1960	ECHO I satellite	(USA)
1960	COURIER I satellite	(USA)
1960	TRANSIT I satellite	(USA)
1961	VENUS I satellite	(USSR)
1961	VOSTOK I satellite	(USSR)
1961	MERCURY-ATLAS 4 satellite	(USA)
1961	OSCAR I satellite	(USA)
1962	TELSTAR I satellite	(USA)

1962	MERCURY-ATLAS 6 satellite	(USA)
1962	OSO I satellite	(USA)
1962	RELAY I satellite	(USA)
1962	MARINER I satellite	(USA)
1962	MARS I satellite	(USSR)
1962	ARIEL I satellite	(UK)
1963	SYNCOM I satellite	(USA)
1964	NIMBUS satellite	(USA)
1964	VOKSHOD I satellite	(USSR)
1964	Packet-switching	(Baron)
1965	PEGASUS I satellite	(USA)
1965	INTELSTAT I satellite	(International)
1965	PROTON I satellite	(USA)
1966	SURVEYOR I satellite	(USA)
1966	ESSA I satellite	(USA)
1966	LUNAR ORBITER I satellite	(USA)
1966	ATS I satellite	(USA)
1966	Optical fibre communications	(Kao, Hockham)
1967	SOYUZ I satellite	(USSR)
1968	IRIS (ESRO I) satellite	(USA)
1969	Internet	(Arpanet)
1969	Aerial matching unit	(Gordon)
1969	SKYNET A satellite	(UK)
1969	AZUR satellite	(Germany)
1970	TUNG-FANG-HUNG satellite	(China)
1970	NATO I satellite	(NATO)
1971	DSCS I satellite	(USA)
1971	SALYUT I satellite	(USSR)
1972	LANDSAT I satellite	(USA)
1973	SKYLAB I satellite	(USA)
1974	Information technology	(Various)
1974	WESTAR I satellite	(USA)
1974	PRESTEL system	(Fedida)
1975	VIKING I satellite	(USA)
1975	RADUGA I satellite	(USSR)
1976	MARISAT I satellite	(USA)
1976	Spread-spectrum techniques	(Dixon *et al*)
1978	Integrated optoelectronics	(Yariv *et al*)
1978	Tamed frequency modulation	(Philips)
1988	Fibre optic transatlantic cable	(DGT, BT, AT&T)
1995	Lasercom	(Thermotrex, Motorola)
1995	Inter-satelite communications	(US Air Force)

4. Components

BATTERIES

1800	Volta's pile	(Volta)
1803	Accumulator	(Ritter)
1839	Magnetohydrodynamic generation	(Faraday)
1839	Fuel cell	(Grove)
1860	Lead-acid cell	(Plante)
1868	Dry cell	(Léclanché)
1870	Clark standard cell	(Clark)

1884	Zinc mercuric-oxide cell	(Clark)
1891	Weston standard cell	(Weston)
1900	Nickel–iron cell	(Edison)
1900	Nickel–cadmium cell	(Junger and Berg)
1954	Solar battery	(Chapin, Fuller *et al*)

CAPACITORS
1745	Leyden jar	(von Kleist *et al*)
1874	Mica capacitors	(Bauer *et al*)
1876	Rolled paper capacitors	(Fitzgerald)
1900	Ceramic capacitors	(Lombardi)
1904	Glass tubular capacitors	(Moscicki)
1956	Solid electrolyte capacitor	(McClean and Power)
1956	Semiconductor diode junction capacitor	(Giacoletto *et al*)

FILTERS
| 1915 | Filters, electromagnetic | (Campbell and Wagner) |

GALVANOMETERS
1820	Electromagnetism (galvanometer)	(Oersted)
1828	Moving coil	(Schweigger)
1828	Astatic	(Nobilli)

GONIOMETERS
| 1907 | Goniometer | (Artom) |

INDUCTORS
1772	Iron dust cores	(Knight)
1909	Ferrites	(Hilpert, Snoek)
1956	Magnetic material (YIG)	(Bertaut and Forrat)
1977	Anisotropic permanent magnet	(Matsuschita Elec.)

MOTORS
1837	Electric motor	(Davenport)
1888	Induction motor	(Tesla)
1902	Synchronous induction motor	(Danielson)

RECTIFIERS
| 1926 | Copper oxide | (Grondahl and Sieger) |

RELAYS
| 1837 | Telegraph bell and signal | (Cooke, Davy *et al*) |
| 1950s | Ferreeds | (Bell Labs) |

RHEOTOME
| 1868 | Waveform plotter | (Lenz) |

RESISTORS
1850	Thermistor	(Faraday)
1885	Moulded carbon composition	(Bradley)
1897	Carbon film	(Gambrell *et al*)
1913	Sputtered metal film	(Swann)
1919	Spiralled metal film	(Kruger)
1925	Cracked carbon	(Siemens and Halske)
1926	Sprayed metal film	(Loewe)
1931	Oxide film	(Littleton)
1957	Nickel–chromium film	(Alderton *et al*)
1958	Field effect varistor	(Bell Labs)
1959	Tantalum film	(Bell Labs)

SWITCHES
1884	Quick break	(Holmes)
1887	Quick make and break	(Holmes)
1950s	Ferreed switch	(Bell Labs)

TRANSFORMERS

1831	Transformer	(Faraday)
1885	Distribution	(Deri)
1885	Power	(Zipernowski *et al*)

WAVEGUIDES

| 1893 | Theory | (Thomson) |
| 1936 | Waveguides | (Southworth *et al*) |

WIRES AND CABLES

1729	Wire conductor	(Gray)
1812	Cable insulation	(Sommering *et al*)
1845	Metallic sheathing	(Wheatstone)
1847	Submarine cable insulation	(Siemens)
1905	Insulated sodium conductor	(Betts)
1933	Polythylene insulation	(ICI)
1949	Microwire	(Ulitovsky)
1965	Wiegand wire	(Wiegand)
1965	Smooth-surfaced wire drawing	(Olsen *et al*)

5. Computers

1642	Calculating machine	(Pascal)
1672	Calculating machine	(Leibniz)
1833	Calculating machine	(Babbage)
1848	Boolean algebra	(Boole)
1854	Calculating machine	(Scheutz)
1889	Tabulating machine	(Hollerith)
1931	Differential analyser	(Bush)
1938	Information theory	(Shannon)
1939	Bell Labs complex computer	(Stibitz *et al*)
1939	Digital computer	(Aitken)
1942	'Velodyne' analyser	(Williams & Uttley)
1943	COLOSSUS	(Newman, Turing *et al*)
1943	ENIAC	(Moore School)
1945	Whirlwind	(MIT)
1945	Computer theory	(von Neuman)
1946	CRT storage computer	(Williams)
1946	ACE	(Turing)
1947	EDVAC	(Penn. University)
1947	UNIVAC	(Eckert, Maunchly)
1948	SEAC	(NBS)
1948	EDSAC	(Wilkes)
1950	Computer graphics	(Burnett)
1950	IBM 650	(IBM)
1950	IBM 701	(IBM)
1950	Hamming code	(Hamming)
1950s	APL language	(Iverson)
1951	Microprogramming	(Wilkes)
1952	SAGE	(IBM, MIT)
1953	IBM 704, 709 and 7090	(IBM)
1956	Transistorised computer	(Bell Labs)
1957	Plated wire memories	(Gianole)
1960	Honeywell 800	(Honeywell)
1960	UNIVAC solid state 80/90	(IBM)

1960	CD 1604	(Control Data Corp.)
1961	Minicomputer	(Englebert)
1965	The mouse	(Digital Equip. Co.)
1969	Semiconductor memories system	(Agusta *et al*)
1969	Magnetic bubble memories	(Bobeck *et al*)
1970	Charge coupled device memories	(Boyle, Smith)
1970	Floppy-disc recorder	(IBM)
1970	UNIX	(Bell, Univ. Cal.)
1971	Microprocessor	(Hoff)
1972	1024 bit random access memory	(Intel)
1973	Logic-state analyser	(House)
1973	Logic-timing analyser	(Moore)
1974	16 bit single chip microprocessor	(National)
1975	4096 bit RAM	(Fairchild)
1975	16 384 bit RAM	(Intel)
1976	One board computer with programmable 1/0	(Intel)
1976	Polysilicon resistor loaded RAM	(Mostek)
1977	CCD analog-to-digital converter	(GE Corporation)
1978	Integrated optoelectronics	(Yariv *et al*)
1978	One megabit bubble memory	(Intel & Texas)
1980	256 K dynamic RAM	(NEC, Toshiba *et al*)
1981	MS-DOS	(Gates)
1984	Digital optical disc	(ATG)
1985	Hard disc card	(PLUS Dev. Corp.)
1985	Digital video recorder	(SONY)
1985	CD-ROM	(Philips)
1985	Windows	(MICROSOFT)
1985	Tactile screen	(Zenith)
1991	Photonic crystals	(Eli Yabionovitch)
1995	Biological memory chip	(Mitsubishi Electric and Santory)
1996	Superfast switch	(Argonne N. Lab)

6. Industrial and Medical

1839	Microfilming	(Dancer)
1843	Facsimile reproduction	(Bain)
1887	Electrocardiograph	(A D Waller)
1908	Geiger counter	(Geiger, Rutherford)
1912	Tungar rectifier	(Langmuir)
1913	Reliability standards	(AIEE)
1914	Ultrasonics (ASDIC, SONAR)	(Langevin)
1914	Thyratrons	(Langmuir)
1916	Reliability, control system	(Bell/Western Elec.)
1918	Induction heating	(Northrup)
1920	Ultra-micrometer	(Whiddington)
1926	Copper oxide rectifier	(Grondahl & Geiger)
1926	Electron microscope	(H Bosch)
1931	CRO cardiograph	(Rijant)
1931	Reliability—quality control charts	(Shewhart)
1933	Ignitron	(Westinghouse)
1937	Xerography	(Carlson)
1940	Cybernetics	(R Weiner)
1943	Reliability—sequential analysis	(Wald)

1943	Magnetic amplifier	(ASIA)
1944	Reliability—sampling inspection tables	(Doge & Romig)
1950s	Ultrasound imaging	(Donald *et al*)
1951	Quality control handbook	(Juran)
1958	Pacemaker	(A Senning)
1960	Computer aided design	(USA Military)
1961	Electronic clock	(Vogel et Cie)
1962	Electronic watch	(Vogel et Cie)
1962	Duane reliability growth theory	(Duane)
1963	Ink jet process	(Sweet)
1963	Electronic calculator	(Bell Punch Co.)
1964	Telemedicine	(Various)
1964	Word processor	(IBM)
1965	Electronic typewriter	(IBM)
1967	Ion beam coating	(Chopra & Randlett)
1971	Electronic digital watch	(Time Computer Corp)
1972	Video games	(Magnavox)
1974	Bar codes	(Dawson)
1975	LASER printer	(IBM)
1977	MRI—magnetic resonance imaging	(G Houndsfield)
1978	All electronic clock face	(Hosiden Elec)
1979	Seven-colour ink-jet printer	(Siemens)
1982	Bubble-jet printing	(CANON)
1986	Scanning tunnelling microscope	(IBM)
1991	Plastic electronics	(Philips)
1995	Glass laser	(Song-Tiong Hoctal, Northwestern Univ.)

7. Microelectronics

(See also 12. Transistors and Semiconductor Devices)

1852	Thin film sputtering process	(Grove)
1913	Sputtered metal film resistors	(Swan)
1940	Thick film circuits	(Centralab)
1949/50	Ion implantation	(Ohl & Shockley)
1952	Semiconductor integrated circuit concept	(Dummer)
1952	Zone melting technique	(Pfann)
1957	Nickel chromium thin film resistors	(Alderton, Ashworth)
1959	Semiconductor integrated circuit patent	(Kilby)
1959	Tantalum thin film circuits	(Bell Labs)
1959	Planar process	(Hoeni)
1960	Epitaxy: vapour phase	(Loor *et al*)
1960	Digital and linear integrated circuits	(Various)
1961	Epitaxy: liquid phase	(Nelson)
1961	Minicomputer	(Digital Equip. Co.)
1962	MOS integrated circuit	(Hofstein & Helman)
1963	Surface acoustic wave devices	(Rowen & Sittig)
1967	Laser trimming of thick film resistors	(Various)
1967	Ion beam coating	(Chopra and Randlett)
1968	C-MOS integrated circuit	(Various)
1968	Aluminium metallisation of ICs	(Noyce)
1969	Collector diffusion isolation	(Bell Labs, Ferranti)

1970	X-ray lithography	(Feder *et al*)
1971	FAMOS integrated circuit	(Frohman-Bentchowsky)
1971	Bumped tape automatic banding	(Lin and Fraenkel)
1971	Liquid crystal study of oxide defects	(Keen)
1972	Microcomputer	(Intel)
1972	1024 bit random access memory	(Intel)
1972	Nitrogen-fired copper wiring	(Grier)
1972	Two-layer resist technique	(Bell Labs)
1972	V-MOS technique	(Rodgers)
1972	Integrated injection logic	(Hart & Slob)
1973	Dry etching technique	(Mitsubishi)
1974	16 bit single chip microprocessor	(National)
1974	Electron beam lithography	(Bell Labs)
1975	Thin film direct bonded copper process	(Burgess *et al*)
1975	LOCMOS integrated circuit	(Philips)
1975	Integrated optical circuits	(Reinhart, Logan)
1975	4096 bit random access memory	(Fairchild)
1975	Silicon anodisation	(Cook)
1976	Microelectronic versatile arrays	(Philips)
1976	One board with programmable 1/0 computer	(Intel)
1976	16 384 bit random access memory	(Intel)
1977	H-MOS	(Intel)
1977	TRIMOS device	(Stanford University)
1977	CCD analog/digital converter	(G E Corporation)
1978	Laser annealed polysilicon	(Texas Instruments)
1978	Integrated Schottky logic	(Philips)
1981	Hydrogenated amorphous silicon films	(Grasso *et al*)
1995	Atomic beam lithography	(Harvard, NI of Standards)
1996	Atomic holography	(NEC Japan)
1996	Electron beam projection System	(Bell Labs)
1996	Surface flat chips	(IBM)
1996	Direct laser writing	(Mikroelektronik Centre)

8. Physics

1780	Galvanic action	(Galvani)
1800	Infra-red radiation	(Herschel)
1807	Ultra-violet radiation	(Ritter)
1808	Atomic theory	(Dalton)
1820	Electro-magnetism	(Oersted)
1821	Thermloelectricity	(Seebeck)
1826	Ohm's law	(Ohm)
1828	Fourier analysis	(Fourier)
1831	Electromagnetic induction	(Faraday)
1832	Self-induction	(Henry)
1834	Electrolysis	(Faraday)
1839	Photovoltaic effect	(Becquerel)
1840	Thermography	(Herschel)
1847	Magnetostriction	(Joule)
1851	Relation between theory of magnetism and electricity	(Kelvin)
1858	Glow discharges	(Plucker)
1878	Cathode rays	(Crookes)
1879	Hall effect	(Hall)

1880	Piezo electricity	(Curies)
1882	Wimshurst machine	(Wimshurst)
1895	X-rays	(Röntgen)
1897	Electron	(Thomson)
1897	Cathode ray oscillograph	(Braun)
1900	Quantum theory	(Planck)
1902	Spontaneous atomic change	(Rutherford & Soddy)
1905	Theory of relativity	(Einstein)
1911	Superconductivity	(Onnes)
1911	Atomic theory	(Rutherford)
1912	Cloud chamber	(Wilson)
1915	Atomic orbit theory	(Bohr)
1918	Atomic transmutation	(Rutherford)
1921	Ferroelectricity	(Vasalek)
1929	Cyclotron	(Laurence)
1930	High field superconductivity	(de Haas & Voogd)
1930	van de Graaf accelerator	(van de Graaf)
1932	Neutron	(Chadwick)
1932	Transmission electron microscope	(Knoll, Ruska)
1932	Cockroft–Walton accelerator	(Cockroft & Walton)
1934	Liquid crystals	(Dreyer)
1934	Trans-uranian atoms	(Fermi)
1934	Scanning election microscope	(Knott *et al*)
1935	Superconducting switching	(Casimir-Jonker *et al*)
1937	Xerography	(Carlson)
1938	Nuclear fission	(Fritsch & Meitner)
1941	Betatron	(Kerst)
1947	Molecular beam epitaxy	(Sosnowski *et al*)
1948	Holography	(Gabor)
1953	MASER	(Townes & Weber)
1955	Infra-red emission from GaSb	(Braunstein)
1955	Cryotron	(Buck)
1956	YIG magnetic materials	(Bertaut, Forrat)
1958	LASER	(Schalow, Townes)
1958	Mossbauer effect	(Mossbauer)
1959	Intrinsic 10μ photoconductor	(Lawson *et al*)
1960	Sub-millimetre photoconductive detector	(Putley)
1961	Transferred electron effect	(Ridley, Watkins)
1961	Transferred electron device	(Hilsum)
1962	Semiconductor laser	(Hall *et al* ; also Nathan & Lasher)
1962	Josephson effect	(Josephson)
1962	LED (Gallium arsenide phosphide)	(Holonyak)
1963	Ion plating	(Mattox)
1963	Gunn diode oscillator	(Gunn)
1963	Surface acoustic wave devices	(Rowen & Sittig)
1964	'IMPATT' diode	(Johnston & de Loach)
1970	X-ray lithography	(Feder *et al*)
1972	X-ray scanner	(EMI)
1972	Automatic crystal growth control	(Bardsley *et al*)
1972	Deep proton-isolated laser	(Dymeut *et al*)
1973	Scanning acoustic microscope	(Quate)
1975	GYROTRON	(Gapanov)
1977	FLAD display system	(Inst. Appl. Solid State Phys.)
1978	Light bubbles	(IBM)

1978	OMIST thyratron	(Nassibian *et al*)
1979	Laser enhanced plating and etching	(IBM)
1979	Amorphous silicon LCD	(RSRE & Dundee Univ.)
1981	Plane-polarised light optical fibre	(Hitachi)
1982	Fission track autoradiography	(AERE)

9. Radar

1924	Radar systems	(Appleton, Briet *et al*)
1937	Radar aiming anti-aircraft guns	(Pollard)
1938	'Gee' navigation	(Dippy)
1938	Klystron	(Hahn & Varian Bros)
1939	Magnetron	(Randall & Boot)
1940	Plan position indicator	(Bowen, Dummer *et al*)
1940	Skiatron	(Rosenthal)
1940	'Oboe' navigation system	(Reeves *et al*)
1941	Radio proximity fuse	(Butement)
1941	'H$_2$S' navigation system	(Dee, Lovell *et al*)
1942	'Velodyne' analyser	(Williams & Uttley)
1942	LORAN	(MIT)
1942	Phantastron circuit	(Williams and Moody)
1942	Sanatron time base circuit	(Williams and Moody)
1942	Miller integrator circuit	(Blumlein)
1943	Ultrasonic radar navigation training device	(Dummer and Smart)
1945	DECCA navigation system	(O'Brien and Schwartz)
1947	Chirp technique	(Bell Labs)
1971	Hologram matrix radar	(Iizuka, Nguyen *et al*)

10. Sound Reproduction

1779	Speech synthesis	(Kratzenstein)
1817	The optophone	(d'Albe)
1860	Microphone, diaphragm type	(Reis)
1876	Telephone	(Bell)
1877	Phonograph	(Edison)
1877	Microphone, carbon	(Edison)
1877	Loudspeakers, moving coil	(Siemens)
1878	Carbon granule microphone	(Hunnings)
1887	Gramophone	(Berliner)
1889	Strowger auto telephone exchange	(Strowger)
1896	Telephone dial	(Keith *et al*)
1898	Magnetic recording (wire)	(Poulsen)
1908	Electronic organ	(Cahill)
1912	Relay auto telephone exchange	(Betulander)
1914	ASDIC	(Langevin)
1915	Acoustic mine	(Wood)
1915	SONAR	(Hunt)
1916	Crossbar telephone exchange	(Roberts and Reynolds)
1917	Microphone, condenser	(Wente)
1919	Crystal microphone	(Nicholson)
1920	Plastic magnetic tape	(Pfleumer)

1924	Reisz carbon microphone	(Neumann)
1925	Loudspeaker, electrostatic	(Various)
1926	Films, sound-on-disc system	(Warner Bros.)
1927	Films, sound-on-film system	(Fox Movietone News)
1930s	Radiophonic sound and music	(Grainger)
1931	Stereophonic sound reproduction	(Blumlein, Bell Labs)
1933	Stereo record	(EMI)
1936	Vocoder	(Bell Labs)
1937	Pulse code modulation	(Reeves)
1948	Films: magnetic recording	(RCA *et al*)
1950s	MODEM	(MIT, Bell Labs)
1950	Floppy discs (patent)	(Nakamata)
1957	Full frequency range loudspeaker	(Walker)
1958	Video tape recorder	(Ampex)
1960	Telephore electronic switching	(Bell Labs)
1961	Tape cassette	(Philips)
1964	Packet switching	(Canon)
1965	Synthesizer	(Moog *et al*)
1967	Audio noise reduction system	(Dolby)
1969	PARCOR speech synthesis	(NTT, Japan)
1972	Video discs	(Philips)
1973	Scanning acoustic microscope	(Quate)
1974	PRESTEL system	(Fedida)
1978	Lightwave powered telephone	Bell Labs)
1978	Laser recording system	(Philips)
1979	satellite echo-cancelling circuit	(Bell Labs)
1979	Compact disc	(Philips)
1982	Camcorder	(Sony)
1987	DAB digital audio broadcasting	(Eureka 147)
1991	Very high density diskette	(Insight, P)
1995	DVD (digital versatile disc)	(International)

11. Television

1884	Nipkow television system	(Nipkow)
1897	Cathode ray oscillograph	(Braun)
1908	Electronic system: theory	(Campbell-Swinton)
1919	Electronic system	(Zworykin)
1923	Iconoscope	(Zworykin)
1925	Mechanical system	(Baird)
1927	Cable TV	(Bell Tel. Co.)
1929	Colour television	(Bell Labs)
1932	Scanning circuit	(Blumlein)
1933	Time base circuit	(Puckle)
1936	Long tailed pair circuit	(Blumlein)
1938	Shadow-mask TV tube	(Flechsig)
1939	Large screen TV projector	(Fischer)
1950	VIDICON TV camera tube	(RCA)
1951	Image animation	(MIT)
1957	PLUMBICON TV camera tube	(Philips)
1958	Colour video recorder	(Ampex)
1965	Virtual reality	(USA Military)
1965	Integrated photodiode arrays	(Weckler)

1968	TRINITRON colour tube	(Sony)
1968	High definition TV	(Nippon)
1970	Video cassette recorders	(Japan, Europe)
1974	PRESTEL system	(Fedida)
1975	VHS recorder	(JVC)
1975	Betamax video recorder	(Sony)
1977	Pocket TV receiver	(Sinclair)
1979	CCD colour TV camera	(Sony)
1980	Large screen colour display	(Mitsubishi)
1987	Compact disc video	(JVC)
1987	Programme control of video recorders	(Matsushita)
1988	Video Walkman	(Sony)
1995	VHS bit stream recorder	(Hitachi, Thomson)

12. Transistors and Semiconductor Devices

(See also 7. Microelectronics)

1917	Crystal pulling process	(Czochralski)
1930	MOS/FET concept	(Lilienfeld)
1935	Field effect transistor	(Heil)
1948	Single crystal fabrication: germanium	(Teal and Little)
1948	Transistor	(Bardeen *et al*)
1949/50	Ion implantation	(Ohl and Shockley)
1950	PIN diode	(Niskizawa)
1950s	Thermo-compression bonding	(Anderson *et al*)
1952	Zone melting technique	(Pfann)
1952	Single crystal fabrication: silicon	(Teal, Bueler)
1952	Alloyed transistor	(RCA)
1953	Surface barrier transistor	(Philco)
1953	Floating zone refining process	(Keck, Emeis *et al*)
1953	Unijunction transistor	(GEC)
1954	Transistor radio set	(Regency)
1954	Silicon solar battery	(Chapin, Fuller *et al*)
1954	Interdigitated transistor	(Fletcher)
1955	Infra-red emission from GaSb	(Braunstein)
1956	Diffusion process	(Fuller, Reis)
1956	Semiconductor diode junction capacitor	(Giacoletto and O'Connell)
1957	Oxide masking process	(Frosch)
1958	Pedestal pulling of silicon	(Dash)
1958	Tunnel diode	(Esaki)
1958	'Technetron' FET	(Teszner)
1958	Field effect varistor	(Bell Labs)
1959	Planar process	(Hoerni)
1960	Light emitting diode (LED)	(Allen & Gibbons)
1960	Epitaxy (vapour phase)	(Loor *et al*)
1961	Epitaxy (liquid phase)	(Nelson)
1962	LED (gallium arsenide phosphide)	(Holonyak)
1963	Gunn diode oscillator	(Gunn)
1963	Silicon-on-sapphire technology	(Various)
1964	IMPATT diode	(Johnston, de Loach)
1964	Transistor modelllng	(Gummel)
1964	Overlay transistor	(RCA)

1965	Self-scanned photodiode arrays	(Weckler)
1966	Nitride-over-oxide semiconductors	(Horn)
1967	TRAPATT diode	(Prager, Chang *et al*)
1968	Amorphous semiconductor switches	(Ovshinsky)
1968	BARRITT diode	(Wright)
1969	Magnetic bubbles	(Bobeck, Fischer *et al*)
1969	Magistor magnetic sensor	(Hudson, IBM)
1970	Charge coupled devices	(Boyle, Smith)
1970	X-ray lithography for bubble devices	(Feder *et al*)
1971	Carrier-domain magnetometer	(Gilbert)
1972	Auto control of crystal growth	(Bardsley *et al*)
1974	CATT triode	(Tu, Cady *et al*)
1976	Amorphous silicon solar cell	(RCA)
1977	FLAD display system	(Inst. Appl. Solid State Phys.)
1978	Laser cold processing semiconductors	(Quantronix)
1979	FLOTOX process	(Intel)
1980	Magnetic avalanche transistor	(IBM)
1980	MCZ silicon crystal growth	(Sony)
1981	Hydroplaning polishing of semaiconductors	(MIT)
1982	Amorphous photosensors	(Sony)
1982	Recrystallisation silicon process	(Texas Instruments)

13. Tubes, Lamps, etc

1855	Cold cathode discharge tube	(Gaugain)
1856	Low pressure discharge tube	(Geissler)
1857	Mercury arc lamp	(Wray)
1878	Carbon filament lamp	(Swan, Stearn *et al*)
1901	Fluorescent lamp	(Cooper-Hewitt)
1904	Two electrode tube	(Fleming)
1906	Three electrode tube	(de Forest)
1910	Neon lamp	(Claude)
1912	Tungar rectifier	(Langmuir)
1914	Thyratron	(Langmuir)
1919	Retarded field microwave oscillator	(Barkhausen, Kurtz)
1919	Housekeeper seal	(Housekeeper)
1922	Negative resistance oscillater	(Gill, Morrell)
1926	Screened grid tube	(Round)
1928	Pentode tube	(Tellegen; Holst)
1931	CRO cardiograph	(Rijant)
1933	Ignitron	(Westinghouse)
1935	Travelling wave microwave oscillator	(Heil)
1935	Multiplier phototube	(Zworykin *et al*)
1936	Cold cathode trigger tube	(Bell Labs)
1937	Polar co-ordinate oscillograph	(von Ardenne *et al*)
1938	Shadow-mask tube	(Flechsig)
1939	Double-beam oscillograph	(Fleming-Williams)
1939	Klystron	(Hahn and Varian Bros)
1939	Magnetron	(Randall and Boot)
1940	Skiatron CRO	(Rosenthal)
1943	Travelling wave tube	(Kompfner *et al*)
1941	Cold cathode stepping tube	(Remington Rand)
1950	VIDICON TV camera tube	(USA)

1956	Vapour cooling of tubes	(Beutheret)
1957	PLUMBICON TV camera tube	(Philips)
1960	FEMITRON microwave ampllfier	(Dyke)
1968	TRINITRON colour tube	(Sony)
1975	Static induction thyristor	(Nishizawa)

Chapter 11

A Concise Description of Each Invention in Date Order

1642   **COMPUTER (Mechanical Calculating Machines)**   **B Pascal (France)**

EDITOR'S NOTE: Although non-electronic this item is included as an essential part of computer history.

The invention of the first mechanical device capable of addition and subtraction in a digital manner has been generally credited to Pascal, who built his first machine in 1642. This claim has been contested on the basis of letters sent to Kepler in 1623 and 1624 by Wilhelm Schickhardt of Tubingen, in which the latter describes the construction of a calaculator. Pascal, who at the age of 19 had wearied of adding long columns of figures in his father's tax office in Rouen, made a number of calculators, some of which are still preserved in museums. His machines had number wheels with parallel, horizontal axes. The positions of these wheels could be observed and sums read through windows in their covers. Numbers were entered by means of horizontal telephone-dial like wheels which were coupled to the number wheels by pin gearing. Most of the number wheels were geared for decimal reckoning but the two wheels on the extreme right had twenty and twelve divisions, respectively for sous and deniers. A carry ratchet coupled each wheel to the next higher place. The stylus-operated pocket adding machines now widely used are descendants of Pascal's machine.

SOURCE: 'The evolution of computing machines and systems' by Serrell, Astrahan, Patterson and Pyre *Proc. IRE* p 1041 (May 1962)

SEE ALSO: 'The inventor of the first desk calculator' by V P Czapla *Computers and Automation* vol 10, p 6 (September 1961)

The Computer from Pascal to von Neuman by H H Goldstine (Princeton, NJ: Princeton University Press) p 7 (1972)

The Origins of Digital Computers edited by B Randell (Berlin: Springer) (1973)

1672   **COMPUTERS (Mechanical Calculating Machines)**   **G W Leibniz (Germany)**

Gottfried Wilhelm Leibniz invented the 'Leibniz Wheel' which enabled him to build a calculating machine which surpassed Pascal's in that it could do not only addition and subtraction fully automatically but also multiplication and division.

SOURCE: *The Computer from Pascal to von Neuman* by H H Goldstine (Princeton, NJ: Princeton University Press) p 7 (1972)

SEE ALSO: *The Origins of Digital Computers* edited by B Randell (Berlin: Springer) (1973)

1729   **WIRE CONDUCTOR**   **S Gray (UK)**

In 1729, Stephen Gray (1696–1736) discovered the difference between electrical insulators and conductors. He found that electricity would flow along wires.

It was well known that some materials would, when rubbed, attract light objects such as pieces of paper. Gray found that some materials held the attractive charge and some did not.

He used a metal ball suspended from a charged rod by means of a thin wire. This 'conducts' the charge to the ball, which thus picks up the pieces of paper.

Gray identified conductors (metallic substances) and insulators (non-metallic substances).

SOURCE: British Library

1745   **CAPACITOR (LEYDEN JAR)**   **von Mushenbrock and Cunaeus (Germany) and von Kleist (Pomerania)**

According to the literature, the Leyden Jar was discovered almost simultaneously by Dean von Kleist of the Cathedral of Camin, Germany, in October 1745 and Peter von Muschenbrock, Professor in the University of Leyden, in January 1746, over 200 years ago. As described by them, it was a glass jar or vial with inner and outer electrodes of various things—water, mercury, metal foil etc. The modern miniature glass dielectric capacitor differs in form and structure from the 200-year-old Leyden Jar, but the principle of operation is the same.

SOURCE: 'History, present status and future developments of electronic components' by P S Darnell *IRE Trans. on Component Parts* p 127/8 (September 1958)

SEE ALSO: *Janus* C Dorsman and C A Crommelin, vol 46, p 274 (1957)

'Observations on the manner in which glass is charged with electric fluid' by E W Gray *Phil. Trans. R. Soc.* vol 77, p 407 (1788)

'Residual charge of the Leyden jar-dielectric properties of various glasses' by J Hopkinson *Phil. Mag.* Part 5, vol 4, p 141 (1877)

1772   **IRON DUST CORES**   **G Knight (UK)**

Iron dust cores consisted of iron filings churned in water, bound with linseed oil, moulded and fired. They were used in a Navy compass. Apparently, Knight was a secretive person and details of his process were not actually published before 1779.

SOURCE: Note from British Science Museum.

SEE ALSO: 'The early history of the permanent magnet' by Andrade *Endeavour* p 27 (January 1958)

B Wilson *Phil. Trans.* vol 69, p 51 (1779) (giving details of the process).

1779   **SPEECH SYNTHESIS**   **Kratzenstein (Russia)**

One of the earliest documented attempts at speech synthesis was made in 1779 when a Russian scientist called Kratzenstein constructed a set of five acoustic resonantors which, when activated by a vibrating reed, produced imitations of the vowels. In 1791 Wolfgang Von Kemplen, a Hungarian, constructed a more elaborate machine which could be made to speak whole words and phrases. It consisted of a large bellows which supplied a stream of air to a reed which, in turn, excited a hand-held rubber tube (resonator). Extra tubes and whistles were added to imitate the nasal and fricative sounds. A much more recent mechanical speech synthesiser was constructed by Reisz in 1937. The motion of the speech articulators was simulated by pressing keys to vary the shape of mechanical vocal tract. It could produce connected speech when operated by a skilled person.

SOURCE: *Signal Processing of Speech* by F J Owens (Basingstoke: MacMillan New Electronics) p 88 (1993)

NOTE: A comprehensive history of speech synthesis may be found in the book by Linggard (1985).

1780 **GALVANIC ACTION** **L Galvani (Italy)**

Luigi Galvani began his studies on the subject of animal electricity in 1780. When performing experiments on nervous excitability in frogs, he saw that violent muscle contractions could be observed if the lumbar nerves of the frog were touched with metal instruments in the presence of distant electrical discharges.

The word 'electricity' was reserved for static electricity and the word 'Galvanism' was proposed by von Humboldt for direct (continuous) current.

SOURCE: 'From torpedo to telemetry' by D W Hill *Electronics & Power* pp 1110–11 (27 November 1975)

1800 **DRY BATTERY** **A Volta (1800), De Luc (1809) and Zamboni (1812) (Italy)**

Volta's invention of the electric battery was announced in a letter to Sir Joseph Banks, the President of the Royal Society and described his 'Volta's Pile'—consisting of copper and zinc discs separated by a moistened cloth electrolyte. These were later improved to consist of paper discs tinned one side, manganese dioxide on the other, stacked to produce 0.75 V between 1 in diameter discs.

Note by Science Museum, London:

Scyffer described experiments with dry cells carried out by Ludicke (1801) Einhof, Ritter (1802) Hachette and Desornes, Biot and many others. Scyffer regarded these as experimental and ascribes the first effective pile to Behrens and to Marechaux but considered the best performance before Zamboni to have been achieved by De Luc in 1809. Zamboni (1812) himself ascribed priority to De Luc since his paper was entitled: 'Descrizione della colonna elettrica del Signore de Luc e considerazione sull analisi de lui Fatta della pile Voltiana'. Work on early dry batteries was, therefore, done from about 1800 to 1812.

SOURCE: 'On the electricity excited by the mere contact of conducting substances of different kinds' *Phil. Trans.* vol 90, p 403 (1800)

SEE ALSO: *A Biographical Dictionary of Scientists* by T I Williams p 535 (London: Adam and Charles Black) (1969) (Volta)

Scyffer *Geschichtliche Dartellung der Galvanisms* pp 135–48 (1848)

1800 **INFRA-RED RADIATION** **W Herschel (UK)**

In 1800, William Herschel, during research into the heating effects of the visible spectrum, discovered that the maximum heating was not within the visible spectrum but just beyond the red range. Herschel concluded that in addition to visible rays the sun emits certain invisible ones. These he called infra-red rays.

SOURCE: *Electronics Engineer's Reference Book* (London: Newnes-Butterworth) chapter 4, p 4-2 (1976)

1801 **ULTRA-VIOLET RADIATION** **J W Ritter (Germany)**

In 1801 the German physicist Ritter made a further discovery. He took a sheet of paper freshly coated with silver chloride and placed it on top of a visible spectrum produced from sunlight falling through a prism. After a while he examined the paper in bright light. It was blackened, and it was blackened most just beyond the violet range of the spectrum. These invisible rays Ritter called ultra-violet rays.

SOURCE: *Electronics Engineer's Reference Book* (London: Newnes-Butterworth) chapter 4, p 4-2 (1976)

1803 **ACCUMULATOR** **J W Ritter (Germany)**

Ritter's charging or secondary pile consists of but one metal, the discs of which are separated by circular pieces of cloth, flannel or cardboard, moistened in a liquid which cannot chemically affect the metal.

When the extremities are put in communication with the poles of an ordinary voltaic pile it becomes electrified and can be substituted for the latter and it will retain the charge.

SOURCE: *Biographical History of Electricity and Magnetism* by Mottelay (London: Charles Griffin & Co) p 381 (1922)

1808   **ATOMIC THEORY**   **J Dalton (UK)**

Dalton conceived the idea that the atoms of different elements were distinguished by differences in their weights. In 1808, he propounded the theory that all chemical elements are composed of minute particles of matter called atoms and that these atoms, as the name implies, cannot be cut up any further. All atoms of one element, he said, were alike but atoms of different elements had different weights. The atom of hydrogen was the lightest atom (1.66×10^{-24} of a gram) and the weights of all other atoms were compared with it.

SOURCE: *New System of Chemical Philosophy* by J Dalton (1808)

SEE ALSO: *A Biographical Dictionary of Scientists* by T I Williams (London: Adam and Charles Black) p 128 (1969)

1812   **UNIVERSAL ELECTRONIC CALCULATOR**   **C Babbage (UK)**
       (DIFFERENCE ENGINE)

Charles Babbage (1791–1871) first conceived the idea of an advanced calculation engine to calculate and print mathematical tables in 1812, as he wanted to eleminate all the sources of inaccuracy associated with compiling mathematical tables by hand. The engine was built in 1824 and assembled in 1832 by Joseph Liement, a skilled toolmaker and draughtsman. It was a decimal digital machine—the value of a number represented by the position of toothed wheels marked with decimal numbers. The front detail of Charles Babbage's Difference Engine is shown in figure 11.1. (See also entry of 1833.)

SOURCE: The Science Museum/Science and Society Picture Library (SCM/CCM/C10011ZH).

1812   **CABLE INSULATION**   **Sommering & Schilling (Germany)**

It was in that year (1812) that Sommering and Schilling conducted a series of experiments in which a soluble material, said to be indiarubber, was first used for insulating wire, following a suggestion made by a Spaniard, named Salva, in 1795 concerning the feasibility of submarine telegraphy. Curiously enough, this first 'cable' developed by Sommering and Schilling was in a sense a power cable as the objective was the detonation of mines. For at least another 50 years, however, practically all development was to be concerned with telegraphy.

SOURCE: 'Electric cables' by S O E Goodall *Proc. IEE* vol 106, Part B, No 25, p 1 (January 1959)

1817   **OPTOPHONE**   **E E Fournier d'Albe (UK)**

The instrument is the invention of Dr E E Fournier d'Albe, and has been developed by Messrs Barr & Stroud, Limited, Glasgow (see figure 11.2). It depends for its action upon a very remarkable property of selenium, a chemical element (discovered in 1817 by the Swedish chemist Berzelius) which, in its grey crystalline form varies greatly in electrical conductivity in accordance with the amount of light to which it is exposed, though the resistance is always high.

Instruments can therefore be constructed that can detect pulsations of light of periods corresponding to those of the vibrations in audible sounds. A properly prepared selenium bridge connected in series with a battery and a telephone receiver and exposed to illumination and eclipse alternating some hundreds of times per second, causes corresponding pulsations of current through the telephone, and produces audible sound of corresponding pitch and quality.

SOURCE: *The British Encyclopaedia* vol 7, p 498 (Gotham Press) (1933)

Figure 11.1. Front detail of Charles Babbage's difference engine (The Science Museum/ Science & Society Picture Library).

Optophone in use

Figure 11.2. The Optophone.

1820   ELECTROMAGNETISM (Galvanometer)   **H C Oersted (Denmark)**

In 1820, Oersted reported the discovery of electromagnetism, and this led him to develop the first galvanometers. It was John Schweigger who constructed the first moving-coil instrument and Nobilli (1828) an Italian physicist, developed a sensitive astatic galvanometer and compared its sensitivity with that of the most sensitive galvanometers then available.

SOURCE: 'From torpedo to telemetry' by D W Hill *Electronics & Power* p 111 (27 November 1975)

SEE ALSO: 'Experiments on the effect of a current of electricity on the magnetic needles' by H C Oersted *Annals of Philosophy* vol 16, p 273 (1820)

1821   THERMOELECTRICITY   **T J Seebeck (Germany)**

The discovery of thermoelectricity is usually attributed to Professor T J Seebeck of Berlin in 1821, although there is some evidence that he might have been anticipated by Dessaignes in 1815. Professor Cummings of Cambridge also discovered the effect independently and published his findings in 1823.

Following Seebeck's work, J C A Pettier completed Seebeck's discovery by showing that the passage of electricity through a junction of two different metals (antimony and copper) could produce a rise in temperature at the junction when passing in one direction and a drop in temperature when passing in the contrary direction.

The introduction of the first successful thermopile in the sense of an array of thermocouples (analogous to the galvanic pile) is attributed to Nobilli. Nobilli's thermopile was subsequently improved by Melloni.

SOURCE: Note from British Science Museum, London (lst and 3rd paragraphs), author (2nd paragraph)

1828   FOURIER ANALYSIS   **J-B-J Fourier (France)**

Jean-Baptiste-Joseph Fourier laid the basis for the analysis of complicated waveforms more than 150 years ago when he showed that any waveform is the sum of single-frequency, or sinusoidal, components. Scientists and engineers have learned that looking at the frequency spectrum of a signal's components provides great insight into the way signals behave with time. The analysis is possible only with a powerful Fourier transform, which maps a time-varying signal into the frequency domain and thus causes the spectral distribution of its sinusoidal components to become visible. See figure 11.3.

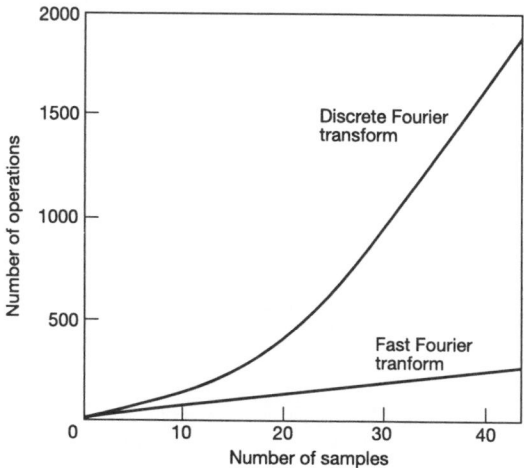

Figure 11.3. Fourier analysis.

Since the Fourier transform requires evaluation of the integral of the input waveform, it can only be performed exactly on a waveshape with a known equation. But in real life, most waveforms are far too complex to be readily defined. The discrete Fourier transform, however, approximates the actual transform by sampling the waveform and operating on each point. It can therefore be computed digitally, but the numbers of operations it requires is so huge—the square of the number samples—that even a reasonable approximation taxes the largest computers.

In 1942, a method was devised for reducing the number of operations from N^2 to $N\ln N$, but as this was before the days of high-speed digital computers and few people cared to grind through the DFT by hand, the work of Danielson and Lanczos lay dormant for nearly twenty years.

Then in the early 1960s interest in the Fourier transform again picked up. J W Tukey worked with J W Cooley on the development of a transform algorithm suitable for the computer, and the result came to be known as the fast Fourier transform. Today, though many variations exist on the original Cooley–Tukey algorithm, any transform that requires $N\ln N$ operations is referred to as the FFT. As the graph shows, the difference in the number of operations required for the DFT and the FFT grows enormously for large numbers of samples.

SOURCE: 'The ubiquitous FET' by R Capace *Electronics* (16 March 1978)

1826 **OHM'S LAW**   **G S Ohm (Germany)**

Ohm was Head of the Department of Mathematics and Physics at the Polytechnic Institute of Cologne when he discovered the law bearing his name.

$$E = IR \qquad R = \frac{E}{I} \qquad I = \frac{E}{R}$$

SEE ALSO: *Die Galvinische Kette Mathematisch Gearbeitet* by G S Ohm (Berlin: Springer) (1827)

1828 **MOVING COIL GALVANOMETER**   **J Schweigger (Germany)**

1828 **ASTATIC GALVANOMETER**   **C L Nobilli (Italy)**

In 1820, Oersted reported the discovery of electromagnetism, and this led him to develop the first galvanometers. It was John Schweigger who constructed the first moving-coil instruments and Nobilli (1828) an Italian physicist, developed a sensitive astatic galvanometer and compared its sensitivity with that of the most sensitive galvanometers then available

SOURCE: 'From torpedo to telemetry' by D W Hill *Electronics & Power* pp 111 (27 November 1975)

SEE ALSO: 'Comparison entre les deux galvanomètres les plus sensibles, la grenouille et le multiplicateur à deux aiguilles, suivie de quelques resultats nouveaux' by C L Nobilli *Chim. et Phys.* vol 43, pp 256–8 RS(1828)

1831 **ELECTROMAGNETIC INDUCTION**   **M Faraday (UK)**

(see also SELF INDUCTION 1832)

In 1831 Faraday wound an iron ring with two coils: one, connected to a voltaic battery, was to create the primary vibration—the iron ring was to concentrate the lateral vibrations from this—and another coil on the opposite side of the ring was to convert these secondary vibrations into another electric current. Thus, on 29 August 1831, Faraday discovered electromagnetic induction.

SOURCE: *A Biographical Dictionary of Scientists* by T I Williams (London: Adam and Charles Black) (1969)p 174

1831 **TRANSFORMER**   **M Faraday (UK)**

A contrivance was used by Michael Faraday in his experiments on electromagnetic induction. This device is described in his diary under the date of 29 August 1831:

'Expts. on the production of Electricity from Magnetism, etc. etc. Have had an iron ring made (soft iron) iron round and 7/8 inch thick and ring 6 inches in external diameter. Wound many coils of copper wire round one half, the coil is being separated by twine and calico there were three lengths of wire each about 24 feet long and they could be connected as one length or used as separate lengths. By trial with a trough each was insulated from the other. Will call this side of the ring A. On the other side but separated by an interval was wound wire in two pieces together amounting to about 60 feet in length, the direction being as with the former coils: this side call B.

Charges a battery of 10 pr. plates 4 inches square. Made the coil on B side one coil and connected its extremities by a copper wire passing to a distance and just over a magnetic needle (3 feet from iron ring). Then connected the ends of one of the pieces on A side with battery; immediately a sensible effort on needle. It oscillated and settled at last in original position. On breaking connection of A side with battery again a disturbance of the needle.

Made all the wires on A side one coil and sent current from battery through the whole. Effect on needle much stronger than before'. A multiwinding transformer and a transformer experiment are described by these words of Faraday's.

SOURCE: 'History, present status and future developments of electronic components' by P S Darnell *IRE Trans. on Component Parts* p 125 (September 1958)

SEE ALSO: *Faraday's Discovery of Electromagnetic Induction* by T Martine (London: Edward Arnold & Co) pp 52–4 (1949)

Note by British Science Museum:

'The word 'transformer' was first used in the electrical sense in 1883 for both static transformers and rotating machines (motor generators). Previous to this 'induction coil' had been used. Faraday was the first to link two electric circuits by a magnetic circuit. The earliest application of the device to transfer power appears to be Jablochkoff's patent No 1996 (date of application 22 May 1877). [1] This was really the first practical 'transformer' and it utilized a piece of apparatus which had existed since 1831.'

1832   **SELF INDUCTION**   **C J Henry (USA)**

(see ELECTROMAGNETIC INDUCTION 1831)

Henry is reported to have discovered the phenomenom of self induction in 1830 but, through his failure to publish, priority was given to Michael Faraday to Henry's great mortification.

REFERENCE: *Joseph Henry* by T Coulson (Princeton, NJ: Princeton University Press) (1950)

SEE ALSO: *A Biographical Dictionary of Scientists* by T I Williams (London: Adam and Charles Black) (1969)p 250

1832   **FIVE-NEEDLE TELEGRAPH SYSTEM**   **Schilling & Cooke (Russia & UK)**

In 1832, Baron Schilling demonstrated a telegraph using five magnetic needles. Had Schilling not died, he might have installed a system for Emperor Nicholas of Russia in 1836. In that same year, William Fothergill Cooke (1808–70) saw Schilling's telegraph, and was so inspired that he determined to invent a system of his own.

SOURCE: British Library

1833   **COMPUTER (Calculating Machines)**   **C Babbage (UK)**

In 1833 Babbage conceived his analytical engine, the first design for a universal automatic calculator. He worked on it with his own money until his death in 1871. Babbage's design had all the elements of a modern general purpose digital computer, namely: memory, control, arithmetic unit and input/output. The memory was to hold 1000 words of 50 digits each, all in counting wheels. Control was to be by means of sequences of Jacquard punched cards. The very important ability to modify the course of a calculation according to the intermediate results obtained—now called conditional branching—was to be incorporated in the form of a procedure for slipping forward or backward a specified number of cards.

[1] Jablochkoff's patent covered a lighting system which includes a mention of a transformer transferring power supply to a number of lamps

SOURCE: 'The evolution of computing machines and systems' by Serrell, Astrahan, Patterson and Pyne *Prog. IRE* p 1042 (May 1962)

SEE ALSO: *The Computer from Pascal to von Neuman* by H H Goldstine (Princeton, NJ: Princeton University Press) p 10 (1972)

'On the mathematical powers of the calculating engine' by Charles Babbage, 26 December 1837 (in Randell's book p 17)

1834   ELECTROLYSIS   M Faraday (UK)

Faraday announced his two laws of electrolysis in 1834 which made explicit the amount of force required; for a given amount of electrical force, chemical substances in the ratio of their chemical equivalent were released at the electrodes of an electrochemical cell. put another way, chemical affinity was electrical force acting on the molecular level.

SOURCE: *A Biographical Dictionary of Scientists* by T I Williams (London: Adam and Charles Black) p 175 (1969)

1837   RELAYS   W F Cooke, C Wheatstone and E Davy (UK)

Telegraph Bell Relay

Telegraph Signal Relay

The first patent was taken out by Edward Davy in 1838 (British Patent No 7719) 'I claim the mode of making telegraph signals or communications from one distant place to another by employement of relays of metallic circuits brought into operation by electric currents'. In 1837 (British Patent No 7390) Cooke and Wheatstone described an electromagnetic relay device for bringing a local battery at the distant station into action to sound an alarm bell there, However, Davy was described by Fakie as working on telegraphy as early as 1836 and entered an opposition to Cooke and Wheatstone's 1837 application for a patent, but the patent was granted. Morse in the USA is credited with a patent in 1840 (US Patent No 1647) which is apparently similar to Davy's patent.

SOURCE: Note by British Science Museum, London.

SEE ALSO: *A History of Electric Telegraphy in the Year 1837* by J J Fakie (London, 1884)

A Biographical Dictionary of Scientists by T I Williams (London: Adam and Charles Black) (Sir W Fothergill Cooke) p 114 (1969)

1837   TELEGRAPHY—MORSE CODE   S B Morse (USA)

While Cooke and Wheatstone of England had proposed electrical telegraph principles, it took the genius of Samuel B Morse to develop the electrical hardware and also recognise the essential elements for a simple code adaptable to his on-off (binary) telegraph system. He first demonstrated his system in 1837–38 and put it into practice in 1844, after he obtained a government grant to connect Baltimore with Washington DC, a distance of 37 miles. This was the first practical development of electrical telecommunications. It met a real need for fast communications and spread rapidly.

SOURCE: 'Telecommunications—the resource not depleted by use. A historical and philosophical resume' by W L Everitt *Proc. IEEE* vol 64, No 9, p 1293 (September 1976)

1837   ELECTRIC MOTOR   T Davenport (USA)

The earliest known examples of a patent for an electric motor is US Patent No 132 granted on 25 February 1837 to Thomas Davenport, of Brandon, Vermont, entitled: 'Improvements in Propelling Machinery by Magnetism and Electromagnetism'.

According to the description contained in the specification, the motor, which is intended to be driven by a 'galvanic battery', is constructed on sound electromagnetic principles.

SOURCE: *Patents for Engineers* by L H A Carr and J C Wood (London: Chapman and Hall) p 87 (1959)

1839   **MICROFILMING**   **Dancer (UK)**

Shortly after the publication of Daguerre's invention of making photographs in 1839, Dancer produced in England photographs of documents of strongly reduced size (1:160) having a side length of about 3 mm. The knowledge of the possibility to produce reduced-size photographs prompted in 1835 the English astronomer John Herschel to suggest, to store documents of general concern (e.g. reference works) in a reduced form, provided the reduction does not involve any hazard for the original documents. The same idea was advanced at the beginning of this century by members of a Belgian library, in order to make old handwritings or prints accessible to many people. Unfortunately, this idea has never materialized.

SOURCE: 'A brief historial review on microfilming' by H Scharffenberg and R Wendell *Jena Review* No 1, p 4 (1976)

1839   **BATTERY (Magnetohydrodynamic)**   **M Faraday (UK)**

The idea of producing electricity from a moving fluid, which is the basis of an MHD generator, was proposed by Faraday in 1839. A conducting fluid is passed between the poles of a magnet and an electromotive force is produced at right angles to the field. This principle is also used in electromagnetic pumps and induction flow meters for conducting fluids. In recent proposals by Kantrowitz and Spron (1959) the working fluid is a conducting gas at high temperatures. The gas moving at a high velocity is passed through a magnetic field at right angles to the direction of florid electrodes placed on opposite sides of the channel extract the power and are connected to the external load.

SOURCE: 'The magnetohydrodynamic generation of power' by K Phillips *AEI Engineering* p 62 (March/April 1964)

SEE ALSO: *Experimental Researches in Electricity* by M Faraday (London, 1839)

'Application of the MHD concept to large scale generation of electric power' by A Kontrowitz and P Sporn (American Electric Power Service Corsoration and AVCO Research Laboratory) (1959)

1839   **PHOTOVOLTAIC EFFECT**   **E Becquerel (France)**

The photovoltaic effect was discovered by Edmond Becquerel as early as 1839. In 1873, Willoughby Smith first observed the photoconductivity of selenium. In 1887, finally, Heinrich Hertz described the photoemissive effect of ultraviolet light on metal electrodes.

SOURCE: 'Beam-deflection and photo devices' by K Schlesinger and E G Ramberg *Proc. IRE* p 991 (May 1962)

SEE ALSO: 'On electric effects under the influence of solar radiation' by E Becquerel *C. R. Acad. Sci. Paris* vol 9, p 561 (1839)

'Effect of light on selenium during the passage of an electric current' by W Smith *American J. Sci.* vol 5, p 301 (1873)

1839?   **BATTERY (Fuel Cell)**   **W R Grove (UK)**

The fuel cell principle—that is, the conversion of chemical energy to electric energy by a path that can avoid the thermodynamic limitation on efficiency imposed by the Carnot relation—has intrigued scientists and engineers for more than a century. In 1839–42, Sir William Grove probably invented the first fuel cell. He used platinum-catalyzed electrodes to combine hydrogen and oxygen so as to produce electricity. It is interesting to note that almost 123 years later the Gemini fuel cell used the same catalyst, though probably in different physical form. One major trouble with Grove's cell was that its voltage fell off badly when an appreciable current drain was put on it. In 1889, Mond and Langer made a hydrogen–oxygen cell with perforated platinum-sheet electrodes, catalysed by platinum black. This cell produced 1.46 watts at 0.73 volt at about 50 percent efficiency. However, it contained

1.3 grams of platinum and required pure hydrogen and oxygen. Thus, its capital cost made it a poor buy as an electric generator. In addition, to complicate things, it ran well only on pure hydrogen and oxygen.

SOURCE: 'Hydrocarbon–air fuel cell systems' by C G Peattie *IEEE Spectrum* p 69 (June 1966)

SEE ALSO: 'On voltaic series in combination of gases by platinum' by W R Grove *Phil. Mag.* vol 14, p 127 (1839)

'On a gaseous voltaic battery' by W R Grove *Phil. Mag.* vol 21, p 417 (1842)

NOTE. The principle of electrochemical fuel cells is by no means new, in fact the first cell was described in 1839 by Sir William Grove. This was a hydrogen fuel device with a sulphuric acid electrolyte and blacked platinum electrodes, generating approximately one volt at a very small current. The first power fuel cell was a 5 kilowatt unit demonstrated in 1959 by the English engineer, F T Bacon, and employed hydrogen–oxygen fuel with an alkaline electrolyte and sintered nickel electrodes. Rights to the development of this cell were obtained by Leesona-Moos Laboratories—a research subsidiary of Pratt & Whitney Aircraft Corp.—from NRDC and a modified version was used in the Apollo spacecraft.

SOURCE: 'Fuel cells and their development in the UK' by W S E Mitchell *Design Electronics* p 34 (February 1966)

1840 THERMOGRAPHY   J Herschel (UK)

Sir William Herschel's famous experiment with thermometers and a prism in 1800 showed the existence of energy beyond the red end of the visible spectrum. As long ago as 1840 his son John demonstrated thermal imaging and saw images in the dark. Considering the sophisticated techniques in use today, John Herschel's methods were both embarrassingly simple and successful. He took a blackened sheet of paper, soaked it in alcohol and focused radiation from a hot source onto the sheet. The infra-red radiation selectively heated parts of the paper, evaporating the alcohol and lightening its colour to form an image.

Variations on this theme took place over the next 100 years but not until 1940, with the pressing need for military systems with real-time infra-red tracking, was significant progress made. Since then optics, detectors, amplifiers, signal processing and displays have improved to the extent that high-performance infra-red systems can now produce television-quality pictures of a scene (at several kilometres' distance) which contains temperature differences of only a fraction of a degree. Thermal resolution, angular resolution and frame time have all been improved by a factor of ten or greater, representing an overall performance improvement of three or more decades.

SOURCE: 'Infra-red imaging systems' by H Blackburn *Systems Technology* (Plessey) No 26, p 15 (June 1977)

1843 WHEATSTONE BRIDGE   C Wheatstone (UK)

Wheatstone described his bridge circuit—which he called the 'differential resistance measurer'—in a comprehensive paper on electric measurements presented to the Royal Society in 1843. That paper would have merited publication in *Proc. IEE*, had there been such a journal then, for the way in which it presented solutions to electrical-engineering problems. The problems were the measurement of electromotive force, current strength and resistance at a time when galvanometers were unstable instruments and there was no sound method of calibrating them. Wheatstone found the theoretical basis for solving these problems in an obscure German publication of 1827 in which G S Ohm showed that there was a simple mathematical relationship linking e.m.f. current and resistance.

The basis of Wheatstone's method of measurements is the use of a calibrated variable resistance to keep the current constant and so avoid any need for calibration of the galvanometers Wheatstone remarked that it was easy to make a calibrated variable resistance (for which he devised the term 'rheostat') because Ohm's work showed that the resistance of a conductor of uniform section was proportional to its length. He showed how to determine the value of an unknown resistance by a simple substitution method. The unknown resistance is included in a circuit with a galvanometer whose reading is noted and then replaced by a rheostat which is adjusted so that the galvanometer reading is the same as before. The scale on the rheostat gives its resistance and hence the value of the unknown resistance.

SOURCE: 'Wheatstone's contribution to electrical engineering' by B P Bowers *Electronics & Power* p 295 (May 1976)

1843 FACSIMILE REPRODUCTION   A Bain (UK)

For the purpose of this review facsimile is considered to be a method by which printed, handwritten and graphic data may be transmitted over communication channels and received in the form of a hard copy. Its origin dates back to 1843, when the Scottish inventor Alexander Bain patented an 'automatic electrochemical recording telegraph'. Next came Frederick Bakewell's cylinder and screw arrangement on which many of the present-day facsimile systems are based.

In Europe facsimile equipment has been commercially available since 1946, In recent years the technology advances in electronics and the drastic fall in semiconductor prices have led to the replacement of bulky separate facsimile transmitters and receivers by small transceivers

SOURCE: 'Facsimile—a review' by J Malster and M J Bowden *The Radio and Electronic Engineer* vol 46, No 2 p 55 (February 1976)

1845 METALLIC SHEATHING OF CABLES   C Wheatstone (UK)

The earliest attempts at metallic sheathing were made in 1845 by Wheatstone, who folded lead strip around the cable core and then joined it with a longitudinal soldered seam. This method bears an interesting similarity to a modern method in which aluminium sheathing is applied using pre–formed strip. The method was superseded by one which involved soldering 50 ft. lengths of lead pipe end to end and subsequently sinking the pipe into contact with the cable core by means of a die. This again has an interesting parallel in a present day method for aluminium sheathing.

In 1879 the first direct extrusion on to cable was made from a Borel press using solid hollow billets pre-heated to 120°C. It was not then considered good practice to recharge with molten lead because of possible thermal damage to the insulation, so the process was limited to a one billet charge. Development of lead sheathed cable has since gone hand in hand with the development of the lead extrusion process. Different types of press have been designed to overcome defects experienced with cable in service but extrusion in one form or another has been universally adopted. By contrast, the present development of aluminium sheathing processes is proceeding along three different lines and it is hard to predict which technique will ultimately prove most successful. The greater difficulty of extruding aluminium as compared with lead has undoubtedly favoured this more varied approach.

SOURCE: 'The metallic sheathing of cables' by A V Garner *AEI Engineering* p 248 (September/October 1962)

1845 CIRCUITS—KIRCHOFF'S LAWS   G R Kirchhoff (Germany)

Two laws that express the behaviour of an electrical network. In 1845 he gave the laws for closed circuits, extending these to general networks (1847) and to solid conductors (1848)

1st law: The total current arriving at any point in an electric network must be zero.

2nd law: The sum of the electromotive forces around any closed circuit is equal to the sum of the IR drops around the circuit.

1847 SUBMARINE CABLE INSULATION   W Siemens (Germany)

Telegraphic instruments had been developed, notably by Schilling, Morse and Cooke and Wheatstone, for the transmission of signals over land lines; and a suitable material for insulating the conductor had already been introduced into Europe. This was gutta percha, the gum from a Malayan tree, exhibited at the Royal Society of Arts in London by Dr Montgomerie in 1843. The electrical and mechanical properties of gutta percha, especially when immersed in sea water, were such that it had no rival for over 70 years as an insulant in submarine cables. Moreover, the Gutta Percha Company had been formed in 1845–6 to exploit its use, and primitive extruders had been constructed for the production of rods and tubes.

Before 1849, many attempts had been made to find a suitable insulant or protection for underground and underwater cables, including tarred rope, glass tubes, split rattan, impregnated cotton and rubber, but none of these lasted long in the sea. In 1847 Werner Siemens used gutta percha for an underground line in Germany, and some was laid in the Port of Kiel in 1848 for the purpose of detonating mines. In America in the same year, Armstrong experimented with gutta percha covered wire in the Hudson River, where Ezra Cornell had previously connected Fort Lee with New York by a rubber insulated line which worked for several months.

SOURCE: *The Story of the Submarine Cable* booklet published by Submarine Cables Limited (AEI, London) p 4 (1960)

1847 MAGNETOSTRICTION   J Joule (UK)

Magnetostriction is a well-known phenomenon in which the mechanical dimension of a magnetic material is altered as the magnetisation is varied, and in which, conversely, the magnetisation is altered as the dimension is changed. Thus, an alternating current applied through a coil wrapped around a specimen can induce mechanical vibrations in it; and alternately, mechanical vibrations set up in such a specimen can transform the mechanical energy into electrical energy in a coil wound around it. Electro-mechanical interactions were observed as early as 1847 with Joules discovery of magnetostriction.

SOURCE: 'Solid state devices other than semiconductors' by B Lax and J G Mavroides *Proc. IRE* p 1014 (May 1962)

SEE ALSO: 'On the effects of magnetism on the dimensions of iron and steel bars' J P Joule *Phil. Mag.* vol 30, p 226 (April 1847)

1848 BOOLEAN ALGEBRA   G Boole (UK)

Formal logic, so necessary for the workings of digital computers could not be satisfactorily explained mathematically before George Boole. In 1848 the English logician published 'The Mathematical Analysis of Logic' and in 1854 'An Investigation of the Laws of Thought', the foundation of what is now symbolic logic. With the theories expounded in these writings, it was possible to express logic in very simple algebraic systems. The equation $X^2 = X$ for every X in the system is basic to Boolean algebra and has only 0 or 1 as an answer in numerical terms. Thus modern computers can make use of this binary system, with their logic parts carrying out binary operations.

SOURCE: *Electronics* p 69 (17 April 1980)

1850 THERMISTOR   M Faraday (UK)

The temperature-sensitive non-linear resistors are known generally as thermistors, a name coined by the Bell Telephone Laboratories (of the USA). They are, however, over 100 years old, for Faraday discovered that silver sulphide possessed a high negative temperature coefficient (although in this case the conduction is ionic and not electronic, and the material therefore suffers from polarisation effects). Uranium oxide was used in Germany, but with this conduction is also ionic and operation is unstable. A magnesium titanate spinet was introduced in 1923, and in the USA about 1912, boron was found to possess negative temperature characteristics. From 1930 onwards the Bell Telephone Laboratories devoted many years of intensive research to the problem and showed that combined oxides of manganese and nickel had valuable properties, They also found that varying the ratio between the manganese and nickel varied these properties The effects of adding small amounts of copper, cobalt and iron were also investigated.

Today, these oxides, treated to become uniphase, are in general use, and are made into beads, rods, blocks, etc.

SOURCE: *Affixed Resistors* 2nd edition by G W A Dummer (London: Pitman) p 147 (1967)

**1851   RELATION BETWEEN THEORY OF MAGNETISM   Lord Kelvin (UK)
AND ELECTRICITY**

Introduced for the first time the vectors later termed magnetic induction and magnetic force by Maxwell.

In a fundamental papers he derived a result expressing the energy of a system of permanent and temporary magnets in terms of a volume integral throughout space.

SOURCE: *A Biographical Dictionary of Scientists* by T I Williams (London: Adam and Charles Black) p 512 (1969)

1852   THIN FILMS (Sputtering Process)   W R Grove (UK)

Although the use of cathodic sputtering as a method for the deposition of thin films predates vacuum evaporation by many years, the latter has been received far more widespread application because evaporation is more convenient for many materials and generally gives high deposition rates. In recent years, however, it has been found that certain materials are more conveniently deposited by sputtering. In some cases it is impossible to deposit materials by any other means.

SOURCE: 'Thin-film circuit technology' by A E Lessor, L I Maissel and R E Thun *IEEE Spectrum* p 73 (April 1964)

SEE ALSO: W R Grove *Phil. Trans. R. Soc. London.* Series B vol 162, p 87 (1852)

1854   COMPUTERS (Calculating Machines)   P G Scheutz (Sweden)

Pehr Georg Scheutz built a difference engine in Stockholm inspired by Babbage's ideas and displayed it in London in 1854 with considerable help from Babbage. The machine had four differences and fourteen places of figures and was capable of printing its own tables (Scheutz was a printer).

SOURCE: *The Computer from Pascal to von Neuman* by H H Goldstine (Princeton, NJ: Princeton University Press) p 7 (1972)

1855   COLD CATHODE DISCHARGE TUBE   J M Gaugain (France)

Experiments with low-pressure glow discharges started very early in the electrical art. Most of them were in small diameter glow tubes, often in the form called Geissler tubes. The first recognition of the fact that such a glow tube having its two electrodes of different size was capable of rectifying the oscillating current from an induction coil appears to have been that of Gaugain in 1855. However, for many years, the only use for these discharges was as light sources. The early glow lamps all required high voltage excitation. With the availability of the rare gases neon and argon and by means of low work function cathodes, glow tubes were developed for low voltage applications.

SOURCE: 'The development of gas discharge tubes' by J D Cobine *Proc. IRE* p 972 (May 1962)

SEE ALSO: J M Gaugain *C. R. Acad. Sci. Paris* vol 40, p 640 (1855)

1856   LOW PRESSURE DISCHARGE TUBES   H Geissler (Germany)

In 1856 Heinrich Geissler, an artist and skilful glass blower of Bonn, Germany, originated the low pressure discharge tubes that were to bear his name. The Geissler tubes were long, small-bore glass tubes, usually shortened by the use of many coils and bends, which were filled with various gases at low pressures and originally excited by high-voltage alternating current. Many beautiful effects could be produced by Geissler tubes filled with different gases and they were often used for decorations. As, for example, a display used to commemorate Queen Victoria's Diamond Jubilee. However, sputtering of the electrodes together with gas clean-up resulted in a short life for the tube. The principal use was for spectral analysis and lecture demonstrations.

SOURCE: 'The development of gas discharge tubes' by J D Cobine *Proc. IRE* p 970 (May 1962)

SEE ALSO: 'The electric-lamp industry' by A A Bright Jr (New York: MacMillan) p 218 *et seq.* (1949)

W DeLaRue, H W Muller and W Spottieswoode *Proc. R. Soc.* vol 23, p 356 (1875)

1857   **MERCURY ARC LAMP**   **J T Wray (UK)**

The first public demonstration of a mercury arc lamp was by Professor J T Wray on the Hungerford Suspension Bridge in London on 3 September 1860. Two British patents were issued to him in 1857. The electric arc was first used commercially for illumination in Paris in 1863. Much later, low-pressure arc 'tubes' were used for illumination. In 1879 John Rapieff described mercury arc lamps in British Patent No 211 but there appears to be no evidence that they were built. Peter Cooper Hewitt showed in public his mercury-arc lamp on 12 April 1901. Georges Glaude, a French inventor, demonstrated the first Neon sign, an improvement of the Geissler tube, at the Grand Palais in Paris in 1910. Developments in luminous tube discharges were made by Moore in 1920. Since these tubes did not have a high light output, they were largely confined to sign applications.

SOURCE: 'The development of gas discharge tubes' by J D Cobine *Proc. IRE* p 970 (May 1962)

SEE ALSO: British Patents issued 1857

1858   **GLOW DISCHARGES**   **J P Pleucker (Germany)**

In 1858 Pleucker investigated experimentally the luminous effects of electric discharge through gases at low pressures. He observed that the glow was deflected in a strong magnetic field.

SOURCE: *A Biographical Dictionary of Scientists* by T I Williams (London: Adam and Charles Black) p 422 (1969)

1860   **MICROPHONE (Diaphragm type)**   **J P Reis (Germany)**

Earliest among microphone diaphragms—perhaps because of its similarity to the eardrum—was a stretched flat membrane (actually a sausage skin) used by Reis to actuate a loose metal-to-metal contact. A stretched flat membrane made of metal or very thin metallized plastic is used in present-day electrostatic microphones. This diaphragm is typically clamped at its periphery by a ring and stretched to any desired tension by a threaded ring.

SOURCE: 'A century of microphones' by B B Bauer *Proc. IRE* p 720 (May 1962)

SEE ALSO: 'Ueber Telephone durch den galvaniscen strom' *Jahnesbericht d. Physikalischen* (Vereins zu Frankfurt am Main, Germany) p 57 (1860–61)

1860   **BATTERY (Secondary)**   **Planté (France)**

The secondary battery business dates back to Plante's discovery of the lead-acid system in 1860.

Secondary or storage cells are electrochemical cells which after discharge can be restored to their original chemical state by passing the current in the reverse direction. Although they have the same set of basic components as primary cells, the anodes and cathodes of secondary cells have a more stringent requirement, in that the electrode reactions have to be reversible. This requirement immediately limits the number of electrode materials available for secondary cells. At present, lead, cadmium, iron and zinc anode materials, and lead dioxide, nickel dioxide and silver oxide cathode materials are the only ones used in commercial secondary cells.

SOURCE: 'Batteries' by C K Morehouse, R Glicksman and G S Lozier *Proc. IRE* p 1474/5 (August 1958)

1865   **RADIO WAVE PROPAGATION**   **J C Maxwell (UK)**

In his first paper on electromagnetism 'On Faraday's Lines of Force' (1855–56) Maxwell set up partial analogies, between electric and magnetic lines of force and the lines of flow of an incompressible fluid. In a series of magnificent papers in 1861–62 he gave a fully developed model of electromagnetic phenomena viewed in the light of the field concept of Michael Faraday of whose validity Maxwell had become fully persuaded by 1858, Adopting the belief of William Thomson (Lord Kelvin) in the rotary nature of magnetism, a magnetic tube of induction was represented by a set of cells rotating about the axis of the tube, interference between the rotations of neighbouring tubes being avoided by rows of intervening cells (in the manner of idle wheels) which corresponded to electric currents. By means of this model Maxwell was able to give an elegant qualitative interpretation of all the known phenomena of electromagnetism. By introducing the notion of elasticity he was then able to give a quantitative description of the propagation of a disturbance in the model. Reinterpreted in terms of the electromagnetic field, this implied that a disturbance in the electromagnetic field should travel with a speed equal to the ratio of the electrodynamic to the electrostatic units of electric force.

SOURCE: *A Biographical Dictionary of Scientists* by T I Williams (London: Adam and Charles Black) p 358 (1969)

SEE ALSO: 'A dynamical theory of the electromagnetic field' by J C Maxwell *Proc. R. Soc. London* vol 13, pp 531–6 (8 December 1864)

1866   TRANSATLANTIC TELEGRAPH CABLE   TC & M Co (UK)

The first successful Atlantic telegraph cables were made and laid in 1866 by The Telegraph Construction and Maintenance Company, then newly formed by the amalgamation of the Gutta Percha Company and Glass, Elliot and Company. The laconic telegram sent on 27 July 1866 by Mr R A Glass, the Managing Director of the Company, from Valentia on the completion of the laying of the first of the two cables by the famous 'Great Eastern', read simply 'All right', and had the more modern two-letter abbreviation been available the message would doubtless have been even shorter.

A few days later came the report that a message of 405 letters from the President of the United States, replying to Queen Victoria, had been sent at a speed of 37 letters per minute. With this performance, modest indeed compared with the modern speed of over 2000 letters per minute, the efforts of the intrepid pioneers were crowned with success and the submarine cable was firmly established as a commercial proposition for oceanic depths.

SOURCE: 'The Story of the Submarine Cable' booklet published by Submarine Cables Limited (AEI) London (1960) p 8

1868   RHEOTOME (Waveform plotter)   H Lenz (Germany)

Heinrich Lenz (of Lenz's Law) developed a segmented commutator or rheotome which could be arranged to sample a periodic waveform at known points of the cycle and thus feed a train of pulses, each corresponding to the amplitude of the waveform at that point, to a slowly responding galvanometer. By plotting the deflections against the time in the circle at which they occurred, the complete waveform could be reconstructed, In 1868 Bernstein used a rheotome to chart the time course of the action potential in a nerve fibre.

By 1876 the capillary electrometer of Marey and Lippman and the string galvanometer of Einthoven were available with a sufficient sensitivity and speed of response to record biolectric events directly. However, preceding Einthoven's studies of the electrocardiogram, Marchand in 1877 and Englemann in 1878 were able to chart the electrocardiogram using a rheotome.

SOURCE: 'From torpedo to telemetry' by D W Hill *Electronics & Power* pp 111 (27 November 1975)

SEE ALSO: 'Des variations electriques, des muscles et due couer an particulier etudiées au moyen de l'electrometre de M Lippmann' E J Marey *C. R. Acad. Sci. Paris* vol 82, pp 975–7 (1876)

ALSO: 'Bettrage zur kenntnis der reizwelle und contractionswelle de herzmuskels' by R Marchand *Plugers Arch. f. d. ges. Physiol.* vol 15, p 511 (1877)

1868   BATTERIES (Leclanché cell)   G Leclanché (France)

The Léclanché dry cell is perhaps the best known cell in common use today. It is widely used in flashlights and other such equipment. This type of cell was originally described by Georges Leclanché in 1868 and has undergone many improvements since that time. Basically, it consists of a nearly pure (99.99 per cent) zinc negative terminal, a carbon positive terminal, and a mixture of ammonium chloride, manganese dioxide, acetylene black, zinc chloride, chrome inhibitor and water. The mixture acts as a depolarising agent to reduce the formation of hydrogen bubbles on the positive electrode as discharge takes place. Improvements which have been made include leak proofing, longer shelf life, pepped-up depolarizers, improved insulation and miniaturisation.

SOURCE: 'Survey of electrochemical batteries' by N D Wheeler *Electro Technology* p 68 (June 1963)

SEE ALSO: Léclanché G *Les Mondes* vol 16, p 532 (1868) (also *C. R. Acad. Sci. Paris* vol 83, p 54 (1876))

1870 BATTERY (Standard Clark Cell)   L Clark (UK)

In about 1870 Latimer Clark, an English engineer and electrician, introduced a new kind of voltaic cell consisting of a positive electrode of mercury covered with a paste of mercurous sulphate and a negative electrode of zinc. The electrolyte was a saturated solution of zinc sulphate. After many determinations Latimer Clark assigned to the cell a mean value of 1.457 volt at 15.5°C. The Clark cell suffered from a very high temperature coefficient of e.m.f. (1200μV/°C). However, in spite of this, the Chicago International Electrical Congress in 1891 adopted the Clark cell together with the silver coulometer in definitions of the ampere and the volt.

SOURCE: 'Standard cells by Muirhead' *Muirhead Technique* vol 18, No 3, p 19 (July 1964)

SEE ALSO: Clark Cell *Proc. R. Soc.* vol XX, p 444 (1872)

Clark Cell *Phil. Trans. R. Soc.* vol CIXIV, p 1 (1874)

NOTE: See also Weston Cell (1891)

1874 CAPACITORS (MICA)   M Bauer (Germany)

Mica sheet as dielectric came into commercial capacitor manufacture only about 1914–18 very largely because not only could it stand up to the mechanical shocks of gunfire better than glass, but it also enabled the size of the capacitors to be reduced substantially for the same effective performance. The drive of war requirements pushed this developed to the fore, although the use of mica as a capacitor dielectric had been 'invented' more than 60 years earlier.

SOURCE: 'Electrical capacitors in our everyday life' by P R Coursey *ERA Journal* No 6, p 10 (January 1959)

SEE ALSO: 'Physical properties of mica' by M Bauer *Z. duet. geol. Ge.* vol 26, p 137 (1874)

'Capacity of mica condensers' by A Zeleny *Phys. Rev.* vol 22, p 651 (1906)

1876 ROLLED PAPER CAPACITOR   D G Fitzgerald (UK)

It appears that the rolled paper capacitor was first covered by a patent filed in 1876 by Fitzgerald, who described:

'The construction of a condenser with layers of paper and conductor (usually tin-foil) alternately interleaved with each other on to a cylinder, and the impregnation of such condenser with paraffin wax after rolling'

SOURCE: 'History, present status and future developments of electronic components' by P S Darnell *IRE Trans. on Component Parts* p 124 (September 1958)

SEE ALSO: 'Improvements in Electrical Condensers or Accumulators' by D G Fitzgerald, British Patent No 3466/1876 (2 September 1876)

'Paper Condensers' by X Boucherot *Eclairage Electrique* (12 February 1898)

'The manufacture of paper condensers' by G F Mansbridge *J Inst. Elec. Engrs.* vol 41, p 535 (May 1908)

'The capacity of paper condensers' by A Zeleny and Andrewes. *Phys. Rev.* vol 27, p 65 (1908)

1876   **TELEPHONE**   **A G Bell (USA)**

It has been a hundred years since a faint but momentous sound was made by Thomas Watson when he plucked a reed of a rudimentary transmitter. But that sound travelled over wire and was heard in another room by Alexander Graham Bell who happened to be holding a similar reed-and-diaphragm apparatus to his ear. This was the first telephone signal.

Later, on 14 February 1876, Bell filed for the now famous patent on the apparatus that he and Watson had been working on—just three hours before one Elisha Gray filed a caveat with the Patent Office, declaring that he was working on a similar device but had not yet perfected it. Had the timing been different, we might now have a 'Ma' Gray instead of Ma Bell.

The first telephone patent was issued to Bell on March 7 1876 three days before the historic moment when the first intelligible human voice was transmitted over the new telephone. Bell, after spilling acid over his clothing, had called out, 'Mr Watson, come here. I want you'. Next year, the first commercial telephone went into service when a Boston banker leased two instruments, each consisting of a simple wooden box that contained both transmitter and receivers The talker had to alternately talk and listen.

According to the records, the seed idea for the Bell System was planted when a merchant named Thomas Sanders made a verbal offer to Bell to finance the telegraphic experiments. They reached a tentative agreement, and shortly afterwards a lawyer named Gardiner G Hubbard made Bell a similar offer. The three put into writing an agreement dated 27 February 1875 and later signed a deed of trusts dated 9 July 1877 forming the Bell Telephone Company, Gardiner G Hubbard, Trustee.

SOURCE: *Electronics* p 91 (11 December 1975)

SEE ALSO: 'Alexander Graham Bell and the invention of the telephone' by J E Flood *Electronics & Power* p 159 (March 1976)

Editor's Note: This issue of *Electronics & Power* is devoted to the early history of the telephone.

'The marriage that almost was' by M F Wolff *IEEE Spectrum* p 41 (February 1976)

'The telephone, its invention and development' by M Woolley *Telecommunication Journal* vol 43, p 175 111/1976

'Bell's great invention: the first 50 years' *Bell Laboratories Record* p 91 (April 1976)

Bell and Gray: 'Contrasts in style, politics and etiquette' by D A Hounshell *Proc. IEEE* vol 64, No 9, p 1305 (September 1976)

1877   **PHONOGRAPH (Gramophone)**   **T A Edison (USA)**

Alexander Graham Bell's invention of the telephone, in 1876, drew attention to the problems of the reproduction of speech. One of those attracted was Thomas Alva Edison, who was all the more interested in the study of sound because he was partially deaf. Sometime in the fall of 1877, Edison wrapped a sheet of tinfoil around a cylinder, set a needle in contact with it, turned a crank to rotate the cylinder, and into a mouthpiece attached to the needle he shouted the nursery rhyme that begins: 'Mary had a little lamb'. After making a few changes, he cranked the cylinder again, and from the horn of the instrument he heard a recognisable reproduction of his voice, Thus was born the first phonograph. The date of invention was later recollected by Edison as 13 August 1877, but there is some question as to how precise that date is. We know, however, that the patent application was filed 24 December 1877, and the patent was granted on 19 February 1878.

SOURCE: 'Disk recording and reproduction' by W S Bachman, B B Bauer and P C Goldmark *Proc. IRE* p 738 (May 1962)

SEE ALSO: US Patent No 200 521 (T A Edison)

1877 MICROPHONE (Carbon)   **T A Edison (USA)**

Among the earliest devices intended for converting vibration into electrical impulses was Reis' loose metal-contact transducer which is reported to have transmitted tones of different frequencies, but not intelligible speech. This latter event seems first to have been achieved by Bell, using a magnetic microphone, on 3 June 1875. However, Bell's microphone proved not to be sufficiently sensitive for telephone work and the experiments of Berliner, Edison, Hughes and others soon thereafter introduced a long era of dominance for the loose-contact carbon transducer. To Edison goes the credit of being the first to design a transducer using granules of carbonised hard coal, still used in present-day microphones.

The carbon granules are made of deep-black 'anthraxylon' coal ground to pass a 60–80 mesh, treated chemically and roasted in several stages under a stream of hydrogen. This drives out volatile matter, washes out extraneous compounds and carbonises the coal. The last step of the process is magnetic and air-stream screening to eliminate iron-bearing and flat-shaped particles.

SOURCE: 'A century of microphones' by B B Bauer *Proc. IRE* p 721 (May 1962)

SEE ALSO: T A Edison: US Patent No 474 230 filed 27 April 1877. Also US Patent No 474 231/2

1877 LOUDSPEAKER (Moving coil type)   **E W Siemens (Germany)**

The motor mechanism consisting of a circular coil located in a radial magnetic field was first disclosed by Siemens. Lodge, Pridham and Jenson and others contributed to the suspension system. However, there were very few developments in loudspeakers in the twenty-seven years following Lodge's disclosure. A breakthrough in the dynamic loudspeaker was made by Rice and Kellogg in 1925. The success of the development was due to their recognition of three physical factors with relation to the action and design of a direct radiator loudspeaker. The first is that the sound-power output of a loudspeaker is the product of the mechanical resistance due to sound radiation and the square of the velocity of the diaphragm. The second is that sound radiation from a small vibrating diaphragm gives rise to a mechanical resistance which is proportional to the square of the frequency. The third is a vibrating system which is mass controlled. It follows then that, if the fundemental resonance occurs below the lowest frequency of interest, the complementary variations of the second and third factors which control the sound output as given by the first factor conspire to provide a uniform response up to the frequency region at which the assumptions begin to fail. This was the contribution of Rice and Kellogg, and it continues to be the basic precept that guides the design of all direct radiator loudpeakers.

SOURCE: 'Loudspeakers' by H F Olson *Proc. IRE* p 730 (May 1962)

SEE ALSO: *Electroacoustics* by F V Hunt (New York: John Wiley and Sons) (1954)

E W Siemens: German Patent No 2355 filed 14 December 1877

O J Lodge: British Patent No 9712 filed 27 April 1898

E S Pridham and P L Jenson: US Patent No 1448 279 filed 28 April 1920

'Notes on the development of a new type of hornless loudspeaker' by C W Rice and E W Kellogg *Trans AIEE* vol 44, pp 461–75 (April 1925)

1878 CATHODE RAYS   **Sir W Crookes (UK)**

In his Bakerian lecture of 1878 and his British Association Lecture of 1879, he announced various striking properties of 'molecular rays' including the casting of shadows, the warming of obstacles and the deflection by a magnet. The title 'Radiant Matter' employed by Crookes in his British Association Lecture of 1879 referred to ordinary matter in a new state in which the mean free path was so large that collisions between molecules could be ignored.

SOURCE: *A Biographical Dictionary of Scientists* by T I Williams (London: Adam and Charles Black) p 120 (1969)

1878   **CARBON FILAMENT INCANDESCENT LAMP**   **J W Swan, C H Stearn, F Topham and C F Cross (UK)**

Swan invented a carbon filament incandescent lamp and a squirting process to make nitro-cellulose fibres for lamp filaments. Stearn was an expert in the production of high vacua and Topham the glass-blower who made the glass globes. Cross discovered the viscose process for the fibres.

SOURCE: 'The Sources of Invention' by J Jewkes, D Sawers and R Stellerman (London: MacMillan) p 59 (1958)

NOTE: T A Edison

Edison's main interest was to perfect the incandescent electric lamp, of which the construction was now being rapidly advanced by Swan. After researches involving the examination of thousands of alternatives, he devised the cotton thread filament which, when joined with bulb patents, made the 'Edison' lamp a commercial success.

SOURCE: *A Biographical Dictionary of Scientists* by T I Williams (London: Adam and Charles Black) p 159 (1969)

1878   **CARBON GRANULE MICROPHONE**   **H Hunnings (UK)**

To Hunnings goes the credit for inventing a carbon transmitter using a multiple loose contact, which soon took the form of a disk electrode projecting into a contact cup containing granulated carbon particles. From that time on, one improvement followed another in a seemingly endless parade.

SOURCE: *Electroacoustics* by F V Hunt (New York: John Wiley and Sons) p 37 (1954)

SEE ALSO: Henry Hunnings (granulated-carbon microphone) British Patent No 3647 dated 16 September 1878; US Patents No 246 512 (filed 14 May 1881) issued 30 August 1881, and No 250 250 (filed 30 September 1881) issued 29 November 1881; both assigned to American Bell Telephone Co.

1879   **DIODE DETECTOR**   **D Hughes (UK)**

In 1879, David Hughes had performed a truly remarkable demonstration of transmitting and detecting a series of recognised pulses of electromagnetic radiation over a distance of some 450 metres along Great Portland Street in London. Hughes was using the as yet undiscovered Hertzian waves, the radio portion of the electromagnetic spectrum. He was using a microphonic detector device which may have functioned like a self-restoring coherer, or, possibly, in a manner closely resembling the rectifying action of crystal detectors which came into use in 1906.

SOURCE: 'The birth pangs of radio' by A R Constable *Proc. on '100 Years of Radio', IEE Conference Publication No 411* p 15 (September 1995)

1879   **HALL EFFECT**   **E H Hall (UK)**

Hall voltage, linear function of magnetic flux.

If a current of particles bearing charges of a single sign and constrained to move in a given direction is subjected to a transverse magnetic field, a potential gradient will exist in a direction perpendicular to both the current and the magnetic field.

SOURCE: *The Encyclopaedia of Physics* 2nd edn, ed R M Besencon (New York: Van Nostrand) p 400 (1974)

1880   **PIEZO ELECTRICITY**   **J Curie and P Curie (France)**

The relation between voltage generated and mechanical pressure on crystallographic materials.

SOURCE: 'Developpement, par pression, de l'electricité polaire dans les cristaux hemiedres a faces inclinées' by J Curie and P Curie *C. R. Acad. Sci. Paris* vol 91, pp 294–5 (July–December 1880)

1882   **WIMSHURST MACHINE**   **J Wimshurst (UK)**

The Wimshurst machine generates electrostatic electricity by friction. It consists of two discs which revolve in opposite directions. The discs are made of an insulating material such as glass and near the peripheries of the discs are mounted small sections of sheet conductor. The charges are generated on the conductor sections by the action of brushes which graze the sectors as they rotate and are picked up by combs. These charges are stored in capacitors and are used to produce a spark between a pair of ball conductors.

REFERENCE: *Text Book of Physics, Part 5: Magnetism and Electricity* by J Duncan and S C Starling (London: MacMillan & Co.) p 960 (1926)

REFERENCE: *Encyclopaedic Dictionary of Electronics and Nuclear Engineering* by R I Sarbacher (London: Pitman) p 1402 (1959)

1884   **NIPKOW TELEVISION SYSTEM**   **P Nipkow**

The first television invention that had practical consequences was the 'electrical telescope', patented by Paul Nipkow in 1884. At the heart of his camera was the now famous Nipkow disk. It had 24 holes equally spaced along a spiral near the periphery of the disk. The image to be transmitted was focused on a small region at the disk's periphery, and the disk was made to spin at 600 revolutions per minute. As the disk rotated, the sequence of holes scanned the image in a straight line. A lens behind the image region collected the sequential light samples and focused them on a single selenium cell. The cell would then produce a succession of currents, each proportional to the intensity of the light on a different element of the image.

At the receiving end, Nipkow proposed using a magneto-optic (Faraday-effect) light modulator to vary the intensity of the reconstructed image. To form the image, a second disk, identical to and rotating synchronously with the one at the transmitter, would be needed.

Nipkow built no hardware—which is probably just as well, because the technology of the time would not have permitted him to build his system; the light modulator alone would have required some 10 watts of control power. His disk, however, was a model for several later television systems that were built, most notably those of British inventor John Logie Baird.

SOURCE: *Electronics* pp 70 and 75 (17 April 1980)

1884   **SWITCH, QUICK BREAK**   **J H Holmes (UK)**

The loose-handle quick-break switch was invented by J H Holmes in 1884. For this device he was granted British Patent No 3256 of 1884, under the title: 'Improvements in or applicable to switches or circuit closers for electrical conducting apparatus'.

SOURCE: *Patents for Engineers* by L H A Carr and J C Wood (London: Chapman and Hall) p 95 (1959)

1884   **BATTERY (Zinc-Mercuric Oxide Cell)**   **C L Clarke (USA)**

The alkaline zinc-mercuric oxide system was first suggested by Clarke in 1884. Although there were a number of additional attempts made over the years to design a practical cell using this system, it was not until early in World War II that a commercially usable mercuric oxide dry cell was invented by Ruben.

SOURCE: 'Batteries' by C K Morehouse, R Glicksman and G S Lozier *Proc. IRE* p 1467 (August 1958)

SEE ALSO: C L Clarke: US Patent No 298 175 (May 6 1884)

'Balanced alkaline dry cells' by S Ruben *Trans. Electrochem. Soc.* vol 92, p 183 (1947)

1885 **TRANSFORMER (Power)**   **C Zipernowski, M Deri and O T Blathy (Hungary)**

The earliest patent covering the construction of the transformer appears to be that of Carl Zipernowski, Max Deri and Otto Titus Blathy, all of Budapest, and since this patent was applied for and granted in this country, the British version can be quoted here.

The date of application was 27 April 1885, the patent being numbered 5201 in that year under the title: 'Improvements in Induction Apparatus for Transforming Electric Currents'.

SOURCE: *Patents for Engineers* by L H A Carr and J C Wood (London: Chapman and Hall) p 89/90 (1959)

1885 **RESISTOR (Moulded Carbon Composition Type)**   **C S Bradley (UK)**

It is of passing interest to note that the earliest moulded-rod composition-type resistor of which the author has been able to trace any record dates back before the days of radio. In 1885, a moulded-composition resistor was patented, comprising a mixture of carbon and rubber heated and moulded to shape and subsequently vulcanized to a hard body.

SOURCE: 'Fixed resistors for use in communication equipment' by P R Coursey. *Proc. IEE* vol 96, Pt. III, p 169 (1949)

SEE ALSO: C S Bradley: British Patent No 8076/1885.

M Slattery: US Patent No 354 275 (1885)

D C Voss: US Patent No 573 558 (1896)

1885 **TRANSFORMER (Distribution)**   **M Deri (Austria)**

For the earliest patent covering the use of the transformer in a distribution system it is necessary to refer to German Patent No 33951 of 1885, since no corresponding patent was applied for in the United Kingdom.

This application was made on 18 February 1885 by Max Deri (in this case described as being of Vienna) under the title (translated) of 'Improvements in the Distribution of Electricity'.

SOURCE: *Patents for Engineers* by L H A Carr and J C Wood (London: Chapman and Hall) p 91 (1959)

1887 **GRAMOPHONE (Phonograph)**   **E Berliner (USA)**

On 4 May 1887, Emile Berliner applied for a patent on what he called a 'Gramophone' to distinguish it from Edison's phonograph of ten years earlier and from Bell's and Tainter's Graphophone of one year earlier. The first figure of the drawings in Berliner's patent shows a record wound on a cylindrical support but by 1888, when he introduced his first model, he had changed to a flat-disk record. The groove in this record had a lateral side-to-side movement, as against the vertical 'hill and dale' system which had been employed by others. This lateral recording process was reminiscent of Leon Scott's phonoantograph, which used a diaphragm and hog bristle to trace a record of sound vibrations on lamp-blacked paper some thirty years earlier.

Berliner also used lamp-black as the recording medium, and combined this method with an etching process which permitted transfer of the original engraving to copper or nickel. Thus Berliner achieved a permanent master recording, and for the first time mass duplication of records was possible. No longer did artists have to repeat each number endless times.

By 1895, Berliner had developed a system utilising many ideas of his own and others: Scott's lateral groove, his own flat disc, and a coating of Bell's and Tainter's wax. The system stood up as the industry standard for half a century, thus Berliner deserves the mantle as the father of disk recording and reproduction.

SOURCE: 'Disk recording and reproduction' by W S Bachman, B B Bauer and P C Goldmark *Proc. IRE* p 738–9 (May 1962)

SEE ALSO: 'The gramophone and the mechanical recording and reproduction of musical sounds' by L N Reddie *J. R. Soc. Arts.* vol LVI, pp 633–49 (8 May 1908)

| 1887 | **ELECTROCARDIOGRAPH** | **A D Waller (UK)** |

The first human electrocardiogram was recorded in 1887 by Augustus Desire Waller (1856–1922), a physiologist from London University, born in Paris.

In 1901 Willem Einthoven, professor of physiology at the University of Leiden in Holland, and former colleague of the physicist and 1908 Nobel Prize-winner Gabriel Lippmann (France), developed the loop galvanometers. This made him the true inventor of the electrocardiograph (ECG), a piece of equipment that weighed 661 lb and required five people to operate it.

Carrying out an electrocardiogram is very straightforward, though interpreting the results is not. It is now possible for a pregnant woman to place an ultrasound probe on her abdomen to monitor her baby's heart. The result can then be sent by the woman by telephone to her doctor, who can then assess the health of the baby, thus saving everyone a great deal of time and effort.

SOURCE: *Inventions and Discoveries 1993* edited by Valerie-Anne Giscard d'Estaing and Mark Young (New York: Facts on File) p 172

| 1887 | **AERIALS (Radio Wave Propagation)** | **H R Hertz (Germany)** |

It remained for Heinrich Hertz to prove the existence of electric waves in space as predicted by Maxwell. The first true antenna appears to have been used by Hertz in his classical experiments at Karlsruhe in 1887, His antenna consisted of two flat metallic plates, 40 cm square, each attached to a rod 30 cm long. The two rods were placed in the same straight line, and were provided at their nearer ends with balls separated by a spark gap about 7-mm long. The spark gap was energised by a Ruhmkorff coil. In order to detect the radiated waves, Hertz employed a receiving circuit consisting of a circular loop of wire broken by a microscopic gap. The radius of the loop was 35 cm which was found by experiment to be the proper size to be in resonance with the oscillator.

SOURCE: 'Early history of the antennas and propagation field until the end of World War I, part I—antennas' by P S Carter and H H Beverage *Proc. IRE* p 680 (May 1962)

SEE ALSO: 'Ueber sehr schnelle elektrische Schwingungen' by H Hertz *Ann. Physikund Chemie* (Wiedeman) NF vol 31, pp 421–448 (15 May 1887)

| 1887 | **SWITCH, QUICK MAKE AND BREAK** | **J H Holmes (UK)** |

Following Holmes' invention of the quick-break switch (see 1884) a later variant provided for a loose-handle operating a spring over a dead centre by means of a toggle action, so that both quick-make and quick-break were obtained. (British Patent No 5648 of 1887.)

SOURCE: *Patents for Engineers* by L H A Carr and J C Wood (London: Chapman and Hall) p 95 (1959)

| 1888 | **INDUCTION MOTOR** | **N Tesla (USA)** |

Tesla's 'master' patent covering the polyphase induction motor was taken out, inter alia, in Great Britain; the British Patent No 6481 of 1888, may therefore be used as a reference.

Its title was 'Improvements relating to the electrical transmission of power and to apparatus therefore' and it was granted in the name of Nikola Tesla of the City and State of New York, USA.

The specification is long and detailed with 18 diagrammatic figures.

SOURCE: *Patents for Engineers* by L H A Carr and J C Wood (London: Chapman and Hall) pp 96–7 (1959)

1889   **COMPUTERS (Tabulating Machinery)**   **H Hollerith (USA)**

Hollerith worked on a machine for tabulating population statistics for the 1890 census in the USA which he patented in 1889. He used a system of holes in a punch card to represent various characteristics: such as male or female, black or white, age, etc. The cards were $6\frac{5}{8}$ by $3\frac{1}{4}$ inches in size, which he chose because it was the size of a dollar bill. Each card contained 288 locations at which holes could be made. The machine is shown in figure 11.4.

Figure 11.4. Hollerith's tabulating machine (The Science Museum/ Science & Society Picture Library).

SOURCE: *The Computer from Pascal to von Neuman* by H H Goldstine (Princeton, NJ: Princeton University Press) p 7 (1972) (Reprinted by permission)

SEE ALSO: 'An electric tabulating system' by H Hollerith, reprinted *The Origins of Digital Computers* edited by B Randell (Berlin: Springer) p 129 (1973)

1889   **STROWGER AUTOMATIC TELEPHONE**   **A B Strowger (USA)**
       EXCHANGE

The man who is generally given the credit of inventing the first practical system of automatic telephony to be used commercially is Almon Brown Strowger, an undertaker from Kansas City, USA. It is said that Strowger's business was losing money because, when people called him, the operators at the telephone exchange would connect the call to other undertakers instead, and he was determined to invent a system that would work without operators. He patented his ideas in 1889, and three years later his equipment was installed in the first public automatic telephone exchange, at La Porte, Indiana, USA.

SOURCE: *The Telephone and the Exchange* by P J Povey (Post Office Publication) p 35 (1974)

SEE ALSO: 'The Story of Telephone Switching' TELONDE, No 2/1977, p 16

1890   **COHERER**   **E Branly (France)**

In Paris, Edouard Branly, physics professor at the Catholic University, observed in 1890 that metal filings, when subjected to 'Hertzian waves' behaved very strangely. Normally, filings do not transmit an electric current because there are air spaces between them but when placed with the range of electromagnetic waves, the filings fuse a little together, enough to offer a conducting path to an electric current. The filings remain a conductor until they are disturbed by shaking or tapping.

Branly called the little glass tube in which he placed his filings a 'coherer'; it was the first form of a 'detector' for electromagnetic waves.

SOURCE: 'A History of Invention' by E Larsen (London: J M Dent & Sons) and (New York: Roy Publishers) p 278 (1971)

| 1891 | **BATTERY (Standard Weston Cell)** | **E Weston (USA)** |

In 1891, Dr Edward Weston, an Anglo-American from New Jersey, filed a patent which was granted in 1893 disclosing a cell in which the electrolyte of the Clark cell was replaced by a saturated solution of cadmium sulphate and the zinc negative electrode was replaced by cadmium amalgam, the depolariser being mercurous sulphate as before. This cell was a definite improvement on the Clark cell since the temperature coefficient of e.m.f. was only about -40 μV/°C. Considerable improvements were made by various workers from 1893 onwards and finally the London Conference held in 1908 authorised the appointment of a special international committee. As a result of the work of this committee, Weston's cadmium cell was assigned the value of 1.01830 international volts at 20°C.

SOURCE: 'Standard cells by Muirhead' *Muirhead Technique* vol 18, No 3, p 19 (July 1964)

SEE ALSO: Weston cell—USA Patent No 494 827 (1893); also British Patent 640 812, 797 381

| 1893 | **WAVEGUIDES** | **J J Thomson (UK)** |

(See also page 140.)

Perhaps the first analysis suggesting the possibility of waves in hollow pipes appeared in 1893 in the book 'Recent Researches in Electricity and Magnetism' by J J Thomson. This book, which was written as a sequel to Maxwell's 'Treatise on Electricity and Magnetism' examined mathematically the hypothetical question of what might result if an electric charge should be released on the interior wall of a closed metal cylinder. Even now, this problem is of considerable interest in connection with resonance in hollow metal chambers. A much more significant analysis, relating particularly to propagation through dielectrically filled pipes, both of circular and rectangular cross section, was published in 1897 by Lord Rayleigh.

SOURCE: 'Survey and history of the progress of the microwave arts' by G C Southworth *Proc. IRE* p 1199 (May 1962)

SEE ALSO: *Recent Researches in Electricity and Magnetism* by J J Thomson p 344 (1893)

'On the passage of electric waves through tubes or the vibrations of dielectric cylinders' Lord Rayleigh *Phil. Mag.* vol 43, p 125 (February 1897)

| 1895 | **X-RAYS** | **W K Röntgen (Germany)** |

On 8 November 1895, while experimenting with a Crookes's tube (see W Crookes) covered with an opaque shield of black cardboard, Röntgen noticed that, when a current passed through the tube, a nearby piece of paper painted with barium platinocyanide fluoresced. In a series of classical papers (1895–7) he described the properties of the new, so-called X-rays, but his attempts to detect their interference by crystals were unsuccessful.

SOURCE: *A Biographical Dictionary of Scientists* by T I Williams (London: Adam and Charles Black) p 448 (1969)

| 1896 | **TELEPHONE DIAL** | **E A Keith, C J Erickson and J Erickson (UK)** |

The early automatic telephone systems of Connolly, Connolly, and McTighe, of Sinclair, and of Strowger, all used push-buttons which the caller was required to press a number of times depending on the number of the telephone he wanted. When automatic exchanges beame larger, and telephone numbers became longer, this type of signalling was no longer practical, and in 1896, E A Keith, C J Erickson, and John Erickson invented the telephone dial.

SOURCE: *The Telephone and the Exchange* by P J Povey (Post Office Publication) p 60 (1974)

1896   WIRELESS TELEGRAPHY   G Marconi (Italy)

On 2 June 1896, Marconi took out in the United Kingdom the first patent for wireless telegraphy based on Hertz's discoveries, though exploiting radiations of a much longer wavelength. His apparatus consisted of a tube-like receiver or 'coherer' connected to an earth and an elevated aerial, its signals were at first transmitted over one hundred yards, a satisfactory demonstration being arranged from the roof of the London General Post Office. Ship to shore communication was established in the following year, when Marconi formed a Wireless Telegraph Company in London for the exploitation of his patents in all countries except Italy; this later developed world-wide affiliations. His first transatlantic signals were made on 12 December 1901 from Poldhu in Cornwall to St John's, Newfoundland, where they were received through an aerial suspended from a kite.

SOURCE: *A Biographical Dictionary of Scientists* by T I Williams (London: Adam and Charles Black) p 352 (1969)

1897   ELECTRON   J J Thomson (UK)

By improving the vacuum of Hertz's cathode ray experiments, Thomson got deflections which, combined with the long-known deflections by a magnet determined the ratio of the charge to the mass of the supposed particles. This ratio e/m was over 1000 times larger than the ratio for hydrogen, the lightest atom known. Thomson considered that this was due to the smallness of the mass and that particles with this small mass were universal constituents of matter, since they were the same whatever the chemical nature of the gas carrying the discharge and the electrodes through which it entered and left. He examined two other cases of the discharge of electricity, namely, those from a hot wire negatively charged and from a negatively charged zinc plate illuminated by ultra-violet light. In both he found charged particles with the same e/m ratio as the cathode rays, and in the second was able to measure the actual charge by condensing drops of water on the particles to form a mist, finding the size of the drops from their rate of fall. The results agreed with the supposed value of the charge on a hydrogen atom as far as this was then known. Thomson called these new light particles 'corpuscles' but later adopted the word 'electron' invented a few years before by J G Stoney for the charge on a hydrogen atom regarded as a natural unit of charge.

SOURCE: *A Biographical Dictionary of Scientists* by T I Williams (London: Adam and Charles Black) p 511 (1969)

SEE ALSO: *The Discovery of the Electron* by D L Anderson (New York: Van Nostrand Reinhold) (1964)

'The first subatomic particle' by O Frisch *New Scientist* p 408 (17 November 1977)

1897   CATHODE RAY OSCILLOGRAPH   F Braun (Germany)

The cathode-ray oscilloscope for the study of the time variation of electron currents was developed by Ferdinand Braun in 1897, the same year in which J J Thomson measured the specific charge of the electron by its deflection in electric and magnetic fields. Ferdinand Braun constructed the first cathode-ray oscillograph at the University of Strasbourg in 1897.

Just like the early X-ray tube, the 'Braun tube' used gas discharge phenomena for the emission and formation of an electron beam. Even after the introduction of a thermionic cathode into cathode-ray tubes by Wehnelt in 1905 an argon atmosphere of about 10^{-3} mm Hg was still retained in commercial oscilloscopes for another 25 years. The effect of ion focusing facilitated the formation of long, filamentary, electron beams.

SOURCE: 'Beam-deflection and photo devices' by K Schlesinger and E G Ramberg *Proc. IRE* p 991, 995 (May 1962)

SEE ALSO: 'On a method for the demonstration and study of currents varying with time'. by F Braun *Wiedemann's Alln.* vol 60, p 552 (1897)

'Cathode rays' by J J Thomson *Phil. Mag.* vol 4, p 293 (1897)

'Ferdinand Braun and the cathode ray tube' by G Shiers *Scientific American* vol 230, p 92 (March 1974)

1897   **RESISTOR (Carbon Film Type)**   **T E Gambrell and A F Harris (UK)**

The earliest type of carbon-film resistor was also used many years before broadcasting and it was apparently forgotten when the greater demand arose. Some of these carbon-film resistors, notably those formed by spraying or otherwise applying the conducting coating and then baking on to a glass filament, were enclosed inside a glass tube with metal end-caps sealed on for terminal connections and with terminals connected to the filament coating by casting into a metal such as type-metal.

SOURCE: 'Fixed resistors for use in communication equipment' by P R Coursey *Proc. IEE* vol 96, pt. III, p 170 (1949)

SEE ALSO: T E Gambrell and A F Harris: British Patent No 25412/1897

1898   **MAGNETIC RECORDING**   **V Poulsen (Denmark)**

No one really knowns how the idea of magnetic recording occurred to Poulsen. During his experiments he found that:

'It would be possible to magnetize a wire to different degrees so close together that sound could be recorded on it by running the current from a microphone through an electromagnet and by either drawing the wire rapidly past the electromagnet or drawing the electromagnet rapidly past the wire.'

This invention had several things to commend it: the wire or tape could be used over and over again by de-magnetizing it and recordings could be played thousands of times without destroying the quality. Poulsen invented this Telegraphone in 1898; with it he won the Grand Prix at the Paris Exposition in 1900. He filed an application for a Danish patent in 1898 and within two years he had filed additional patent applications in the United States and most European countries. These early patents suggested, as recording media, steel wires and tapes and discs of material coated with magnetisable metallic dust, though he himself used only steel wire and tape in his machinery. The basic principles enunciated by Poulsen are still applied in all types of modern magnetic recorders.

SOURCE: *The Source of Invention* by J Jewkes, D Sawers and R Stillerman (London: MacMillan & Co.) p 326 (1958)

SEE ALSO: 'The development of the magnetic tape recorder' *Engineer* (18 March 1949)

1900   **CAPACITORS (Ceramic)**   **L Lombardi (Italy)**

Ceramic materials have been used for many years as electrical insulators. They are able to withstand severe working conditions because they are vitrified by firing at temperatures of the order of 1200°C. Being completely inert they will withstand their rated working voltage indefinitely and retain their shape and physical characteristics under normal conditions.

SOURCE: *Fixed Capacitors* (2nd edn) by G W A Dummer (London: Pitman) p 115 (1956)

SEE ALSO: 'An improved process for manufacturing thin homogenous plates, more particularly applicable for use in electrical condensers' L Lombardi, British Patent No 9133, filed May 17 1900.

'Permittivity of titania' by W Schmidt *Ann. Phys. Lpz.* vol 4, p 959 (1902)

'Dielectric losses and breakdown strength of porcelain' by F Beldi *Brown Boveri Rev.* vol 18, p 172 (May 1931)

'Insulating materials of the steatite group' by E Schonberg *Elektrotech. Z.* vol 54, p 545 (June 1933)

Tubular metallized ceramic capacitors. Porzellanfabrikkahla, Germany. British Patent No 440951/1934

| 1900 | **BATTERY (Nickel–Iron Cell)** | **T A Edison (USA)** |

The nickel–iron–alkaline batteries as they exist today are essentially the same as discovered by Edison around 1900 and marketed in 1908. The negative electrode consists of pockets of active material grouped together to form a plate, while the positive plate is an assembly of perforated nickel tubes filled with nickel hydroxide and nickel.

SOURCE: 'Batteries' by C K Morehouse, R Glicksman and G S Lozier *Proc. IRE* p 1478 (August 1958)

SEE ALSO: 'The Edison nickel–iron–alkaline cell' by F C Anderson *J. Electrochem. Soc.* vol 99, p 244C (1952)

| 1900 | **QUANTUM THEORY** | **M Planck (Germany)** |

Max Planck had propounded in 1900 the theory that energy is not emitted in a continuous flow, but always in a collection of packets of finite size. He was driven to accept this view in order to explain the distribution of the energy found in the light from hot bodies such as the sun. His conception struck at the principle of continuity as the comprehensive basis of nature.

SOURCE: *Science at War* by J G Crowther and R Whiddington (London: HMSO) (1947) p 124

| 1900 | **BATTERY (Nickel–Cadmium Cell)** | **Junger & Berg (Sweden)** |

The nickel–cadmium battery discovered by Junger and Berg about 1900 is closely related to the nickel–iron Edison battery. Both are mechanically rugged and will withstand electrochemical abuse in that they can be overcharged, overdischarged, or stand idle in a discharged condition. The nickel–cadmium battery differs from the Edison battery in the use of cadmium anodes in place of iron.

SOURCE: 'Batteries' by C K Morehouse, R Glicksman and G S Lozier *Proc. IRE* p 1478 (August 1958)

SEE ALSO: *Storage Batteries* by G W Vinal 4th edn (New York: Wiley) (1955)

| 1901 | **RADIO: HEAVISIDE/KENNELLY LAYER** | **O Heaviside (UK) and A Kennedy (USA)** |

When Marconi first succeeded in establishing radio communication around the curvature of the earth in 1901, Oliver Heaviside in England and Arthur Kennelly in the USA proposed that the phenomena might be due to the existence of an ionized layer in the upper atmostphere surrounding the earth which could reflect radio waves. However, another explanation was that the waves were detectable because of diffraction around the earth's curvature, a phenomenon which would be more observable, the longer the wavelength used. Another explanation could be that of the bending of the wavefront which would occur if the dielectric constant of the atmosphere should progressively change with altitude due to the effects of temperature, density and moisture.

The observation of the waves of high frequency drew renewed attention to the proposals of Heaviside and Kennelly. Their hypothetical layer was dubbed the Heaviside–Kennelly layer. Breit and Tuve in 1926, definitely confirmed the existence of several such layers by sounding the atmosphere with short pulses sent from a transmitter on earth which were reflected from the layers back to a receiver adjacent to the transmitter. The time delay identified the height of each reflection and determined physical characteristics of the nature of the reflection and the degree of penetration as a function of frequency, The Breit–Tuve experiments were a predecessor of later radar techniques. The vagaries observed by the earlier experimenters began to be explainable and even predictable. The region of the several layers was renamed the 'Ionosphere' and the individual layers designated the D, E, F_1 and F_2 layers, the D layer being the lowest.

SOURCE: 'Telecommunications—the resource not depleted by use. A historical and philosophical resume' by W L Levitt *Proc. IEEE* vol 64, No 9, p 1297 (September 1976)

SEE ALSO: 'A test of the existence of the conducting layers' by G Breit and M Tuve *Phys. Rev.* vol 28, p 554 (September 1926)

1901   **FLUORESCENT LAMP**   **P Cooper-Hewitt (USA)**

The first low-pressure mercury discharge lamp was introduced at the beginning of the twentieth century by Peter Cooper-Hewitt, the American individual inventor. Sir Humphrey Davy had discovered the effect of a discharge through mercury early in the nineteenth century. Cooper-Hewitt's lamp was inefficient by modern standards, though better than contemporary incandescent lamps; it also produced the characteristic bluelight of the mercury discharge lamp. It differed from the modern fluorescent lamp in being designed to produce visible radiation, not ultra-violet. In 1901 he used rhodamine dye, which fluoresces red, to improve the light's colour, but the rhodamine deteriorated too rapidly for this to be a success. The Moore and neon discharge lamps, introduced at much the same time as the Cooper-Hewitt, also contributed to the development of the fluorescent lamp. D McFarlan Moore, an American individual inventor, was the first to apply to hot cathode used on it, and he also anticipated Wehnelt in constructing lasting electrodes. Georges Claude's introduction of the neon tube, and the desire to modify its colour, stimulated interest in fluorescent powders and in means of employing them with a lamp.

SOURCE: *The Sources of Invention* by J Jewkes, D Sawers and R Stellerman (London: MacMillan) pp 298/9 (1958)

SEE ALSO: 'Lighting by luminescence' by A Claude *Light and Lighting* (3 June 1939)

1902   **INDUCTION MOTOR (SYNCHRONOUS)**   **E Danielson (Sweden)**

Danielson's synchronous induction motor is an excellent example of the sudden appearance of an invention which was complete at its first inception.

It was not patented in Great Britain, but the full information is available in United States Patent No. 694092 of 25 February 1902 granted to Ernst Danielson of Westeras, Sweden

After relating the advantages of the over-magnetised synchronous motors for eliminating lag (of current) and referring to the difficulty of starting such machines, the specification proceeds:

'The invention consists, briefly, in combination with an ordinary induction-motor and a suitable resistance for connecting to the secondary element of said motor, a source of continuous electric currents (and) a switch arrangement so connected that the secondary part of the motor may by means of said switch arrangement either be connected to the said resistance or to the said source of continuous currents'.

With reference to the self-synchronizing action, the specification states:

'When the exciting current is supplied, the motor is changed from an asynchronous to a synchronous one, provided, however, that the exciting-current is strong enough to pull the motor in step.'

The drawing and diagram of connections (showing a three-phase-secondary winding) which are attached to the specification are completely up-to-date, and were it not for the somewhat archaic outlines of the machines, might have been taken from a present-day text book or manufacturer's pamphlet.

SOURCE: *Patents for Engineers* by L H A Carr and J C Wood (London: Chapman and Hall) (1959) pp 97/8

1902   **SPONTANEOUS ATOMIC CHANGE**   **E Rutherford and F Soddy (UK)**

Very swift and brilliant analysis of the phenomena led Rutherford and Soddy to announce in 1902, only six years after the original discovery of radioactivity, their theory of the spontaneous disintegration of atoms. They asserted that atoms, the very foundation of matter and nature, were exploding, and not according to any rule, but merely by chance. Einstein has said that he 'could not believe that the Almighty had organised the world according to the throwing of dice'. It needed a bold spirit to adopt chance as the first principle in the explanation of the transmutation of the fundamental atoms of matter.

Rutherford forthwith bent his full genius to the task. Within nine years he had made the rays reveal the general structure of the atom. He proved that atoms are not little hard balls, but very spacious structures, consisting of a few distant electrons circulating round a relatively heavy nucleus, like the planets round the sun of a miniature solar system. Nearly all the mass of the atom was concentrated in the nucleus, which carried a positive electric charge exactly balancing the sum of the negative electric charges carried by the circulating electrons.

SOURCE: *Science at War* by J G Crowther and R Whiddington (London: HMSO) (1947) p 123

1904   TWO ELECTRODE TUBE   J A Fleming (UK)

Fleming received his early education in London, where he attended London University; later he spent four years at Cambridge, and he was appointed 'electrician' to the Edison Electric Light Company in 1882. During a visit to the United States in 1884 he visited Edison to discuss electric lighting problems, and it is of particular moment to the story of the valve that, during the visit, Edison demonstrated a discovery he had made a year before the Edison effect. Using a carbon-filament lamp in which a metal plate had been sealed, Edison found that when the plate was connected through a galvanometer to the positive terminal of the filament a current flowed, but no current flowed when connection was made to the negative terminal. He used the device to regulate the supply voltage in power stations. Fleming was very interested in the phenomenon, and, on his return home, carried out researches in which he showed that 'the space between the filament and the metal plate is a one-way street for electricity'.

It was on 16 November 1904 that Professor J A Fleming (1849–1945) described in British Patent Specification No 24850 a two-electrode valve for the rectification of high-frequency alternating currents.

SOURCE: 'Fleming and de Forest—an appreciation' by Captain C F Booth *IEE Pub. Thermionic Valves 1904–1954* (London: IEE) p 1 (1955)

1904   CAPACITORS: GLASS (TUBULAR)   I Moscicki (UK)

The glass dielectric capacitor was later manufactured in tubular form known as the Moscicki tube— which provided the only form of capacitor available for Marconi's early experiments in practical wireless communication; and in a modified form, using flat glass plates interleaved with zinc sheets and all immersed in oil, continued to provide the condensers for the spark wireless transmitting apparatus up to and including part at least of the 1914–18 war period.

SOURCE: 'Electrical capacitors in our everyday life' by P R Coursey *ERA Journal* No 6, p 10 (January 1959)

SEE ALSO: I Moscicki 'Improvements in electric condensers', British Patent No 1307, filed 18 January 1904

1905   INSULATED SODIUM CONDUCTOR   A G Betts (USA)

The concept of using sodium as an electrical conductor is hardly a new one. Back in 1901, a basic Swiss patent was issued for this purpose. In 1905 and 1906, French and American patents, respectively, were issued to Anson G Betts, who recognised the favourable economics of the metal. His patent specified that 'the sodium be enclosed—prefereably hermetically—by a sheathing of substantially nonoxidizable reinforcing material.'

In 1927, H H Dow and R H Boundy of the Dow Chemical Company began to experiment with sodium in steel pipes, and in 1930, Dow constructed a line 10 cm in diameter by 260 m long, by joining together 6-metre lengths of sodium-filled steel pipe. This uninsulated conductor operated at Dow's plant in Midland, Mich. for about ten years at currents ranging from 500 to 4000 amperes direct current. The results of this experimental work have been published in considerable detail.

SOURCE: 'Insulated sodium conductors—a future trend' by L E Humphrey, R C Hess, G I Addis, A E Ruprecht, P H Ware, E J Steeve, J A Schneider, I F Matthysse and E M Scoran *IEEE Spectrum* p 73 (November 1966)

SEE ALSO: 'Sodium in pipe successful as electrical conductor' by R H Boundy *Elec. World.* vol 100, No 26, p 852 (24 December 1932)

'A 4000-ampere sodium conductor' by R H Boundy *Trans. Electrochemical Soc* pp 151–60 (September 1932)

Cooling Electrical Machines and Cables by T De Koning (The Hague, Netherlands: Groothertoginne) pp 202–32 (1955)

1905   **THEORY OF RELATIVITY**   **A Einstein (USA)**

The brilliant theory of relativity was proposed by Einstein in 1905 to explain the observed fact that the speed of light is constant under all conditions. The theory of relativity was extended to all the motions of nature, and it began to be evident that many apparently different things were the same fundamental thing seen from different points of view. Space and time, which according to the old ideas seemed utterly different, were in the light of relativity seen to be two different aspects of one underlying unity.

From this line of development, Einstein showed that mass and energy were one of these pairs of interchangeable aspects. Mass could be conceived as congealed energy; and he even calculated how much energy there would be in a unit of mass. He found that it would be enormous. When energy was released, mass disappeared. The annihilation of mass would be accompanied by a vast and proportionate output of energy. If a mass of one ounce of matter could be transformed into energy, it would produce enough to turn nearly one million tons of water into steam.

SOURCE: *Science at War* by J G Crowther and R Whiddington (London: HMSO) (1947) p 125

1906   **CRYSTAL DETECTOR (CARBORUNDUM)**   **H C Dunwoody (USA)**

Silicon shares with carborundum the distinction of being used in one of the first crystal rectifiers. The use of carborundum was discovered by an associate of de Forest, Gen. Henry H C Dunwoody, who applied for a patent on 23 March 1906. It saved the day for de Forest's company, which had just been enjoined from infringing on the electrolytic detector of Reginald Fessenden after a lengthy legal battle. De Forest's Audion, only recently invented, was not yet commercially practicable. Ironically, it was Pickard who helped de Forest's company use its newly invented carborundum, showing the company the proper form of holder and battery bias for good results, He was never compensated for this consulting work.

SOURCE: 'The crystal detector' by A Douglas *IEEE Spectrum* p 66 (April 1981)

1906   **RADIO BROADCASTING**   **R Fessenden (USA)**

The first documented successful broadcasting of speech and music was conducted by Dr. Reginald Fessenden at Brant Rock, Mass., on Christmas Eve, 1906, utilizing a 50 kc radio-frequency alternator which produced about 1 kw of power and which was built by the General Electric Co., under the direction of Dr E F W Alexanderson. Modulation was accomplished by means of a microphone which is belived to have been water-cooled and which was connected in the antenna circuit. Clear reception was obtained at many locations including ships at sea.

Subsequently, Dr Lee de Forest conducted experimental broadcasting in 1907 from his laboratory in New York City, in 1908 from the Eiffel Tower in Paris, and in 1910 from the Metropolitan Opera House in New York City. These experiments, which were conducted with arc transmitters of about 500 watts power, modulated by microphones in the antenna-ground system, while successful, were handicapped by the high noise level inherent in arc transmitters.

SOURCE: 'AM and FM Broadcasting' by R F Guy *Proc. IRE* p 811 (May 1962)

SEE ALSO: 'History of radio to 1926' by G L Archer (New York: The American Historical Society) (1938)

1906 **THREE-ELECTRODE TUBE**   **L de Forest (USA)**

While Fleming was developing his two-electrode valve, Dr Lee de Forest was working in the United States on somewhat similar lines, and on 25 October 1906, de Forest applied for a patent for a three-electrode valve a triode—as a device for amplifying feeble electric currents, the amplification being achieved by using a voltage on the intermediate electrode (grid) to control the plate current. A few months later de Forest extended the patent to cover the use of the valve as a detector. The introduction of the third electrode to provide an amplifier as compared with the two-electrode rectifier very greatly extended the potential applications of the thermionic valve, and much credit is due to de Forest for his achievement.

Unfortunately, the invention of the triode led to considerable bitterness and litigation involving Fleming and de Forest, the former insisting to the end of his long life that de Forest's work was dependent on his own two-electrode valve. On the other hand, de Forest has always maintained that he was not aware of Fleming's patent before taking out his own. Initially the American courts held that de Forest's addition of the grid was dependent on Fleming's work. The story of the patent litigation did not end until 1943, when the United States Supreme Court decided that the original Fleming patent had always been invalid.

SOURCE: 'Fleming and de Forest—an appreciation' by Captain C F Booth *IEE Pub. Thermionic Valves 1904–1954* (London: IEE) p 2 (1955)

1907 **GONIOMETER**   **A Artom (Italy)**

The direction of aircraft on long-distance routes can be plotted more accurately than ever before, thanks to the invention of the radio-goniometer by Italian Alessandro Artom. Now navigation by radio waves is possible, making the old, cumbersome slide-rule calculations obsolete.

SOURCE: *The Timetable of Technology* (London: Michael Joseph and Marshall Editions) p 26 (November 1982)

1907 **CRYSTAL DETECTOR (PERIKON)**   **G W Pickard (USA)**

In addition to the silicon detectors, Wireless Specialty sold detectors using other minerals discovered by Pickard in November 1907: Pyron (iron pyrite) and Perikon (zincite in contact with chalcopyrite). The name Perikon was coined from PERfect pIcKard cONtact. Each mineral had its own field: Perikon was most sensitive, but had to be readjusted often, whereas silicon was very stable and able to withstand heavy static discharges.

SOURCE: 'The crystal detector' by A Douglas *IEEE Spectrum* p 66 (April 1981)

1908 **ELECTRONIC ORGAN**   **T Cahill (USA)**

Beginning with the 1900s there were many attempts to offer electric or electronic substitutes for organs, but none enjoyed any degree of commercial success. They were based on photo-optics, magnetic or electrostatic prerecordings, vacuum tube or neon lamp oscillators, or amplified blown reeds, etc. One of these, called the Telharmonium, was invented by Thadius Cahill and demonstrated in 1908. The size of a small power-generating station, it consisted of almost a hundred alternator generators for all the frequencies of the scale. Then through a console of switches, synthesised musical signals were transmitted over telephone lines without benefit of amplifiers.

In 1935 Mr Laurens Hammond, based on his synchronous electric clock, invented the first commercially successful mass-produced electric organ that started an industry. Since then, many manufacturers all over the world have joined this industry, offering instruments in a variety of sizes and prices that have transformed modern music

SOURCE: *Electronics Engineer's Reference Book* (London: Newnes-Butterworth) chap. 17, p 17-2 (1976)

SEE ALSO: T Cahill: US Patent No 1295691 (25 February 1919)

L Hammond: US Patent No 1956350 (24 April 1934)

1908 GEIGER COUNTER   E Rutherford and H Geiger (UK)

About this time, experiments were being conducted with the collection of current by a positive wire and negative cylinder arrangement that were to have a profound effect on science. These experiments were reported in a paper by Rutherford and Geiger, which showed that the number of charges of an ionizing event could be multiplied several thousand times by the ionising action of electrons in the high field region near the wire. This was the start of what for a time were called Gieger–Muller counter tubes and now simply Geiger counter tubes. The technique of the proportional type of counter was established in 1928, and the following year, schemes for determining the coincidence of ionising events were presented. Thus direction, scattering, absorption, etch types of experiments were possible and the modern era of cosmic-ray and nuclear research developed.

SOURCE: 'The development of gas discharge tubes' by J D Cobine *Proc. IRE* p 971 (May 1962)

SEE ALSO: E Rutherford and H Geiger *Proc. R. Soc.* A, vol 81, p 612 (1908)

1908 TELEVISION (Electronic)   A A Campbell-Swinton (UK)

Boris Rosing, of the St. Petersburg Technological Institute, seems to have been the first physicist who thought of using Braun's tube for the reception of images. As early as 1907 he suggested a system of remote electric vision, with a Nipkow disc for scanning the scene to be transmitted and a cathode-ray tube as the receiver. At about the same time the English inventor, A A Campbell-Swinton, also proposed a system of electronic television, but with cathode-ray tubes for transmission as well as for reception. He published his ideas in the scientific magazine *Nature* in 1908, and elaborated them again in 1911 and 1920, explaining that the image transmitted in this way could be split up into, and reassembled from, about 400 000 points of different light value within $\frac{1}{25}$ of a second.

SOURCE: *A History of Invention* by E Larsen (London: J M Dent & Sons) (New York: Roy Publishers) p 323 (1971)

1909 FERRITES (HF)   G Hilpert (Germany) and J L Snoek (Holland)

The first proposal for high-frequency application was made in Germany in 1909 by Hilpert, who first synthesised such ferrites. However, the practical development and exploration of these materials did not come about until Snoek from the Philips Laboratories in Holland carried out extensive investigations on the high-frequency properties of such materials as manganese and nickel ferrite well into the UHF region.

SOURCE: 'Solid-state devices other than semiconductors' by B Lax and J G Mavroides *Proc IRE* p 1012 (May 1962)

SEE ALSO: 'Genetische und konstitutive Zusammenhange in den magnetischen Eigenschaften bei Ferriten und Eisenoxyden' by G Hilpert *Berichte deutsch chemisch, Gesell* vol 42, p 2248–61 (1909)

New Developments in Ferromagnetic Materials by J L Snoek (New York: Elsevier) (1947)

1910 NEON LAMP   G Claude (France)

The majority of cold cathode tubes use a gas filling of which neon is the principal constituent. Development of the modern cold cathode tube may thus be said to date from 1898, the year in which Sir William Ramsay discovered neon. Ten years later Georges Claude began to isolate substantial quantities of a helium-neon mixture and in 1910 Claude exhibited two 38-ft. neon tubes. These were the precursors of the familiar neon sign and decorative lighting tube of today, in which the light comes principally from the long positive column.

Filament lamps and electric power were both expensive at that time. There was accordingly a strong incentive to develop a cheap and robust lamp of low power consumption and suitable for use on domestic supply voltages. Professor H E Watson has described the work which led to the appearance of the domestic neon lamp, first in Germany in 1918 and later in Holland. Once sputtering had been substantially reduced, the design of a 'beehive' neon for 220 V supplies presented few problems. For 110 V supplies, however, an alloy or activated cathode was required in order sufficiently to reduce the cathode fall. The successful development of such a tube by Philips in Holland stimulated renewed activity in the USA and eventually, in 1929, the General Electric Co., produced a miniature neon indicator.

SOURCE: 'A survey of cold cathode discharge tybes' by D M Neale *The Radio and Electronic Engineering* p 87 (February 1964)

SEE ALSO: 'The development of the neon glow lamp (1911–61)' by H E Watson *Nature* vol 191, No 4793, p 1040–1 (9 September 1961)

1911   **ATOMIC THEORY**   **Lord Rutherford (UK)**

The general model of the atom was first proposed by Rutherford in 1911 and consisted of a nucleus of protons and neutrons, about which rotated electrons in orbits. Such a system appears not unlike our solar system in a scale relative to the size of the components.

SOURCE: 'Semiconductor electronics—1. Solid-state physics' *Electro-Technology* p 95 (October 1960)

1911   **SUPERCONDUCTIVITY**   **K Onnes (Netherlands)**

The discovery of superconductivity dates back to 1911 when Kamerlingh Onnes was carrying out a systematic investigation at Leiden of the electrical resistivity of metals in the range of low temperatures opened up by his recent liquefaction of helium in 1908. In the course of these measurements he observed a sudden drop in the resistance of mercury at 415 K, and was unable to detect any remaining resistance below that temperature; in short, the metal had become superconducting.

SOURCE: *Materials for Conductive and Resistive Functions* by G W A Dummer (New York: Hayden Book Co.) p 121

SEE ALSO: 1911 Onnes H K Commun. Phys. Lab., University of Leiden, No 119b ('Further experiments with liquid helium'); No 120b ('The resistance of pure mercury at helium temperatures. Further experiments with liquid helium'); No 122b ('Disappearance of the electrical resistance of mercury at helium temperatures'); No 124c

1913 Onnes H K Commun. Phys. Lab. University of Leiden, No 133b, Suppl. No 34

1914 Onnes H K Commun. Phys. Lab., University of Leiden, No 139f ('Appearance of galvanic resistance in supra conductors when brought into a magnetic field')

1912   **TUNGAR RECTIFIER**   **I Langmuir (USA)**

During the spring and summer of 1912, Dr. Irving Langmuir of General Electric became interested and active in the development of the gas-filled, tungsten-filament, incandescent lamp. These early laboratory lamps were constructed with heavy-coiled tungsten-spiral filaments that operated at low voltage. The bulb space around the filament was relatively small so that it would operate at high temperature and a few drops of mercury were placed in the bulb. There was usually another portion of the bulb where the mercury vapour condensed and ran back into the filament portion.

These lamps were frequently tested by operation from a 110 V dc line through a series resistance. As one of the objects of these tests was to study the life and characteristics of the lamp at very high temperatures, there were frequent burnouts. It was noticed that, when the filament sometimes burned out during operations, an arc formed at the break.

In all of this early work the primary interest was in the arc as a source of light and apparently no tests on the arc as a rectifier were made at that time. However, one of the men sketched a tube with a hot filament and a separate anode plus liquid mercury.

In the early part of 1915 the possible need for a garage battery charger was considered and during 1916 it became a going concern.

SOURCE: 'Early history of industrial electronics' by W C White *Proc. IRE* p 1130 (May 1962)

1912   **CLOUD CHAMBER (For Revealing the Ionisation**   **C T R Wilson (UK)**
Tracks of Radioactive Particles)

His intention was to produce an artificial cloud by the adiabatic expansion of moist air. For condensation to occur it was then believed that each droplet required a nucleus of dust. However, Wilson showed that even in the complete absence of dust particles some condensation was possible, and that it was greatly facilitated by exposure to X-rays. He was led to conclude that charged atoms (or ions) were the necessary nuclei (1896–7). After much labour he produced, in 1911, his Cloud Chamber, in which the paths of single charged particles showed up as trails of minute water droplets.

The Wilson Cloud Chamber, sometimes modified or refined, has been an indispensable tool of modern physics ever since.

SOURCE: *A Biographical Dictionary of Scientists* by T I Williams (London: Adam and Charles Black) p 564 (1969)

1912   **RELAY AUTOMATIC TELEPHONE EXCHANGE**   **G A Betulander and N G Palmgren (Sweden)**

The idea of using relays as switching circuits in a telephone exchange was by no means new; it was first conceived in 1912 by G A Betulander and N G Palmgren of Sweden who also developed the crossbar selector. Although his name is less well known, Betulander is to the relay exchange what Strowger is to the selector exchange.

SOURCE: *The Telephone and the Exchange* by P J Povey (Post Office Publication) p 84 (1974)

1912   **RADIO (Ionospheric propagation)**   **W H Eccles (UK)**

Following Marconi's success in 1901 in transmitting signals across the Atlantic, Kennelly and Heaviside postulated the existence of a conducting (ionized) layer in the earth's upper atmosphere and suggested that such a layer might cause the waves to follow the curvature of the earth. After it became clear that diffraction could not explain the substantial field strengths actually received at great distance, increased attention was directed to this proposal of an ionised region.

The theory of radio-wave propagation through the ionosphere is based on work by Eccles in 1912, on the ionizing effect of solar radiation, and on the effective refractive index of an ionized medium. Larmor in 1924 re-examined the work of Eccles and others and ascribed the major part of the refractive effect to the presence of free electrons in large numbers. The Eccles–Larmor theory, as later extended and developed by Appleton, Hartree and others to include the effect of anisotropy due to the earth's magnetic field, is now considered the basic theory of radio-wave propagation in the ionosphere. This work was later extended by Booker and others to cover oblique propagation in a nonhomogeneous ionosphere.

SOURCE: 'Radio-wave propagation between World Wars I and II' by S S Attwood *Proc. IRE* p 689 (May 1962)

SEE ALSO: 'On the diurnal variations of the electric waves occurring in nature and on the propagation of electric waves round the bend of the earth' by W H Eccles *Proc. R. Soc.* A vol 87, pp 79–99 (June 1912)

1912   **CIRCUITRY (Regenerative)**   **de Forest, E H Armstrong and I Langmuir (USA) and A Meissner (Germany)**

The regenerative circuit was invented in 1912 by de Forest, Armstrong and Langmuir in the United States and by Meissner in Germany. After 20 years of litigation, the United States Supreme Court finally decided in de Forest's favour.

In the regenerative detector circuit RF energy is fed back from the anode circuit to the grid circuit to give positive feedback at the carrier frequency, thereby increasing the sensitivity of the circuit.

Regenerative receivers marked a big step forward in providing greatly increased sensitivity. Inherently they provided large amplification of small signals and small amplification of large signals. By 1922 they had reached the high point in their development and had almost entirely superseded crystal sets.

SOURCE: 'The development of the art of radio receiving from the early 1920s to the present' by W O Swinyard *Proc. IRE* p 794 (May 1962)

SEE ALSO: 'Some recent developments of regenerative circuits' by E H Armstrong *Proc IRE* vol 10, pp 244–60 (August 1922)

'The regenerative circuit' by E H Armstrong *Proc. Radio Club of America* (April 1915)

1912	**CIRCUITRY(Heterodyne and Superheterodyne) (see figure 5.2)**

H M Fessenden and E H Armstrong (USA)

Professor Fessenden, in his search for an improved receiver, invented the heterodyne system in 1912. Previous receivers had merely acted as valves, detecting by turning a direct current on and off in amounts proportional to the received signal. In contrast, the heterodyne system operated through the joint action of the received signal and a local wave generated at the receiving station. Combination of these two alternating currents resulted in an audio beat-note, the difference frequency between the two waves. Although Fessenden's local oscillator was an arc source, very bulky and troublesome, it was nevertheless the forerunner of superheterodyne and single-banded reception.

The next advance in double-detection technique involved amplification of the beat-note or intermediate frequency. Several parallel developments took place in the United States and in Europe. It is difficult to name an inventor since the superheterodyne system as a basic idea seemed to appear from several sources at about the same time. The works of J H Hammon, A Meissner, Lucian Levy, E F Alexanderson, and E H Armstrong stand out. Armstrong fully appreciated the problem and obtained a patent in 1920 that was of major importance in the practical application of the superheterodyne system.

SOURCE: 'Radio receivers—past and present' by C Buff *Proc. IRE* p 887 (May 1967)

SEE ALSO: 'The superheterodyne—its origin, development and some recent improvements' by E H Armstrong *Proc. IRE* vol 12, p 549 (October 1924)

1913	**RELIABILITY (Standards)**	**AIEE (USA)**

At the risk of some controversy, it can be stated that the first step in the new field of reliability can be traced to discussion generated by the AIEE Standard No. 1 published in 1913. The title of this document was 'General Principles upon which Temperature Limits are Based in the Rating of Electric Machines and Apparatus'. This Standard started the first cycle of appreciation for the chemical composition of electrical materials. During the subsequent development of the electrical and electronic field, there has been a cyclic recurrence of this emphasis which can be recognised today as 'analysis for the basic mechanism of failure in materials used in highly reliable electronic parts'. However, in all fairness, it should be stated that the element of time-dependent degradation was not a part of the referenced first document. This element did not become a formally recognized factor until the great upsurge period for reliability in the 1950s.

SOURCE: 'The reliability and quality control field from its inception to the present' by C M Ryerson *Proc. IRE* p 1323 (May 1962)

1913	**ATOMIC ORBIT THEORY**	**N Bohr (Denmark)**

In 1913 Bohr proposed certain changes in the earlier atomic concept. These were: (1) there are stable orbits in which the electrons do not radiate; (2) in these orbits the angular momentum is an integral number times some constant; (3) if an electron changes from one orbit to another, energy is emitted or absorbed corresponding to the difference in energy between those orbits. Thus, various stable orbits correspond to various permissible energy levels.

SOURCE: 'Semiconductor electronics—1. Solid-state physics' *Electro-Technology* p 95 (October 1960)

1913   **RESISTORS (Thin Metal Film)**   **W F G Swann (UK)**

Since this is a review article it is necessary to refer to some of the earlier papers on the subject of thin films. A typical example is the paper that Dr W F G Swann presented to the British Association meeting over sixty years ago in 1913, which illustrates how little our ideas have changed in this period. Swann measured the resistivity of sputtered platinum films as a function of the sputtering time and hence of the film thickness; he also recorded the temperature coefficient of resistance of these films between liquid nitrogen temperature and 100°C and observed the abnormally high resistivity associated with very thin films and also a negative temperature coefficient of resistance in the thinnest films.

SOURCE: 'Resistive thin films and thin-film resistors—history, science and technology' by J A Bennett *Electronic Components* p 737 (September 1964)

SEE ALSO: W F G Swann *Phil. Mag.* vol 28 p 467 (1914)

1914   **THYRATRONS**   **I Langmuir (USA)**

In 1914 Dr Langmuir first suggested a method of controlling the arc in a mercury-pool tube by means of a grid. He showed how a grid voltage could be used to control the starting of the main arc in each rectifying cycle. Thus, the average arc current through the tube was controlled when an ac anode voltage was used.

In 1922 Toulon, a French scientist, improved on this method of control by varying the phase of the grid voltage with respect to the anode voltage rather than to its amplitude. Thus, the arc could be made to start at any point in the anode voltage cycle and this resulted in a very practical and convenient method of controlling the average value of a rectified anode current.

In 1936 Dr A W Hull developed the idea of operating a hot-cathode diode in a low pressure of an inert gas or vapour. As a result, the space charge effect was eliminated and the voltage drop of the discharge dropped to a low and more or less constant value of 5 to 10 volts. At this low arc drop voltage, the positive-ion bombardment of the cathode is not destructive.

SOURCE: 'Early history of industrial electronics' by W C G White *Proc. IRE* p 1132 (May 1962)

1914   **ASDIC**   **(UK)**

The British traced their sonar system to World War I, when a piezoelectric oscillator was used to emit the sound waves in a system called the ASDIC, named for the Allied Submarine Detection Investigation Committee), set up in 1917. At the outbreak of World War II, Britain outfitted its fleet with ASDICs that had the quartz-steel transducer. In the US, meanwhile, work on the magnetostriction-tube transducer was being done under contract by the Submarine Signal Company of Newport, RI.

SOURCE: *Electronics* p 174 (17 April 1980)

1915   **ACOUSTIC MINE**   **Wood (UK)**

The acoustic mine was invented by Wood, later Deputy Superintendent of the Admiralty Research Laboratory, in the war of 1914–18.

The first encounter with a German acoustic mine in the recent war occurred in the Firth of Forth on 31 August 1940, when one was exploded by a motor boat. Then the cruiser 'Galatea' reported that twice in a week mines had exploded well ahead of her. Destroyers began to explode mines half a mile away. One suggestion was that the phenomenon was due to unexploded bombs. Even at the end of September, some experts doubted the existence of acoustic mines.

The mine contained a reed tuned to vibrate on a frequency of 240 per second. The noise was picked up and communicated by the reed to a carbon microphone. The later forms of mine contained counters which operate like a telephone exchange. The mine will not go off until it is called up, so to speak, for, say, the seventh time. The first six ships will pass over it safely and it will explode under the seventh. The mines contained clocks which could keep them disarmed for many days, until a fixed date.

SOURCE: *Science at War* by J G Crowther and R Whiddington (London: HMSO) p 172 (1947)

1915   **SONAR (Submerged Submarine Detection by**   **P Langevin (France)**
        Ultra-Sonics)
(See also chapter 6.)

The detection of submerged submarines by sound has been particularly highly developed by British scientists. The method has been evolved from a suggestion made in 1912, after the sinking of the liner Titanic by collision with an iceberg. A British engineer named Richardson suggested that icebergs might be detected by the echo of a pulse of sound waves emitted from the approaching ship. The use of short sound waves which can be readily 'beamed' like a searchlight is necessary in order to secure precise definition of the direction of the berg. In 1914 the only known practical way of obtaining such short or supersonic waves was from oscillators made of mica and caused to vibrate by electrical stresses.

An Allied Committee was at this time formed to develop anti-submarine technique. It included Dr R W Boyle of Canada, Professor W H (later Sir William) Bragg, Professor P Langevin of France and Rutherford. Langevin succeeded in 1915 in producing ultra-sonic waves by applying the piezo-electric effect discovered by Jacques and Pierre Curie. When quartz crystals are cut in the appropriate way electrification expands or contracts them; and conversely when mechanically stretched or compressed they produce an electric charge. Thus an alternating electric potential applied to a quartz crystal causes it to vibrate and these vibrations are of the type that produces sound waves. Hence by suspending a quartz crystal in water, and applying to it a pulse of alternating current of appropriate frequency, the crystal will be made to vibrate and communicate to the water a pulse of sound waves of the desired length. This will be transmitted through the water and be reflected by any obstacle. If such an obstacle is of sufficient size an echo can be received on the quartz which sent out the transmission. The echo creates a weak pulse of alternating current which can be detected by electrical instruments. In the spring of 1918, scientists at the Admiralty Experimental Station at Harwich succeeded with an apparatus of this kind in securing super-sonic echoes from a British submarine at a range of a few hundred yards.

SOURCE: *Science at War* by J G Crowther and R Whiddington (London: HMSO) p 153 (1947)

SEE ALSO: 'Uses of ultrasonics in radio, radar and sonar systems' by W P Mason *Proc. IRE* p 1374 (May 1962)

Electroacoustics by F V Hunt (Cambridge, Mass.: Harvard University Press) chapter 1 (1954)

1915   **FILTERS (ELECTROMAGNETIC)**   **G Campbell & K W Wagner (UK)**

In 1831 Michael Faraday formulated the law of electromagnetic induction and self-induction. Some 84 years later, in 1915, G Campbell and K W Wagner utilized Faraday's law in their invention of the first electromagnetic or LC wave filter. Significant advances in filter theory and technology then followed rapidly, until today, filters have so permeated electronic technology that it is hard to conceive of a modern world without them. Consumer, industrial, and military electronic systems all require some kind of signal filter, and, in the past, LC filter networks provided one of the most efficient and economical methods of implementing them.

SOURCE: 'Inductorless filters: a survey' by G S Moschytz *IEEE Spectrum* p 30 (August 1970)

SEE ALSO: 'The golden anniversary of electric wave filters' by A I Zverev *IEEE Spectrum* vol 3, pp 129–31 (March 1966)

'Introduction to filters' by A I Zverev *Electro-Technol (New York)* pp 63-90 (June 1964)

1915   **RADIO (Single Sideband Communication)**   **J R Carson (USA)**

The military and international companies with a need to communicate abroad leaned heavily on radio, and by the 1950s designers were hunting for ways to ease spectrum crowding. Single-sideband transmission appeared to hold great promise.

Like so many other radio techniques, it had been developed originally for use in telephone systems. In 1915, J R Carson of AT & T's Development and Research Dept. had proved that only one sideband was needed for transmitting intelligence. Three years later the first commercial application of single-sideband showed that it was possible to use this technique to increase channel capacity. By the mid-1930s single-sideband transmission at high radio frequencies had proved successful. In the late 1950s, after its first major application in the Strategic Air Command became known, single-sideband really caught on.

SOURCE: 'Communications' by J H Gilder *Electronic Design* vol 24, p 96 (22 November 1972)

1916 **TELEX** **Markrum Co (USA)**

The first teleprinter, making it possible to send written messages through telephone lines, was invented in 1916 by Markrum Co. of Chicago. The system became operational in 1928 and was extended at a national level by the Bell Laboratories in 1931 under the name telex from teleprinter exchange.

SOURCE: *The Book of Inventions and Discoveries* Associate Editor Valerie-Anne Giscard d'Estaing (UK: Queen Anne Press, Macdonald & Co.) (1990) p 243

1916 **RELIABILITY (Control)** **Bell/Western Electric (USA)**

Perhaps one of the first clear-cut examples of what would now be called a reliability-control programme occurred in 1916. The Western Electric Company and the Bell Telephone Laboratories co-operated in a planned programme to produce good performing and trouble-free telephone equipment for public use. The following elements of a good reliability programme were involved:

(1) An ably planned, forward-looking R & D programme considering the system needs.

(2) Development programmes leading to mature designs verified by design reviews and engineering model tests.

(3) Part improvement projects using test-to-failure methods, consideration of tolerances and performance under stress loading.

(4) Standardization and simplification in design.

(5) Product evaluation under field application conditions for prototypes and pilot production models.

(6) Quality control during manufacture to assure achieving the reliability inherent in the design.

(7) Feedback from the field to provide information for the designers.

SOURCE: 'The reliability and quality control field from its inception to the present' by G M Ryerson *Proc. IRE* p 1323 (May 1962)

1916 **'CROSSBAR' TELEPHONE EXCHANGE** **J G Roberts and J N Reynolds (USA)**

The first 'crossbar' based on an idea conceived as long ago as 1901, by Homer J Roberts, of the USA, but the idea needed many years of development work before it could be transformed into a practical system. This was done by John G Roberts, and John N Reynolds of the USA, who patented the system in 1916, and by Gotthelf A Betulander and Nils G Palmgren of Sweden who designed the first satisfactory mechanism. The first fully operational crossbar exchange was installed in Sweden in 1926. By the late 1930s, crossbar had been developed to the point where it could offer several advantages over Strowger, although it was more expensive. At that time, crossbar, like Strowger, was a step-by-step system but it used a totally different kind of selector.

SOURCE. *The Telephone and the Exchange* by P J Covey (Post Office Publication) p 75 (1974)

1917 **CONDENSER MICROPHONE** **E C Wente (USA)**

One electrostatic device chat has earned a secure place for itself in the transducer art is the condenser microphone. Wente's 'uniformly sensitive instrument' of 1917 probably represents the first transducer design in which sensitivity was deliberately traded for uniformity of response, and it was almost certainly the first in which electronic amplification was relied on to gain back the ground lost by eschewing resonance.

SOURCE: *Electroacoustics* by F V Hunt (Cambridge, Mass': Harvard University Press) p 169 (1954)

SEE ALSO: 'The sensitivity and recision of the electrostatic transmitter for measuring sound intensities' by E C Wente *Phys. Res.* vol 19, p 498 (May 1922)

1917   CRYSTAL PULLING TECHNIQUE   J Czochralski

Most of the single crystal silicon used for p-n junction devices is produced by one of three techniques. The oldest of these is the pulling process commonly called the Czochralski technique whereby the crystal is grown from a charge of Si melted in a crucible (usually quartz). Historically, the next development was the float-zone technique of crystal growth, involving the movement of a narrow molten zone up or down a vertical rod of initially polycrystalline silicon. This technique has been widely used because it eliminates the problem of crucible contamination. The third crystal growth technique, which has only relatively recently been commercially exploited, is a crucibleless pulling technique whereby the molten region is formed on the end of a thick silicon rod and the crystal is seeded and pulled from this molten pool in a manner combining, to some extent, features of Czochralski and float-zone growth.

SOURCE: 'Microinhomogeneity problems in silicon' by H F John, J W Faust and R Stickler *IEEE Trans. on Parts, Materials and Packaging* vol PMP-2, No 3, p 51 (September 1966)

SEE ALSO: 'Measuring the velocity of crystallisation of metals' by J Czochralski *Z. Phys. Chem.* vol 92, p 219 (April 1917)

1918   ALEXANDERSON ALTERNATOR   E F W Alexanderson (USA)

Alexanderson tested special Swedish iron strips 1.5 mills thick in strong magnetic fields, and found the iron capable of satisfactory operation at 100 kHz, so he designed the alternator with an iron core. However Fessenden rejected the design and insisted on the use of a wooden core as he was sure that iron would melt in a strong magnetic field at 100 kHz.

By mid-1906 General Electric had built an alternator with a wooden core as specified, and Fessenden used it for his tests from Brant Rock, Mass. on Christmas Eve, 1906. He succeeded in broadcasting both speech and music, and the transmission was heard as far away as Norfolk, Virginia. However Dr Alexanderson did not give up his idea of using an iron cored armature, and was able to obtain authority from GE to build a model alternator to his own design. When this machine was demonstrated to Fessenden he was convinced of its potential and placed an order for two 100 kW alternators using iron cores.

By 1915 a 50 kW, 50 kHz experimental alternator was being tested and Dr Alexanderson was able to modulate it with voice, using a DeForest Audion valve to control a magnetic amplifier. Dr Alexanderson was also responsible for the design of a multiple tuned antenna for use with the alternator.

By 1917 the 50 kW alternator was ready to be tested in the American Marconi station at New Brunswick, NJ, but by this time America had entered the war and all radio stations were taken over by the US government.

SOURCE: 'The Alexanderson Alternator, a 'near-perfect' system of W/T transmission' by K Weedon *IEE Conf. Publ. 411* p 69 (September 1995)

SEE ALSO: 'Transatlantic radio communication' by E F W Alexanderson *Proc. IEEE* vol 72, no 5, p 626 (May 1984)

1918   INDUCTION HEATING (High Frequency)   E F Northrup (USA)

The production of heat by induced currents was recognised as early as 1880. The heating of transformer cores due to eddy currents was, of course, understood at an early date. Probably the real engineering and developmental pioneer in induction heating with frequencies above the power range was Professor

E F Northrup of Princeton University. In 1918 he built practical 'furnaces' for frequencies above 10 000 cycles and powers as high as 60 kW using a spark gas oscillator.

SOURCE: 'Early history of industrial electronics' by W C White *Proc. IRE* p 1134 (May 1962)

1918   **RADIO (Ground Wave Propagation)**   **G N Watson (UK)**

At all frequencies, electromagnetic energy is diffracted into the geometric shadow of an obstacle, the amount increasing with decreasing frequency. In the radio-frequency range, diffraction yields significant and usable signal strengths in the shadow zone beyond the geometric field strength in the shadow zone of the earth was attacked by many mathematicians over 15 years with resultant divergent answers. The first significant break-through was produced by Watson in 1918, who showed that waves radiated by an antenna on the surface of a perfectly conducting sphere would be attenuated exponentially at great distances. The numerical values of field strength predicted by Watson under these conditions proved to be far lower than known experimental values. This discrepancy promoted increased interest in the ionosphere as a mechanism that might explain the wide divergence in the mathematical and experimental values of field strength, particularly in the kilocycle range of frequencies.

SOURCE: 'Radio-wave propagation between World Wars I and II' by S S Attwood *Proc. IRE* p 688 (May 1962)

SEE ALSO: 'The diffraction of electric waves by the earth' by G N Watson *Proc. R. Soc. A* vol 95, pp 83–89 (October 1918) and vol 96, pp 546–63 (July 1919)

1918   **THE DYNATRON**   **A W Hull (USA)**

The dynatron possesses substantial advantages over other types of oscillator that are available for the same purposes and, in additions has applications for which other types are not available. For example, a two-terminal oscillatory circuit, with one terminal 'earthy' can be maintained in oscillation, there being no necessity for tappings, or auxiliary circuit elements. The oscillators are in general more stable in frequency than those maintained by triodes and are under more precise control.

SOURCE: 'Applications of the dynatron' by M G Scroggie *The Wireless Engineer* vol X, no 121, p 527 (October 1933)

SEE ALSO: 'The dynatron' by A W Hull *Proc. IRE* vol 6, pp 5–35 (February 1918)

'The dynatron detector' by Hull, Kennelly and Elder *Proc. IRE* p 320 (October 1922)

1918   **ATOMIC TRANSMUTATION**   **Lord Rutherford (UK)**

When certain radioactive atoms explode, they release some of their energy by communicating it to the atomic fragments flung out in the explosion. Some of these fragments consist of nuclei of atoms of helium. They are flung out with immense speed and hence high energy of movement. Why not direct these against the citadel of an atom, against its nucleus. Rutherford was preoccupied with this idea when the war of 1914 began. He was called by the British Government to assist in the scientific struggle against the German submarines and was very busy with this work. But he kept the attacks on the atom going. While the Germans were shelling Verdun, Rutherford was bombarding the nucleus in his laboratory at Manchester, when he could find the time.

In 1918 he failed to appear at the meeting of an important war science committee. When one of his colleagues asked him why he had failed to appear, he said that he had just got definite evidence that it might be possible to disintegrate an atom at will, and that if this proved to be true 'it was far more important than the war'.

Immediately after the conclusion of the war, while still in Manchester, he completed this investigation and in 1919 published a conclusive proof that atoms of nitrogen could be transmuted by bombardment with atomic projectiles from natural radioactive substances; and later, in Cambridge, he extended his investigations to the study of the artificial disintegration of the atoms of many light elements. He was assisted in this especially by Chadwick, who had been one of his most brilliant pupils and collaborators at Manchester.

SOURCE: *Science at War* by J G Crowther and R Whiddington (London: HMSO) p 126 (1947)

1918 MULTIVIBRATOR CIRCUIT   H Abraham & E Bloch (France)

The multivibrator, described by Abraham and Bloch in 1918, is a very important circuit, and is in use for many purposes (see figure 11.5). In its original form, however, it is more commonly used for the generation of pulses than as a time base or discharger circuit.

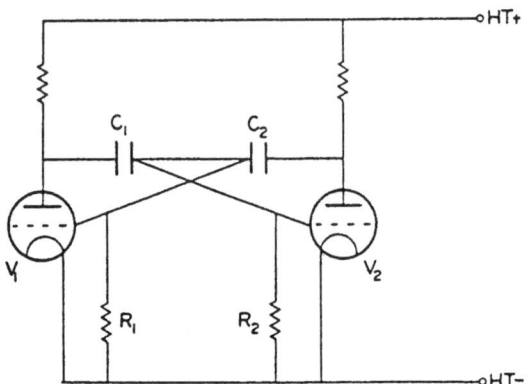

Figure 11.5. Multivibrator circuit.

The multivibrator consists of two valves, each having the anode coupled to the grid of the other via a condenser, and with a leak resistance to earth. The method of operation of the circuit is as follows: when the anode current in one valve increases, it passes a negative signal to the grid of the other, due to the increased voltage drop across the anode load.. The anode current of the second valve is thus reduced, which results in the grid of the first valve becoming more positive and increases the anode current of this valve still further. It will thus be apparent that the effect is a cumulative one and that the anode current in the first valve rapidly reaches a maximum while the anode current in the second is cut off. This circuit condition then remains until the charge in the condenser C_1 has leaked away through R_2 sufficiently to permit anode current to flow once more in the second valve. When this occurs the cumulative effect again takes place, but in the reverse direction. The multivibrator, therefore, has two unstable limiting conditions which occur when either of the valves is at cut-off and the other at zero grid potential.

SOURCE: *Time Bases* by O S Puckle (London: Chapman & Hall) p 25 (1944)

SEE ALSO: 'Notice sur les Lampes-valves à 3 Electrodes et leurs applications' Publication No 27 of the French Ministere de la Geurre (April 1918)

1918 CIRCUITRY (Neutrodyne)   L A Hazeltine (USA)

The neutrodyne circuit was invented in 1918 by Hazeltine. Basically this was a tuned radio-frequency (TRF) amplifier which employed a specific type of neutralisation. A current obtained from the plate circuit was fed back into the grid circuit in the proper magnitude and phase to balance out, or neutralise, the effect of grid-to-plate capacitance inside the tube, and thus it achieved stability and prevented oscillation.

SOURCE: 'The development of the art of radio receiving from the early 1920s to the present' by W O Swinyard *Proc. IRE* p 794 (May 1962)

SEE ALSO: 'Tuned radio frequency amplification with neutralisation of capacity coupling' by L A Hazeltine *Proc. Radio Club of America* vol 2 (March 1923)

1918   NOISE (Shot Effect)   **W Schottky (Germany)**

The first realisation that unwanted random noise was a factor to contend with in the field of communications came during World War I when attempts were being made to design high-gain vacuum-tube amplifiers. It was soon found that there was a limit to the number of stages which could be cascaded in the quest for high gain due to an unacceptably high background noise which masked the weak signals being amplified. In his classic paper Schottky first explained one of these effects and formulated the random component in the plate current of a vacuum tube.

Schottky ascribed the random fluctuations in the plate current to the fact that this current is composed not of a continuous but rather of a sequence of discrete increments of charge carried by each electron arriving at the plate at random times. The average rate of charge arrival constitutes the dc component of the plate current on which is superimposed a flucuation component as each discrete charge arrives. He referred to this phenomenon as 'schroteffekt' or 'shot-effect'.

SOURCE: 'Noise and random processes' by J R Ragazzini and S S L Chang *Proc. IRE* p 1146 (May 1962)

SEE ALSO: 'Theory of shot effect' by W Schottky *Ann. Phys.* vol 57, p 541 (December 1918)

1919   TELEVISION (Electronic System)   **V Zworykin (USA)**

1927   TELEVISION (Electronic System)   **P Farnsworth (USA)**

It was already recognised that a complete cathode-ray system provided the answer, but no experimental work was started and the idea lay dormant until, after going to America in 1919, Zworykin joined the Westinghouse Company. His major difficulty centred on the electronic camera tube. He, like Campbell-Swinton, had conceived the idea of charge storage by 1919. Zworykin lacked the necessary funds to carry his ideas forward into practical form at the time he conceived them, and it was several years before he could concentrate on their elaboration. Westinghouse took him on to their research staff but their laboratory devoted itself mainly to radio research, and, since Zworykin was given no freedom to pursue his ideas on television, he resigned to join a development company in Kansas. Returning to Westinghouse in 1923, he drew up an agreement whereby he retained the rights to the television inventions he had disclosed in 1919, while Westinghouse acquired the exclusive option to purchase the patents at a later date.

Philo Farnsworth was essentially an individual inventor who, though fortunate enough to find substantial financial backing, always retained his autonomy in research. Largely self-taught, he appears at an early age to have conceived a completely electronic system. He was of the type which prefers to work on a small scale with relatively simple equipment. Working in laboratories in Los Angeles and later in San Francisco he was able to demonstrate a complete electronic system in 1927, when he filed his first patent application, including his image dissector tube, which constitutes his most important inventive contribution After long-drawn-out patent interference proceedings, Farnsworth and Zworykin each received basic patents on their different systems of television transmission.

SOURCE: *The Sources of Invention* by J Jewkes, D Sawers and R Stillerman (London: MacMillan & Co.) p 385/6 (1958)

SEE ALSO: 'The history of TV' by G R M Garratt and A H Mumford *Proc. IEE; Part III A, Television* vol 99 (1952)

1919   RETARDED FIELD MICRO-WAVE OSCILLATOR   **H Barkhausen & K Kurtz (Germany)**

In 1919 it was noted by Barkhausen and Kurtz, during tests for the presence of gas in transmitting valves in which the grid was held at a high positive potential and the anode at a negative one, that oscillation could be maintained in a circuit connected between grid and anode, or between other pairs of electrodes.

The explanation was given that electrons are accelerated from the cathode to the positive grid through which some of them pass. These are then retarded in the grid anode space and turn back to the grid, some of them again passing through, and being reflected at the cathode to repeat the behaviour.

SOURCE: 'Microwave valves: A survey of evolution, principles of operation and basic characteristics' by C H Dix and W E Willshaw *J. Brit. IRE* p 578 (August 1960)

SEE ALSO: 'The shortest waves obtainable with vacuum tubes' by H Barkhausen and K Kurtz *Z. Phys.* vol 21, pp 1–6 (1920)

1919   CIRCUITRY (Flip-Flop Circuits)   Eccles and Jordan (USA)

In digital circuitry probably the most important single device is the bistaple circuit or 'flip-flop'. Used today in innumerable modifications the original form of this circuit was developed in 1919 by Eccles and Jordan. During the late 1930s flip-flops in tandem connection began to be used by nuclear physicists to scale down the counting rate from Geiger–Muller counters by factors of $2n$. These devices were all essentially binary counters in which the last 'carry' pulse was used as the output to a mechanical register. In 1944, Potter described a scale-of-ten counter in which feedback caused a binary counter to skip six of its possible 16 states. A neon lamp indicating the state of each flip-flop provided a binary-coded readout for each decade. In 1946, Grosdoff described another binary counter with feedback to provide a scale-often circuit. Grosdoff's circuit featured a direct readout obtained by a matrix of resistors arranged to light one of ten neon lamps for each stablestate of the counter. By connecting such decades in tandem direct reading decimal counting registers of arbitrarily large capacity were achieved and the way was cleared for the widespread application of high speed electronic counting.

SOURCE: 'Digital display of measurements in instrumentation' by B M Oliver *Proc. IRE* p 1170 (May 1962)

SEE ALSO: 'A highly accurate continuously variable frequency control system' by C H Vincent *Instruments and Methods* vol 7, p 325 (1960)

1919   VALVES—HOUSKEEPER SEAL   W G Houskeeper (USA)

The breakthrough came in 1919 when W G Houskeeper patented a method of sealing base metals through glass. Before this the leads brought through the glass had to be wires or ribbons. With this invention it was possible to make large diameter seals using metals such as high purity copper which has high conductivity, and is easily worked into any desired shape. The expansion of copper is $165 \times 10^{-7}/°C$ and that of a common glass is $52 \times 10^{-7}/°C$ so a matched seal is not possible. The technique developed by Houskeeper was to taper the metal to a feather edge and then seal the glass to the thin edge of the taper. The copper is cleaned, oxidised and then sealed to the glass. The thin copper is sufficiently flexible to equalise the difference in expansion and contraction between the glass and metal.

With this development it became possible to make the anode part of the vacuum envelope and cool its outer surface directly. By 1925 valves with water cooled anodes were in commercial production and they mark the beginning of the era of high-power broadcast transmitters.

SOURCE: *Electronics Engineer's Reference Book* (London: Newnes-Butterworth) chap. 7, pp 7–31 (1976)

SEE ALSO: 'The art of sealing base metals through glass' by W G Houskeeper *J. Am. IEE* vol 41, p 870 (1923)

1919   CRYSTAL MICROPHONE   Nicholson (USA)

The crystal microphone, destined to be widely used in home tape recorders and public address systems, is made in the USA by Nicholson. The mike works on the piezo-electric principle, by which small voltages are produced on the surface of a solid such as a crystal. Sound quality is good, and costs low, but the mike reacts adversely to heat, humidity and rough handling.

SOURCE: *The Timetable of Technology* (London: Michael Joseph, Marshall Editions) p 61 (November 1982)

1919   **'MILLER' TIME BASE CIRCUIT**   **J M Miller (USA)**

There is a family of time bases (see figure 11.6) which utilize the device known as a Miller Capacitance Valve, i.e, a valve circuit with a condenser C connected directly between the grid and the anode. The insertion of this condenser results in feed-back from the anode of the valve to its grid. When such a valve circuit is provided with a resistive anode load the effect on the grid circuit is substantially the same as if a capacitance of $C(A + 1)$ in series with a resistance equal to I/g had been connected between the grid and the cathode, where A is the gain of the valve stage and g is the mutual conductance in amperes per volt. This is usually known as the 'Miller Effect' (see figure 11.6).

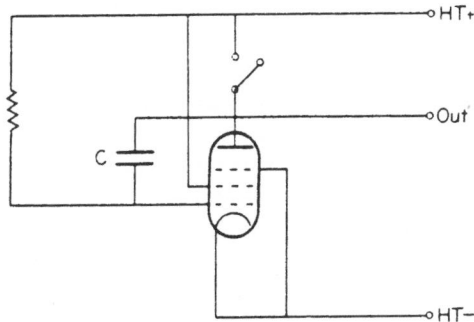

Figure 11.6. Miller time base circuit.

SOURCE: *Time Bases* by O S Puckle (2nd edn) (London: Chapman & Hall) p 158 (1951)

SEE ALSO: 'Dependence of the input impedance of a three-electrode vacuum tube upon the load in the plate circuit' by J M Miller *Sci. Papers of the Bureau of Standards* vol 15, p 367 (1919)

1919   **RESISTOR (Metal-Film Type)**   **F Kruger (Germany)**

Proposals for the production of metallic-film resistors that will possess a high degree of stability date back over 25 years, but mostly they could be prepared only with comparatively low resistance values. By spiralling the coating in a manner similar to that used for high-stability resistors, it has been found possible to increase the resistance of resistors made with gold films up to 200 000 ohms.

SOURCE: 'Fixed resistors for use in communication equipment' by P R Coursey *Proc. IEE* vol 96, pt. III, p 173 (1949)

SEE ALSO: F Kruger: British Patent No 157909/1919

1920   **PLASTIC MAGNETIC TAPE**   **Dr Pfleumer (Germany)**

Magnetic tape recording becomes possible outside the laboratory. The reason is the introduction, by Austrian researcher Dr Pfleumer, of plastic tapes in place of steel wires or tapes. The magnetophones marketed by the Germans in the 1930s will incorporate these plastic tapes.

SOURCE: *The Timetable of Technology* (London: Michael Joseph, Marshall Editions) p 65 (November 1982)

1920   **ULTRA-MICROMETER**   **R Whiddington (UK)**

Few would suspect that one of the first authentic disclosures of practical electronic instrumentation was published in the November 1920 issue of The Philosophical Magazine by R Whiddington MA, DSc, Cavendish Professor of Physics in the University of Leeds.

The title of Professor Whiddington's communication is 'The Ultra-Micrometer; an Application of the Thermionic Valve to the Measurement of Very Small Distances'. The summary of the paper reads: 'If a circuit consisting of a parallel-plate condenser and inductance be maintained in oscillation by means of a thermionic valve, a small change in distance apart of the plates produces a change in the frequency of the oscillations which can be accurately determined by methods described. It is shown that changes so small as 1/200 millionth of an inch can easily be detected. The name 'ultra-micrometer' is tentatively suggested for the apparatus'.

SOURCE: Letter: 'A pioneer of electronics' by F G Diver *The Radio and Electronic Engineer* vol 45, No 11, p 687 (November 1975)

SEE ALSO: 'The ultra-micrometer—an application of the therrnionic valve to the measurement of very small distances' by R Whiddington *Phil. Mag.* Series 62, vol 40, p 634 (November 1920)

1921   CRYSTAL CONTROL OF FREQUENCY   W G Cady (USA)

While working with a quartz crystal connected in the circuit of a self-excited vacuum-tube oscillator, Cady discovered that the frequency of self-oscillation could be stabilized over a small range by the vibration of the quartz crystal. In retrospect it appeared that a good many other experiments had made use of crystals associated with vacuum-tube circuits with which they might have observed this stabilizing action; but most of them sought to avoid the 'anomalous' effects that occurred in the neighbourhood of resonance in the crystal, and it remained for Cady to discover this remarkable stabilizing action of a high-Q resonator. Cady extended these results by applying two pairs of terminals to the crystal and connecting these as a feedback path for a three-tube amplifier in such a way that the circuit would oscillate but only at the resonance frequency of the crystal.

Cady described and demonstrated his crystal circuits (January 1923) to Professor G W Pierce of Harvard University. Pierce took up the study of such crystal-oscillator circuits immediately, and within a few months his experiments had led him to the invention of several improved forms of crystal oscillator in which a two-terminal crystal could be made to control uniquely the frequency of oscillation in a single-tube circuit.

SOURCE: *Electroacoustics* by F V Hunt (Cambridge, Mass.: Harvard University Press) p 53, 55 (1954)

SEE ALSO: 'The piezo-electric resonator' by W G Cady *Phys. Rev.* A, vol 17, p 531 (April 1921)

Proc. Inst. Radio Engrs. vol 10, pp 83–114 (April 1922); also *Piezoelectricity* (New York and London: McGraw-Hill Book Co.) (1946)

Walter G Cady (crystal resonator) US Patent No 1450 246 (filed 28 January 1920) issued 3 April 1923; also (crystal stabilization and a 3-tube oscillator controlled by a 4-terminal crystal) US Patent No 1472 583 (filed 28 May 1921) issued 30 October 1923.

1921   FERROELECTRICITY   J Valasek

Ferroelectricity was first discovered by Valasek in 1921 in Rochelle salt, which is piezoelectric above its Curie point; however, the most important ferroelectric material today is barium titanate, which was discovered in 1942 and developed shortly thereafter at MIT for ferroelectric applications. Since then, other materials of the Perovskite structure and of other structures have been found to be ferroelectric.

SOURCE: 'Solid state devices other than semiconductors' by B Lax and J G Mavroides *Proc. IRE* p 1014 (May 1962)

SEE ALSO: 'Piezoelectric and allied phenomena in Rochelle salt' by J Valasek *Phys. Rev.* vol 17, p 475 (April 1921)

ALSO: 'High dielectric constant ceramics' by A von Hippel, R E Breckenridge, F E Chesley and L Tisya *Ind. Engrg. Chem.* vol 38, p 1097 (November 1946)

1921   SHORT WAVE RADIO   Amateurs (USA and Europe)

One of the main developments in radio after 1918 was the discovery of the usefulness of the shorter wave-bands. It was generally considered that wavelengths below 200 metres were useless except for short distance transmission, though cases were known of long ranges being obtained on short waves.

These were regarded as freaks, however, and wavelengths below 200 metres were, after 1918, allocated to amateurs who encouraged by these 'freak' results, arranged trial broadcasts from America to England in December 1921 on 200 metres. Their success showed that short-wave low-power broadcasts could be heard over long distances

SOURCE: *Sources of Invention* by J Jewkes, D Sawers and R Stellerman (London: MacMillan) p 353 (1958)

1922   **CIRCUITRY (Superregeneration)**   **E H Armstrong (USA)**

Superregeneration was discovered by E H Arrmstrong during the defense of his patent case for the regenerative receiver. In a superregenerative receiver, sustained oscillations are squelched by periodic variation of the effective resistance of the input resonant circuit. Oscillations periodically build up in a circuit resonant at the signal frequency. Sustained oscillations are prevented by periodic application to the grid of the superregenerative tube of a signal that damps the oscillations. The quenching frequency is usually between 20 000 and 100 000 cps. The superregenerative detector, because of its broad tuning and considerable sensitivity, was practical and popular earlier when unstable modulated oscillators were used as transmitters at frequencies above 30 Mc.

SOURCE: 'Radio receivers—past and present' by C Buff *Proc. IRE* p 887 (May 1962)

SEE ALSO: 'Super-regenerative receivers' by J R Whitehead (Cambridge: Cambridge University Press) (1950)

1922   **NEGATIVE RESISTANCE OSCILLATOR**   **E W B Gill and J H Morell (UK)**

Following the work of Barkhausen and Kurtz (1919) which for the first time reported the generation of electrical oscillations depending primarily on the oscillatory motion of electrons in a vacuum, and not on the excitation of oscillatory currents in a tuned circuit, and this discovery might be said to represent the starting point of the whole field of modern microwave valves.

Subsequent work on this device called generally the 'retarding field' generator, by Gill and Morrell, reported in 1922 and subsequently by many others, showed that this simple type of operation had many variants including one in which with adequate emission, a negative resistance could be provided by the tube over a frequency band.

SOURCE: 'Microwave valves: a survey of evolution, principles of operation and basic characteristics' by C H Dix and W E Willshaw *J. Brit. IRE* p 578 (August 1960)

1923   **'SQUEGGER' CIRCUIT**   **E V Appleton, J F Herd and R A Watson Watt (UK)**

The first hard valve time base was of the transformer-coupled type and was developed about 1923 by Appleton, Herd and Watson Watt. The circuit was known as a 'squegging oscillator' or 'squegger' (see figure 11.7).

The circuit consists of a transformer-coupled valve V_1, oscillating fairly violently at a radio frequency and having a condenser C_1 in the cathode-grid circuit. At each positive peak of potential at the grid, current flows from the transformer secondary winding through the valve V_1 from grid to cathode and into the condenser C_1 which thus accumulates a charge such that the mean grid potential becomes increasingly negative. When the negative potential reaches the cut-off bias potential of the valve, the anode current and, hence, also the alternating potential at the grid, is cut off and remains so until the charge on the condenser leaks away sufficiently through the diode to permit the resumption of oscillation. While the anode current is cut off, the condenser C_1 loses its charge via V_2 and the grid makes a potential excursion towards zero volts. Upon this excursion there is superimposed a damped oscillatory motion which is the second half-cycle of the oscillation.

SOURCE: *Time Bases* by O S Puckle (2nd edn) (London: Chapman & Hall) (1951)

SEE ALSO: E V Appleton, J F Herd and R A Watson Watt: British Patent No 235254.

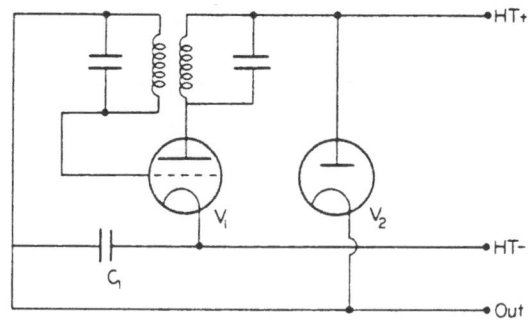

Figure 11.7. Squegging oscillator circuit.

1923   **ICONOSCOPE**   **V K Zworykin (USA)**

The combination of electron beam scanning and storage was first proposed and carried into practice by V K Zworykin in 1923—at a time preceding any other electronic pickup system. The target was an aluminium film oxidised on one side, which was photosensitised with cesium vapour and faced a metal grill serving as collector for photoelectronics. The metal side, which served as a signal plate, was scanned by a high-speed electron beam, which penetrated through the oxide layer, forming a temporary conducting path permitting the locally stored charge to flow off through the signal plate.

While the tube as described was capable of transmitting only rudimentary patterns, it became the ancestor of the extended line of storage camera tubes which dominate all phases of television picture transmission. Their common features are electron beam scanning and a picture target with small transverse conductivity capable of storing charge released in response to light.

The first practical storage camera tube was the iconoscope. Here, the picture was projected on a mosaic of photosensitive elements capacitatively coupled to a signal plate. The mosaic was then scanned by a high-velocity beam, restoring the mosaic to a uniform potential and releasing photoelectrically stored charge for forming the picture signal. The secondary-emission ratio was greater than unity, so that the equilibrium potential under the beam was close to that of the electrodes facing the mosaic.

SOURCE: 'Beam-deflection and photo devices' by K Schlesinger and E G Ramberg *Proc. IRE* p 993 (May 1962)

SEE ALSO: V K Zworykin: US Patent No 2141 059 (20 December 1938) filed 29 December 1923

'The Iconoscope—a modern version of the electric eye' by V K Zworykin *Proc. IRE* vol 22, p 16 (January 1934)

1924   **REISZ TRANSVERSE-CURRENT CARBON**   **G Neumann (Germany)**
       MICROPHONE

A very significant invention in the microphone field was the Reisz transverse-current microphone, which became the standard BBC microphone from 1926 to 1935—the familiar white marble octagonal microphone. It was invented by Georg Neumann, who worked for the Reisz Company in Germany at the time, about 1924.

A block of marble or other insulating material had two deep recesses cut in it, these being joined by a shallow surface trough. Each recess had a carbon electrode in it, connected to a terminal at the rear. The block was covered by a thin diaphragm of mica or paper and the space between this and the trough-plus-recesses was filled with fine granulated carbon.

The diaphragm was very light, and well damped by the carbon granules, giving the microphone a performance very much superior to that of other carbon microphones, such as those used in telephones.

SOURCE: Communication from P J Baxandall, Malvern (22 July 1982)

SEE ALSO: *BBC Engineering 1922–72* p 42 (BBC Publications) (1972)

'Georg Neumann—in memoriam' *J. Audio Eng. Soc.* p 708 (October 1976)

1924 LINEAR SAW-TOOTH TIME BASE CIRCUIT R Anson (UK)

The earliest attempt at the construction of a linear saw-tooth time base is believed to be due to Anson, and was developed about the year 1924. The author believes that this instrument appeared as a result of the development of the Anson relay in which a neon tube was used for signal-shaping purposes in conjunction with telegraph receivers. The neon tube is a two-electrode valve, filled with neon gas at a low pressure, in which neither of the electrodes is heated. When a potential, which is dependent upon the gas pressure, the proximity of the electrodes, the material of which the electrodes are made and their surface condition is applied between the electrodes the gas becomes ionized. This potential is normally about 130 volts. When the potential falls to about 100 volts the gas deionizes.

The neon time base consists of a condenser charged through a hlgh resistance and discharged by the neon tube when the charge upon the condenser reaches the striking voltage of the tube.

SOURCE: *Time Bases* by O S Puckle (London: Chapman & Hall) p 13 (1944)

SEE ALSO: R St G Anson: British Patent No 214754

1924/5 RADAR E Appleton, G Briet, R A Watson Watt SFR, GEMA *et al* (primarily UK)

The first use found for the reflecting properties of radio waves was in measuring the height of the Heaviside layer. This was done in Britain in 1924 by Sir Edward Appleton and M F Barnett, and in the USA in 1925 by Dr Gregory Briet and Dr Merle A Tuve of the Carnegie Institute. Breit and Tuve were the first to apply the pulse principle.

By 1939, Germany, Great Britain, Holland and the USA all possessed military radar apparatus, while the first peaceful application had been made in France in 1935. In France and Germany the work was done by the scientists of radio companies: in Britain and the USA by scientists in government research stations.

Scientists of the Societé Francaise Radioelectrique began to study the use of metric and decimetric radio waves to detect obstacles in 1934, first with a view to saving life at sea and then for military uses. An 'obstacle detector' working on decimetric waves and employing magnetrons and the pulse principle was fitted to the liner Normandie in 1935; it appears to have been successful and equipment was installed to detect ships entering and leaving the harbour at Le Havre in 1936. Work was meanwhile done on equipment for detecting aircraft, although no radar warning system had been installed by 1939.

German work, begun before 1935, was carried on under contracts from the Navy, by a new firm GEMA, and later by other firms. As a result, by 1939 a large number of radar sets were in operation for detecting aircraft. Though the German work began at least as early as that in other countries and had reached a very similar level of development by 1939, it lagged behind after the outbreak of war. German policy was based on the assumption that the war would be short, and consequently less effort was put into such basic work as radar than in Britain or the USA.

Interest in the possibility of using radio waves to detect aircraft arose in Britain about 1934. The reflection of radio waves from aircraft had been observed in 1931 and 1933; on the latter occasion the possibilities for aircraft detection were carefully analysed. H E Wimperis, Director of Scientific Research at the Air Ministry and his assistant, A P Rowe suggested that the country should increase its efforts to develop a method of detecting aircraft at a distance and an investigating committee of three scientists, Sir Henry Tizard, Professor A V Hill and Professor Patrick Blackett, was set up. It was soon realised that radio beam would be the ideal alternative to the existing inadequate acoustic warning equipment.

Robert A Watson-Watt (now Sir Robert) who was a lecturer in physics at University College, Dundee, before he began a research career in the government laboratories, played the major role in developing practical radar equipment in Britain. He was superintendent of the Radio Division of the National

Physical Laboratories at the time the pressure for improved air defence reached its peak. He felt confident that radio waves could be employed to detect aircraft. His two memoranda of February 1935 described his suggested means for so using them; after a demonstration of the echoes produced by aircraft from the BBC Daventry short-wave station, the Tizard Committee recommended that work on the lines suggested by Watson-Watt should be started. Working with six assistants, of whom A F Wilkins was the principal, Watson-Watt developed the first practical radar equipment for the detection of aircraft on the Suffolk coast in the summer of 1935. The main problems he solved were the construction of a high-power transmitter, the modulation of it with short pulses, the development of receivers to handle the pulses and of suitable transmitting and receiving aerials The performance of the first equipment was considered promising enough for the Air Ministry to build a chain of five radar stations.

The development of radar had meanwhile been proceeding independently in the United States. Military interest began after L A Hyland, an associate of A H Taylor, discovered accidentally in 1930 that aircraft cause interference in radio waves and Leo Young successfully applied the pulse apparatus to this. Despite the fact that radar looked so promising, the Navy was reluctant to spend any significant amount on it, but, through the persistent efforts of Harold G Bowen, chief of the Naval Laboratory, $100 000 was allocated for radar research. Robert M Page, head of the research section of the Naval Laboratory's radio divisions developed some of the first modern radar equipment. In 1938, two years after successful laboratory demonstrations of the equipment, the American Navy finally fitted radar devices to some of its ships.

After 1940 Great Britain and the United States co-operated in radar development.

SOURCE: *The Sources of Invention* by J Jewkes, D Sawers and R Stillerman (London: MacMillan & Co) pp 346, 347 and 348 (1958)

1925   RESISTOR (Cracked-Carbon Type)   Siemens and Halske (Germany)

It was undoubtedly in Germany that the first practical use was made of a cracked-carbon film in lieu of metal to form a highly stable resistance coating and resistors of this general type were manufactured by several firms in that country for a number of years before the war. Amongst these, the Siemens and Halske organisation seems to have produced the largest quantities, so that this type of resistor commonly came to be known as the 'Seimens resistor'.

One of the earliest disclosures of the cracking of hydrocarbon vapour to produce a hard carbon layer is contained in the Seibt patent of 1930 and in the Stemag patent of the same year. The Siemens and Halske patent relating to these resistors is dated March 1932, but the fundamental method is already there referred to as 'well known!'.

SOURCE: 'Fixed resistors for use in communication equipment' by P R Coursey *Proc. IEE* vol 96, Pt. III pp 174–5 (1949)

SEE ALSO: Siemens and Halske: Akt. Ges. German Patent No 438429/1925

Siemens and Halske: Akt. Ges.British Patent No 387150 (1932)

C A Hartman: German Patent No 438 429 (1925)

1925   ELECTROSTATIC LOUDSPEAKER   (Various)

The electrostatic loudspeaker failed to gain wide commercial acceptance, in spite of extensive development activity devoted to it during 1920–35, for the very sound reason that several serious shortcomings still adhered to its design. Either the diaphragm or the air gap itself had usually been relied on to provide the protective insulation against electrical breakdown, but this protection was often inadequate and limits were thereby imposed on the voltages that could be used and on the specific power output. Close spacings, a film of trapped air, stiff diaphragm materials, and vulnerability to harmonic distortion combined to restrict to very small amplitudes both the allowable and the attainable diaphragm motion: and as a consequence, large active areas had to be employed in order to radiate useful amounts of sound power, especially at low frequencies. But when large areas were employed, the sound radiation was much too highly directional at high frequencies. Several of the patents cited below bear on one or another of these features, and it is now apparent that an integration of such improvements would have made it possible to overcome almost—but not quite—every one of these performance handicaps. Occurring singly as they did, however, no one of these good ideas was able by itself to rescue the electrostatic units from the burden of their other shortcomings. Taken together, however, with the newly available diaphragm materials and with the important addition of one or two new ideas, the modern form of electrostatic loudspeaker can so completely surmount these former handicaps that it merits careful consideration as a potential competitor for the moving-coil loudspeaker in many applications.

SOURCE: *Electroacoustics* by F V Hunt (Cambridge, Mass.: Harvard University Press) p 173, 174 (1954)

SEE ALSO: For example, Colin Kyle, US Patents No 1644 387 (filed 4 October 1927;) issued 4 October 1927, and No 1746 540 (filed 25 May 1927) issued 11 February 1930: Ernst Klar (Berlin) German Patent No 611 783 (filed 22 May 1926) issued 5 April 1935, and US Patent No 1813 555 (filed 21 May 1927, renewed 14 November 1930) issued 7 July 1931 (insulating spacers, perforated plate coated with a dielectric); Hans Vogt (Berlin), more than a score of contemporary and relevant German patents, for example, German Patents No 583 769 (filed 25 December 1926) issued 9 September 1933 and No 601 117 (filed 17 May 1928) issued 8 August 1934, and US Patent No 1881 107 (filed 15 September 1928) issued 4 October 1932 (tightly stretched diaphragm between perforated rigid electrodes): Edward W Kellogg (GEC) US Patent No 1983 377 (filed 27 September 1929) issued 4 December 1934 (sectionalised diaphragm with inductances for impedance correction); William Colvin, Jr. US Patent No 2000 437 (filed 19 February 1931) issued 7 May 1935 (woven-wire electrodes); D E L Shorter, British Patent No 537 931 (filed 21 February 1940, complete spec. 23 January 1941, accepted 14 July 1941) (diaphragm segmentation with external dividing networks for improving directivity and impedance).

'Wide range electrostatic loudspeakers' by P J Walker *Wireless World* pt. 1, p 208 (May 1955), pt. 2, p 265 (June 1955), pt. 3, p 381 (August 1955)

1925   **SHORT WAVE COMMERIAL RADIO**   **L J W van Boetzelaer (Holland)**
          COMMUNICATION

On 23 April 1925, an experiment began in Hilversum which turned out to have far-reaching consquences for the link between The Netherlands and the Dutch East Indies, now Indonesia, causing a huge long-wave transmitter to become obsolete only months after it came into operation.

At the Nederlandsche Seintoestallen Fabriek, now PTI, a young research engineer, L Jan Was van Boetzelaer had just received a new water cooled 4 kW transmitting triode from the Philips factory at Eindhoven, which had a grid-anode capacity low enough to permit oscillating at 11.5 MHz (!). Working in a humble wooden shed, Mr van Boetzelaer, after many difficulties, managed to get the primitive transmitter 'on the air'.

To see how far this transmitter could be heard, it was arranged that the steam ship 'Prins der Nederlanden', sailing to the East, would listen in on a daily schedule. With admirable perseverance, van Boetselaer operated the morse key until late at night, hoping to be read. The ship's reactions, coming in slowly via coastal stations, were favourable. Then, in a bold mood, the diligent operator invited Malabar to send a cable if they happened to read the transmissions. The telegram sent back caused great excitement in Hilversum. It flatly stated that reception on 26 metres had been loud and clear since the beginning of the experiment.

SOURCE: *Philips Telecommunication Review* vol 33, No 4, p 191 (December 1975)

1925   **IONOSPHERE LAYER**   **E Appleton (UK)**

At Cambridge University, British physicist Edward Appleton lays the foundation for the development of radar: he measures the height of the ionosphere and finds that radio waves are reflected from the upper atmosphere to a height of 310 miles (500 km) above ground level.

SOURCE: *The Timetable of Technology* (London: Michael Joseph, Marshall Editions) p 78 (November 1982)

1925 TELEVISION (Mechanical Scanning)   J L Baird (UK)

When Baird, in 1923, decided to devote his untried inventive genius to the development of a practical television scheme, the problem seemed to him to be comparatively simple. Two optical exploring devices rotating in synchronism, a light-sensitive cell and a controlled varying light source capable of rapid variations in light flux were all that were required, 'and these appeared to be already, to use a Patent-Office term, known to the art'. Baird, however realised the difficult nature of the problem. 'The only ominous cloud on the horizon', he wrote 'was that in spite of the apparent simplicity of the task no-one had produced television'.

Baird's principal contemporaries in this challenge were C F Jenkins of the USA and D von Mihaly of Hungary. Other inventors were patenting their ideas on television at this time (1923) but only Jenkins, Mihaly and Baird and a few others were pursuing a practical study of the problem based on the utilisation of mechanical scanners.

SOURCE: 'The first demonstration of television' by R W Burns *Electronics & Power* p 953 (9 October 1975)

1925 NOISE (Johnson)   J B Johnson (UK)

In addition to fluctuation effects produced by vacuum tubes, it was found that random noise signals were generated in metallic resistors made of homogeneous materials. These effects were found to be temperature dependent and are known as thermal noise. A number of basic contributions to the understanding of thermal noise were made in the 1920s and 1930s among which was the outstanding paper by Johnson in 1925. The source of thermal conductor noise was traced to the random excitation of the electron gas in the conductor in consequence of its existence in an environment of thermally-agitated molecules. The effect is similar to the Brownian movement of particles suspended in a llquid in which the thermally-agitated molecules of the liquid collide with the suspended particle and impart to it a certain amount of energy. Since the particle is cohesive, collision with any one of its molecules sets the entire particle in motion thereby resulting in random movements observable under a microscope.

SOURCE: 'Noise and random processes' by J R Ragazzini and S S L Chang *Proc. IRE* p 1147/8 (May 1962)

SEE ALSO: 'Thermal agitation of electricity in conductors' by J B Johnson *Phys. Rev.* vol 32, 2nd series, p 97 (July 1928)

1926 SCREENED GRID TUBE   H J Round (UK)

By the end of the 1914–18 war the triode was the only tube in common use as a detector and amplifier and generator of high-frequency oscillations. Broadcasting was, in fact, started with the triode, although some of its limitations were, by that time, well recognised. One major limitation arose from the inherent electrostatic capacitance between the grid and the anode, within the valve itself. This gave rise to a coupling between grid and anode circuits which resulted in uncontrollable, and therefore undersirable, reaction between the output circuit and the input circuit. The introduction of a screed grid, between the control grid and the anode, to reduce this inter-electrode coupling, was first suggested by A W Hull. Schottky had earlier suggested a four-electrode tube, but his suggestion of the introduction of an additional grid had been to secure an increase of amplification factor. Hull's suggestion, by contrast, was directed to the reduction of grid-anode capacitance. But it remained for H J Round to bring the screen-grid valve into practical use in 1926.

Other, and later, versions of the screen-grid valve were developed, the screen grid being provided with one or more skirts which extended to the walls of the container bulb. In one form the grid, and in another the anode, was brought out uniquely at the top of the tube, the other electrodes being brought out from the base. In this way the undesired capacitance between the control grid and the anode could be reduced to 0.001 or $0.01\mu\mu$F

SOURCE: 'Thermionic devices from the development of the triode up to 1939' by Sir Edward Appleton *IEE Pub. Thermionic Valves 1904–54* (London: IEE) p 22–3 (1955)

1926   **COPPER OXIDE RECTIFIER**   **L O Grondahl and P H Geiger**
                                                           (USA)

In the course of an investigation of copper oxide formed on a piece of copper, during which current was passed through the oxide in a direction at right angles to the surface of separation, it was observed that the resistance of the combination was less when the current flowed from the oxide to the copper than when it flowed in the reverse direction. In the first unit, the ratio of the resistances in the two directions was about 3 to 1. The phenomenon was so different in nature from anything that had been observed in other known types of rectifiers that an intensive study and experimental investigation was undertaken during which it became more and more evident that the new device has characteristics which make it very probable that it will find general application as a rectifier.

SOURCE: 'A new electronic rectifier' by L O Grondahl and P H Geiger *Proc. AIEE Winter Convention, New York (February 7–11, 1927)* p 357

1926   **CIRCUITRY (Automatic Volume Control)**   **H A Wheeler (USA)**

In 1925 the stage was set for the invention of a practical automatic volume-control circuit, and on 2 January 1926, Wheeler invented his diode AVC and linear diode-detector circuit. This circuit was first incorporated in the Philco Model 95 receiver which he designed at the Hazeltine laboratory and which was announced about September 1929.

Full AVC bias voltage was applied to the first two RF tubes, and, to prevent distortion in the third RF stage, half AVC bias voltage was applied to that stage. The automatic volume-control action was sufficiently gradual to permit accurate tuning by ear, and it was unnecessary to touch the volume control once it was adjusted

SOURCE: 'The development of the art of radio receiving from the early 1920's to the present' by W O Swinyard *Proc. IRE* p 795 (May 1962)

SEE ALSO: 'Automatic volume control for radio receiving sets' by H A Wheeler *Proc. IRE* vol 16, p 30 (January 1928)

1926   **FILM SOUND RECORDING (Sound-on-Disc System)   Warner Bros (USA)**

As a result of lagging interest in the motion pictures by the public, Warner Bros. in 1926 decided to test the popularity of sound picturese To minimize the cost of the venture, this studio arranged to have Western Electric develop the necessary equipment to synchronise disk-recording machines with cameras that were housed in booths to suppress the camera noise. Arrangements were made with the Victor Talking Machine Company to do the recording in their facilities and with their personnel. The Victor Talking Machine Company was a Western Electric licensee, and their studios vere equipped with Western Electric recording equipment.

Western Electric developed motor drives for theatre projectors and disc turntables.   These were mechanically connected to the same constant speed motor system. Essentially standard public address system amplifiers and loudspeakers were used. The first picture produced was 'Don Juan'. In October 1927 'The Jazz Singer' followed and was a success.

The public reaction was so enthusiastic that the large theatre chains wanted equipment immediately to play the pictures. Western Electric agreed to lease equipment to them. As a result of the success of the first pictures, Warner Bros. installed disk-recording equipment in their studios. This system was called Vitaphone. It was destined to be supplanted by systems that recorded the sound as photographic images on the same film the picture was printed on. Having demonstrated the popularity of sound pictures and developed the equipment, the industry proceeded with great speed to convert studios for sound-picture production.

SOURCE: 'Film recording and reproduction' by M C Batsel and G L Dimmick *Proc. IRE* p 745 (May 1962)

1926   **FIXED RESISTOR (Sprayed Metal Film)**   **S Loewe (Germany)**

The basic idea involved is the very old art of decorating chinaware with precious metals. In Germany in 1926, Loewe developed a resistive film by atomizing a liquid solution of platinum resinate by forcing compressed air through it and applying the spray to an insulating base. Heating the film thus formed reduces it to the metal.

SOURCE: 'Resistors—a survey of the evolution of the field' J Marsten *Proc. IRE* p 922 (May 1962)

SEE ALSO: S Loewe: German Patent No 591 735 (1926) and US Patent No 1717 712 (1926)

1926   **TRANSITRON OSCILLATOR**   **B van der Pol (Holland)**

One of the earliest forms of single-valve trigger circuit was described by van der Pol in 1926. This circuit employs a tetrode in which the screen and grid have a resistance–capacitance coupling, the grid being fed from the high potential source via a high resistance, which forms part of the coupling network.

The term 'transitron' has recently come into use to denote any circuit which employs a single pentode valve in such a way that amplification is possible without the phase reversal. The name was originally coined by Brunetti, who defined it as a 'retarding-field negative-transconductance device'. Generally, these circuits are extremely versatile, and one may produce a sinusoidal, sawtooth or square wave output from one oscillator by means of very simple switching. It is equally easy to convert the continuous oscillator into a flip-flop by changing the bias potential. When the circuit is arranged as a relaxation oscillator or as a flip-flop it is very valuable as a switching device, and has many applications to time bases and control circuits.

SOURCE: *Time Bases* by O S Puckle (2nd edn) (London: Chapman & Hall) pp 56, 57 (1951)

SEE ALSO: 'On relaxation oscillations' by B van der Pol *Phil. Mag.* vol 2, p 978 (1926)

'The transitron oscillator' by C Brunetti and E Weiss *Proc. IRE* vol 27, p 88 (1939)

1926   **YAGI AERIAL**   **H Yagi (Japan)**

Dr H Yagi studied under the direction of Professor Dr G H Barkhausen at the Dresden Technische Hochschule from 1913 to 1914, under the direction of Professor Dr J A Fleming at University College, London from 1914 to 1915, and under the direction of Professor Dr G W Pierce at Harvard University, Cambridge, Massachusetts from 1915 to 1916.

In 1926, during his career as a professor at Tohoku University, he invented the VHF directive 'Yagi antenna' which was widely put into practical use for domestic television reception. Some years later, as a result of this invention, the Academy of Technical Science in Copenhagen awarded him the Valdemar Poulsen Gold Medal for his outstanding contributions to radio technique.

SOURCE: 'Death of Dr Hidetsugu Yagi' *Telecommunication Journal* vol 43, p 372 (V/1976)

1926   **ELECTRON MICROSCOPE**   **H Busch (Germany)**

In 1926 Hans Busch (Germany) laid the theoretical foundations for the electron microscope. In 1928 two of his fellow countrymen, Max Knoll and Ernst Ruska, from the Technische Hochschule in Berlin, carried out experiments based on his research that led to the development of the first operational electron microscope in 1933. It was perfected by Rulska, who, with Heinrich Rohrer and Gerd Binnig, was awarded the Nobel Prize for Physics in 1986 for the invention of the tunnel effect miscroscope (see below).

SOURCE: *Inventions and Discoveries 1993* edited by Valerie-Anne Giscard d'Estaing and Mark Young (New York: Facts on File) p 172

1927   **CABLE TELEVISION**   **Bell Telephone Co (USA)**

The first cable television transmission was carried out in the United States by the Bell Telephone Company in 1927. The experiment took place between Washington and New York. A Nipkow scanning disc was used for the transmission and another for the reception.

This cable technique was then taken up again for the purposes of reaching those areas without access to traditional Hertzian transmission.

In 1949 a small town in Oregon in the United States had bad reception of programmes transmitted from Seattle on account of the mountains which surround it. It was decided that a large aerial would be installed on high ground. From there a cable network transmitted programmes, without any risk of parasitic oscillation.

It was not until the 1960s that cable television experienced real growth in the United States and Canada. Today 22 million Americans are subscribers to different cable systems (paid by subscription), 2.5 million of which are Disney Channel subscribers.

The development of optical fibres in place of traditional coaxial cable (invented by the Americans Affel and Espensched in 1929) makes it possible to go from passive viewing of programmes to active audience participation, that is to say, users are able to choose their programmes and to participate directly in the contents of the programmes themselves (quick surveys, questionnaires, games etc).

The fibre optic technique makes it possible to transmit through the same cables not only television programmes, but also radio programmes, telecommunications and data material.

Thanks to direct television satellites, Europe was able to benefit from 120 channels in 1993.

SOURCE: *The Book of Inventions and Discoveries* Associate Editor Valerie-Anne Giscard d'Estaing (UK: Queen Anne Press, Macdonald & Co.) (1990) p 243

| 1927 | **NEGATIVE FEEDBACK AMPLIFIER** | **H S Black (USA)** |

In one of the most fundamental discoveries in the history of communications, H S Black in 1927 at Bell Laboratories found that by feeding part of an amplifiers output back into its input (negative feedback), it was possible by sacrificing some amplification to achieve stable operation at low distortion.

SOURCE: *Mission Communications—the Story of Bell Laboratories* by Prescott C Mabon (Murray Hill, NJ: Bell Laboratories Inc.) p 171–2 (1975)

| 1927 | **FILM SOUND RECORDING (Sound-on-Film System)** | **Fox Movietone News (USA)** |

The first commercially successful photographic sound recording system (Fox Movietone News) used a variable intensity method of modulating a beam of light to expose the film negative. The gas-filled lamp known as the Aeo-light had an oxide coated cathode, and its intensity could be modulated over a considerable range by varying the anode voltage, at audio frequencies, between 200 and 400 volts. The Aeo-light was mounted in a tube which entered the camera at the back. Directly against the film was a light restricting slit which passed a beam about a tenth inch long and 0.001 inch high, placed between the picture and the sprocket holes. The Aeo-light could produce a sufficiently high intensity to expose the sensitive negative films used for picture taking. The system worked quite well for news photography, where the sound and pictures were taken simultaneously on the same camera.

SOURCE: 'Film recording and reproduction' by M Batsel and G L Dimmick *Proc. IRE* p 746 (May 1962)

| 1928 | **PENTODE TUBE** | **Tellegen and Hoist (Holland)** |

The substantial suppression of secondary emission in a tetrode is not an easy matter, particularly where it is desired to operate with high anode and screen potentials, and so by far the most common method of suppressing secondary emission is by way of the inclusion of a suppressor grid, between screen grid and anode, as in the pentode invented by Tellegen and Hoist of the Philips Company, in Holland. The suppressor grid is maintained at the filament potential. Pentodes, before 1939[1] had become extremely popular for both high- and low-frequency amplifications.

SOURCE: 'Thermionic devices from the development of the triode up to 1939' by Sir Edward Appleton *IEE Pub. Thermionic Valves 1904–54* (London: IEE) p 23–4 (1955)

1928   FREQUENCY STANDARDS (Quartz Clocks)   J W Horton & W A Marrison (USA)

The tuning fork was developed to a point at which it gave a stability of 1 part in 10^7 per week and could have been improved still further. By this time, however, the first quartz clock had been made by Horton and Marrison and it seemed clear that quartz possessed many advantages. One fundamental advantage was the higher frequency of quartz vibrations. Frequencies of many millions of cycles per second were already being used for radio transmissions, and it was not very convenient to measure them in terms of a standard having such a low value as 1 kc.

SOURCE: 'Frequency and time standards' L Essen *Proc. IRE* p 1159 (May 1962)

SEE ALSO: 'Precision determination of frequency' by J W Horton and W A Marrison *Proc. IRE* vol 16, p 137 (February 1928)

1928   RADIO (Diversity Reception)   H A Beverage, H O Peterson and J
**   B Moore (USA)**

Because of the turbulence and abrupt changes encountered in the HF medium, special attention had to be given to improved means of reception as it became apparent that transmitter power increases alone were not sufficient.

Among the significant techniques developed, one of the most important is diversity reception, wherein a considerable improvement is obtained due to the statistical independence in the fading characteristics of two or more paths. Diversity reception is basic, improving the reception of any type of modulation at any frequency. H A Beverage, H O Peterson and J B Moore described and developed a triple-space-diversity system for on–off telegraph reception around 1928. This employed three antennas spaced about 1000 feet apart. The rectified outputs of three separate receivers were combined across a common load resistor and the voltage across the resistor keyed a local tone generator. As long as the voltage was above a certain minimum from any receiver, a properly keyed tone signal was reproduced.

SOURCE: 'Radio receiver—past and present' by C Buff *Proc. IRE* p 888 (May 1962)

SEE ALSO: 'Diversity receiving system of RCA Communications Inc. for radiotelegraphy' H H Beverage and H O Peterson *Proc. IRE* vol 19, pp 531–61 (April 1931)

1929   COLOUR TV   Bell Laboratories (USA)

Colour comes to television. With spectacularly good results, the first transmission is beamed between Washington and New York. The 50-line system used by the Bell Telephone Laboratories transmits the three primary colours—red, blue and green—along three separate channels. Later in the year, the basis of modern colour TV is laid down when several colour signals are transmitted over a single channel.

SOURCE: *The Timetable of Technology* (London: Michael Joseph, Marshall Editions) p 86 (November 1982)

1929   CYCLOTRON   E O Laurence (USA)

Laurence used a curved path for the particles, so that the particles could circulate continuously, travelling long distances in a relatively small volume and using the same accelerating system over and over again. An electrically charged particle entering a magnetic field directed at right angles to the motion of the particle, proceeds to move in a circle with constant speed; as the particle speed is increased, the radius of the circle in which the particle moves also increases. Further acceleration occurs at each revolution.

SOURCE: *The Sources of Invention* by J Jewkes, D Sawers and R Stellerman (London: MacMillan) pp 290/1 (1958)

SEE ALSO: 'Atomic slingshot' by Howard Blakeslee *Science Digest* (April 1949)

'Maestro of the atom' by L A Schuler *Scientific American* (August 1940)

1929   MICROWAVE COMMUNICATIONS   A G Clavier (France)

In 1920 Barkhausen positive-grid oscillator provided a means for the efficient generation of 40-cm waves. This revived the interest in the centimeter waves. In 1929 Andre G Clavier, then associated with Laboratoire Central de Telecommunications in Paris, started an experimental project to challenge the then accepted principle that wire or cable circuits should be used in preference to radio whenever physically possible.

In 1930 a link was started between two terminals in New Jersey using 10-ft. parabolic antennas. Just as testing started, the project was transferred back to France. On 31 March 1931, Clavier and his associates demonstrated that microwave transmission provided a new order of economy, quality, dependability and flexibility in communications over a 40-km path between Calais and Dover. The circuit provided both telephone and teleprinter service using 17–6cm waves transmitted in a 4° beam by means of a parabolic reflector 3 metres in diameter with a power output of a fraction of a watt. Andre Clavier went on to establish the first commercial microwave radio link in 1933 from Lympne, England, to St. Inglevert, France.

SOURCE: 'Microwave communications' by J H Vogelman *Proc. IRE* p 907 (May 1962)

1930   TRANSISTOR (MOSFET Concept)   J Lilienfeld (Germany)

A 1930 patent was issued to Julius Lilienfeld of the University of Leipzig for a device that could be compared to today's MOSFET, or insulated gate field-effect transistor. The device was reported to provide a means of obtaining amplification in a thin film of copper sulfide. However, a working device was probably never built, since the low mobility of holes in the material and other factors would seem to preclude any amplification.

SOURCE: 'Solid state devices' *Electronic Design* vol 24, p 72 (23 November 1972)

1930   VAN DE GRAAF ACCELERATOR   R J van de Graaf (USA)

For nuclear structure research, constant-potential accelerators use the electrostatic belt generator invented by van de Graaf. About 5.5 million volts were insulated in air between two large generators in 1930 (equipment now in the Boston Museum of Science).

SOURCE: *The Encyclopedia of Physics* (2nd edn) Editor R M Besancon (New York: Van Nostrand Reinhold & Litton Educational Pub. Inc.) p 13 (1974)

SEE ALSO: 'Electrostatic generators for the acceleration of charged particles' by R J van de Graaf, J G Trump and W W Bruechner *Rep. Prog. Phys.* vol 11, p 1 (1948)

1930s   METEOR SCATTER (BURST) SYSTEMS   Schanker *et al* (USA)

Billions of meteors enter the Earth's atmosphere every day and burn up at altitudes typically from 80 km to 120 km; this range of altitudes includes also the ionised E layer. When the meteors vaporise, they create an ionised trail which re-radiates incident radio waves quite efficiently in the upper HF and lower VHF ranges. This effect was discovered during activities aimed at predicting 'sporadic' E layer reflections. It was eventually realised that one of the 'sporadic' effects was due to irregular meteor trail reflections. Schanker estimates that this discovery was made in the 1930s; it is well known that large scale experiments to probe the ionosphere were taking place internationally.

Because meteor scatter paths between two terminals are open only for short periods, ranging from milliseconds to 1 or 2 seconds, high speed transponding (hand-shaking) is essential before information can be passed. Thus, systems suitable for general use had to wait for the production of high speed, cheap microprocessors and modern memory chips and have developed rapidly since 1982.

A typical meteor scatter burst system can provide information exceeding 75 b/s (sometimes much greater) over each 24 hour period at distances up to 2000 km.

SOURCE: Private Communication from J R Guest, Malvern Wells, UK

SEE ALSO: *Meteor Burst Communication* by Jacob Z Shanker (Boston: Artech House)

'The Canadian JANET system' by Davis G W L *et al Proc. IRE* (December 1995)

1930   **HIGH FIELD SUPERCONDUCTIVITY**   **W J de Haas and J Voogd (Netherlands)**

The story of high field superconductivity began as long ago as 1930, when de Haas and Voogd discovered that resistance was restored in the Pb–Bi eutectic only by magnetic fields as high as 16 000 to 20 000 gauss at 4.2 K. The eutectic alloy had a transition temperature of about 8.8 K. This discovery immediately suggested to its authors the old idea of making a superconducting solenoid with wire of this material, but because of the very low critical current densities that they observed, the idea was soon dropped and in fact lay fallow for over 20 years. Then in 1955 Yntema described a superconducting solenoid wound with niobium wire which produced fields up to 7000 gauss, but this received little attention. Autler (1960) made a similar coil producing 4300 gauss but, the subject did not really take off until the discovery in 1961 by Kunzler and his co-workers of the remarkable current-carrying properties of the intermetallic compounds Nb_3Sn. This material had been found to have the very high critical temperature of 18.0 K by Matthias *et al* (1954) and indeed this is still the highest known transition temperature of any material, give or take a few tenths of a degree.

SOURCE: *Materials for Conductive and Resistitive Functions* by G W A Dummer (New York: Hayden Book Co) p 141

SEE ALSO: W J de Haas and J Voogd *Commun. Phys. Lab. University of Leiden* No 208b (1930)

'Superconducting winding for electromagnetics' by G B Yntema *Phys. Rev.* vol 98, p 1197 (1955)

'Superconducting electromagnetics' by S H Autler *Rev. Sci. Instrum.* vol 31, p 369 (1960)

'Superconductivity in Nb_3Sn at high current density in a magnetic field of 88 kgauss' by J E Kunzler, E Beuhler F S L Hsu and J H Wernick *Phys. Rev. Lett.* vol 6, p 890 (1961)

1930s   **RADIOPHONIC SOUND—MUSIC**   **P Grainger (Australia)**

These techniques first came into real use during the 1950s with the maturation of the magnetic tape recorder although as long ago as the 1930s Percy Grainger, the Australian composer of 'Country Gardens' fame, had produced a brief composition based on pure frequencies for the Theremin, an early electronic sound generator. The beginnings were however with 'musique concrète', pioneered in Europe, although as the name suggests the 'music' was made through the manipulation of pre-recorded natural sounds and was in fact orchestrated noise. With further study it became apparent that if more 'musical' sounds were used as the raw material, i.e. sounds with a more ordered harmonic structure, greater malleability was achieved as the timbre changes encountered during pitch changes, due to differing tape speeds on playback, still bore some audible relationship to each other.

SOURCE: *Electronics Engineer's Reference Book* (London: Newnes-Butterworth) ch 17, p 17-16 (1976)

1931   **FIXED RESISTOR (Oxide Film)**   **J T Littleton (USA)**

The seed for this important contribution was provided by Littleton (1931) who developed an iridized, conducting tin-oxide coating for glass insulators. Its resistivity was sufficiently low to equalise potential across the insulator, thereby reducing corona effect, but too high for use in conventional resistors. Mochel modified this film by the addition of antimony oxide which stabilized its electrical properties. By varying the tin–antimony proportions, negative or positive temperature coefficients are obtained.

SOURCE: 'Resistors—a survey of the evolution of the field' J Marston *Proc. IRE* p 922 (May 1962)

SEE ALSO: J T Littleton: US Patent No 2228 795 (1931)

J M Mochel: US Patent No 2564 707 (1947) Reissue 25 556

1931   **STEREOPHONIC SOUND REPRODUCTION**   **A D Blumlein (UK) and Bell Labs.**
                                                      (USA)

Stereophonic reproduction *per se* was pioneered almost simultaneously by Blumlein in Great Britain and at the Bell Telephone Laboratories. Blumlein's contributions are presumed to be described in his patents. He showed a complete system applicable to sound-on-disc motion pictures, including microphone arrays utilising bidirectional as well as omnidirectional microphones, transmission circuits, and disc recording systems utilising simultaneous lateral and vertical recording. Economic difficulties are believed to have prevented completion and commercial exploitation of these systems.

The recognised early systems approach to large audience stereophonic reproduction was a public demonstration of Bell Telephone Laboratories equipment under the guidance of Dr Harvey Fletcher on 27 April 1933. The Philadelphia Orchestra was in the Academy of Music in Philadelphia and it was reproduced in Constitution Hall, Washington, DC.

SOURCE: 'The history of steophonic sound reproduction' by J K Hilliard *Proc. IRE* p 776 (May 1962)

SEE ALSO: A D Blumlein: British Patent No 394 325 (14 December 1931)

Also US Patent No 2093 540

'Perfect transmission and reproduction of symphonic music in auditory perspective' F B Jowett *et al Bell Telephone Quart.* vol 12, p 150 (July 1933)

1931   **CRO CARDIOGRAPH**   **P Rijlant (Belgium)**

The first workers to use the cathode-ray tube in bio-electrical work were Gasser and Erlangers, in their work on the action potentials of nerves. At the present time, many workers in bio-electrical fields, realising its unique properties, are adapting the cathode-ray tube to their own particular problems; a general account will be found in Holzer's book. Among the first to adapt it to electrocardiographic work was Rijlant, of Brussels (1931 *et seq*) who has published many papers on the subject. Schmitz, in Germany, and Matthews (1933) in this country, were also among the first to publish electrocardiograms recorded by the cathode-ray oscillograph (CRO). Matthews (1934) was able to show that Rijlant's electrocardiograms were inaccurate, and to point out that the new waves (P_2, T_2, etc) described by him were really caused by deficiencies in his amplifer.

SOURCE: 'The examination and recording of the human electro-cardiogram by means of the cathode-ray oscillograph' by D Robertson *J. IEE* vol 81, p 497 (1937)

SEE ALSO: 'Some observations on the adaptation of the cathode ray oscillograph to the recording of bio-electrical phenomena with special reference to the electrocardiogram by D Robertson *Proc. R. Soc. of Medicine (Section of Physical Medicine)* vol 29, p 593 (1936)

Cathode Ray Oscillography in Biology and Medicine by W Holzer (Vienna: Maudrich) (1936)

'The cathode ray oscillogram of the human heart' by P Rijlant *Comptes Rendues des Séances de la Societé de Biologie* vol 109, p 42 (1932)

1931   **COMPUTERS (Differential Analyser)**   **V Bush (USA)**

Early analogue computer for solving differential equations.

SOURCE: 'The differential analyser—a new machine for solving differential equations' by V Bush *Journal of the Franklin Institute* vol 212, p 477 (1931)

SEE ALSO: *The Computer from Pascal to von Neumann* by H H Goldstine (Princeton, NJ: Princeton University Press) p 88 (1969)

1931   **RELIABILITY—QUALITY CONTROL CHARTS**   **W A Shewhart (USA)**

After languishing in libraries for several years, the work of Dodge and Romig in acceptance sampling and the work of Shewhart on control charts finally was brought to light during World War II through the nationwide training programmes sponsored by the Office of Production Research and Development of the War Production Board.

Although the underlying concepts were developed by scientific investigators and statisticians in the preceding decades, the genius of Dodge, Romig and Shewhart lay in their recognition of basic principles as an aid to solving practical problems, and their ability to recognise and formulate a systematic approach.

SOURCE: 'Treating real data with respect' by J A Henry *Quality Progress* p 18 (March 1976)

SEE ALSO: *Economic Control of Quality of Manufactured Product* by W A Shewhart (New York: Van Nostrand) (1931)

1932 NEUTRON   J Chadwick (UK)

Some years before, the German physicists Bothe and Becker had discovered some very penetrating radiations obtained by bombarding the light metal beryllium with particles shot out of polonium. They assumed that the radiations were of a wave nature. The French physicists Frederic Joliot and his wife Irene Curie—whose mother Marie Curie had discovered polonium and named it after her native country—studied these radiations and made a very striking experiment. They found that if a piece of paraffin wax was placed in front of them then the amount of radiation seemed to be increased, and not decreased by the inter-position of the wax. Chadwick, by further experiment and interpretation, was able to prove that this paradox could be resolved if the radiations were not waves, but a new kind of atomic particle without any electric charge. Thus he discovered the neutron.

SOURCE: *Science at War* by J G Crowther and R Whiddington (London: HMSO) p 127 (1947)

1932 COCKCROFT–WALTON ACCELERATOR   J D Cockcroft and E D S Walton
(Atom-Smasher)   (UK)

Cockcroft was an electrical engineer from Manchester. He had graduated as an engineer at the Manchester College of Technology and joined the engineering firm of Metropolitan-Vickers Electrical Company Ltd. After spending four years in the army in the war of 1914–18 he returned to his firm. He engaged in advanced study with Professor Miles Walker and was presently awarded a post-graduate scholarship to continue his studies at Cambridge. His engineering knowledge fitted him to devise powerful electrical apparatus, and he attacked the problem of devising an electrical machine by which an electrical field of several hundred thousand volts could be applied to atomic particles, so that they could be given a very high speed and energy, like those thrown out naturally by radium. Cockcroft and his colleague Walton succeeded in disintegrating lithium with electrically accelerated protons in 1932, shortly after Chadwick's discovery of the neutron. This was a great advance, because electrical machines could be developed, and large streams of atomic projectiles could be produced at will.

Cockcroft used protons, the nuclei of hydrogen atoms, in his first experiments. Each proton released about sixty times as much energy as it possessed itself. But the number of protons accelerated was relatively insignificant, and the amount of energy used in producing the accelerating field was much greater than the total amount released in the atomic disintegrations. As a machine, Cockcroft's atom-smasher was very inefficient in the engineer's sense.

SOURCE: *Science at War* by J G Crowther and R Whiddington (London: HMSO) p 128 (1947)

1932 CIRCUITRY (Energy Conserving Scanning Circuit)   A D Blumlein (UK)

As with most of the diagrams, figure 11.8(a) is taken from the Patent Specification, and shows the basic features of the line scan circuit which is now universal in television receivers, although not brought into a common use until 1946. Figure 11.8(b) shows the method of operation, involving three separate regimes during the cycle.

It is now such a well-known circuit that it will not be described in detail, but it is interesting to compare it with the single LC circuit which is all that is necessary for a sinusoidal waveform. whereas to handle the saw-tooth waveform it is necessary to provide also the switches in the form of the valve and diode as shown. The element of symmetry mentioned earlier can be seen here.

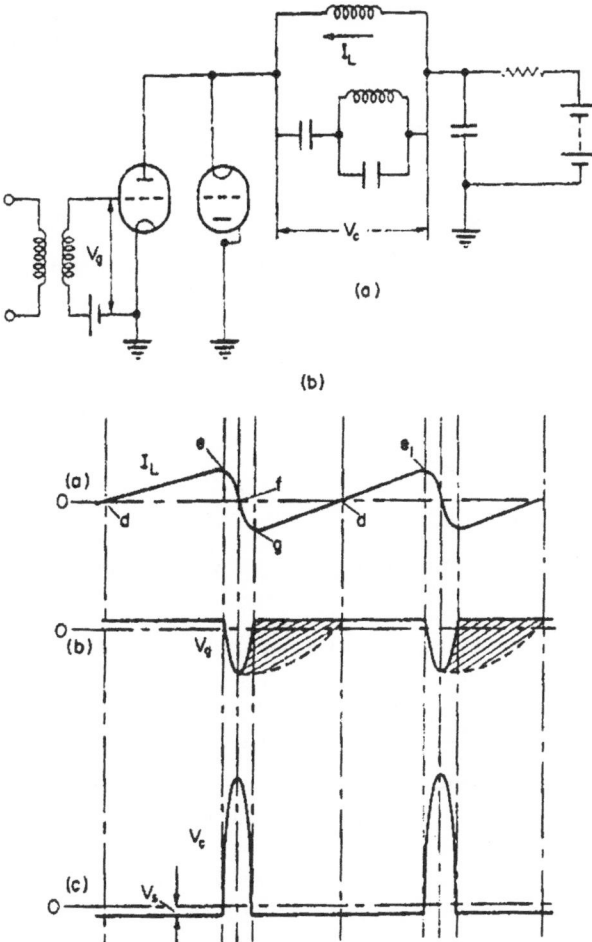

Figure 11.8. Energy conserving scanner circuit. (a) Basic features; (b) method of operation.

SOURCE: 'The world of Alan Blumlein' *British Kinematography Sound and Television* vol 50, No 7, p 209 (July 1968)

SEE ALSO: British Patent Specification No 400 976 (1932)

1932   **TRANSMISSION ELECTRON MICROSCOPE**   **M Knoll & E Ruska (Germany)**

The first electron microscope was built at the Technical University of Berlin early in 1931. It had two electro-magnetic lenses in series and achieved a modest magnification of 17. Improvements were made later. A condenser lens was added and an iron shield with a narrow gap built around the magnetic lens. Ruska in 1934 was able to demonstrate a magnification of 12 000.

SOURCE: *The Encyclopaedia of Physics* (2nd edn) Editor R M Besancon (New York: Van Nostrand Reinhold & Litton Educational Pub. Inc.) p 2 (1974)

SEE ALSO: 'Uber Fortschritte im Bau und in der Leistung des Magnetischen Elektronemikroskops' by E Ruska *Z. Phys.* vol 87, (9 & 10), p 580 (1934)

'Origin of the electron microscope' by M M Freundlich *Science* vol 142, (3589) p 185 (1963)

1933   **STEREO RECORD**   **EMI (UK)**

The first stereophonic records were produced in Great Britain by EMI (Electric and Musical Industries) in 1933. The research, directed by the physicist Alan Dower Blumlein, culminated in the recording of stereo 78s. The work of Blumlein and EMI remained experimental untl 1958, when the American company Audio Fidelity and the British companies Pye and Decca issued the first commercial stereo records thanks to numerous technical advances.

SOURCE: *Inventions and Discoveries 1993* edited by Valerie-Anne Giscard d'Estaing and Mark Young (New York: Facts on File) p 138

1933   FREQUENCY-MODULATION   E H Armstrong (USA)

The credit for promoting FM as a broadcast service goes to E H Armstrong. For years he had been seeking a way to reduce static, and finally he turned his attention to FM. Toward the end of 1933 he had perfected a system of wide-band frequency modulation which seemed to overcome natural and many forms of man-made static. In this system the carrier was frequency modulated ± 75 kc by audio components up to 15 kc.

SOURCE: 'The development of the art of radio receiving from the early 1920's to the present' by W O Swinyard *Proc. IRE* p 797 (May 1962)

SEE ALSO: 'Frequency modulation' by S W Seeley *RCA Rev.* vol 5, p 468 (April 1941)

Selected Papers on Frequency Modulation edited by J Klapper (New York: Dover Publications Ltd.) (1970)

NOTE The actual invention of frequency modulation goes back to 1902 US Patent 785 303. D Ehret (Endeavor Review April 1978)

1933   POLYETHYLENE INSULATION   ICI (UK)

An outstanding event in the cable world in recent years was the discovery of polyethylene in 1933 by Imperial Chemical Industries Ltd. At first only minute quantities could be produced, but by 1937 a small amount was made available for experimental use. The opportunity was at once taken and, after extended research, a mile of submarine cable insulated with Telcothene, a synthetic material based on polyethylene, was made by Submarine Cables Ltd. in 1939.

SOURCE: 'The story of the submarine cable' Booklet published by Submarine Cables Ltd (AEI) London p 13 (1960)

1933   HARD VALVE TIME BASE CIRCUIT   O S Puckle (UK)

In 1933, O S Puckle developed a time base which employs a variation of the multivibrator as a condenser charging medium. This raised the maximum repetition frequency, as compared with that obtainable from a thyratron time base, from about 40 kc/s up to a maximum of about 1 Mc/s.

SOURCE: *Time Bases* by O S Puckle (London: Chapman & Hall) p 30 (1944)

SEE ALSO: 'A time base employing hard valves' O S Buckle British Patent 419198; also *Journal of the Television Society* vol 2, p 147 (1936)

1933   RADIO ASTRONOMY   K G Jansky (USA)

While looking for the sources of static in overseas radio signals, K G Jansky in 1933 discovered radio energy coming from the stars—thus launching the science of radio astronomy.

SOURCE: *Mission Communications—the Story of Bell Laboratories* by Prescott C Mabon (Murray Hill, NJ: Bell Laboratories Inc.) p 170 (1975)

1933   'IGNITRON' (Mercury-Arc Rectifier)   Westinghouse (USA)

In 1933 the Westinghouse Company announced its Ignitron. Its potential value was at once recognised and an active developmental programme soon commercialised it extensively. Progress in making and applying Ignitrons was rapid. By the end of 1934 a welding control unit using glass Ignitrons was installed in a customer's shop.

SOURCE: 'Early history of industrial-electronics' W C White *Proc. IRE* p 1133 (May 1962)

1934 **FREQUENCY STANDARDS (Atomic Clocks)**   **C E Cleeton and N A Williams (USA)**

In 1934 Cleeton and Williams at Michigan University excited a spectral line of ammonia at a frequency of 23 870 Mcs by a source of radio waves generated in the laboratory. The source used by them for exciting the transitions was a magnetron which generated a fairly wide band of frequencies and the ammonia was at atmospheric pressure, at which only a very broad resonance effect is observed. The width was mainly due to the effect of collisions and this can be reduced by reducing the pressure.

SOURCE: 'Frequency and time standards' L Essen *Proc. IRE* p 1161 (May 1962)

SEE ALSO: 'Electromagnetic waves of 1.1 cm wavelength and the absorption spectrum of ammonia' C E Cleeton and N A Williams *Phys. Rev.* vol 45, p 234–7 (February 1934)

1934 **TRANS-URANIAN ATOMS**   **E Fermi (Italy)**

In 1934, Professor Enrico Fermi in Rome poured out a bewildering series of discoveries by systematically bombarding atoms of all the elements with neutrons. He found that several dozen of them could be transmuted by neutrons, and he obtained particularly interesting results from uranium. This is the most complicated of the ninety-two different kinds of chemical atoms found on the earth. These can be placed in an order of complication, depending on the number of electric charges on the nucleus. The first in the series is hydrogen, with one positive charge, and therefore known as Atom No 1, and the last is uranium, with ninety-two charges, and therefore known as Atom No 92. It is not surprising that Atom No 92 should be naturally radio-active. It might well be too complicated to be stable. It is in fact an ancestor of radium, whose atomic number is 88.

Fermi found that the bombarded uranium produced numerous atoms with chemical properties quite different from uranium. He concluded that he had made new atoms, more complicated than uranium atoms, and supposed that these must be 'trans-uranian' atoms, Nos. 93, 94 etc. He seemed to have made a new series of atoms hitherto not found on the earth.

SOURCE: *Science at War* by J G Crowther and R Whiddington (London: HMSO) p 129 (1947)

1934 **LIQUID-CRYSTALS**   **J Dreyer (UK)**

Although the liquid-crystal state was first noted in 1889, it was not until around 1934 that serious consideration was given to these electro-optical devices, in the Marconi laboratories, in England. John Dreyer found that their orderly molecular arrangement could be used to orient dye molecules for making polarisers—a method still used even though his work was done in the 1940s and patented in 1950. The present explosion in liquid-crystal research began in the 1960s when the RCA laboratories in the United States began to investigate them. Its course of development can be charted by counting the US and British patents that have been granted—one each year in 1936, 1946, 1950, 1951, 1963, 1965 and 1967. Then, suddenly, seven in 1968, 11 in 1969, at least two in 1970 and more than 11 in 1971. Few companies claim as long-lived an association with the subject as Marconi and RCA: most have been in the field for two years or less.

SOURCE: 'The fluid state of liquid-crystals' by M Tobias *New Scientist* p 651 (14 December 1972)

1935 **SUPERCONDUCTING SWITCH**   **Casimir-Jonker and W J de Haas (Netherlands)**

The idea of using the superconductive transition to switch a small resistance into and out of a circuit at will seems to have occurred at about the same time in the laboratories at Leiden and Toronto in 1935. Casimir-Jonker and de Haas (1935) developed an apparatus to detect the first trace of resistance in a superconducting specimen, using a sensitive magnetometer to observe the change of field external to the cryostat when the current decayed in a super conducting circuit in series with the specimen. A superconducting lead solenoid around the specimen was used to restore its resistance, and a resistance as small as 3×10^{-11} ohm produced a decay of current rapid enough to be detected.

At Toronto, Grayson Smith and Tarr (1935) used a moving-coil magnetometer inside the cryostat itself, consisting of fixed lead field coils and a moving copper coil. A short section of lead wire in series with the field coils acted as a superconducting switch and could be driven normal by means of a current in a copper solenoid. The apparatus was used for measuring small persistent currents, again by observing their decay. Used in this way as a super-conduct ing galvanometers it was capable of detecting currents as small as 10^{-4} amp in a circuit of self-inductance of 5×10^{-4} henry.

SOURCE: *Materials for Conductive and Resistive Functions* by G W A Dummer (New York: Hayden Book Co.) p 134

SEE ALSO: 1935: Casimir-Jonker J M and de Haas W J *Physica* vol 2, p 935

1935: H Grayson Smith and F G A Tarr *Transaction of the Royal Society of Canada* vol 29, p 23

1935   **TRAVELLING WAVE MICROWAVE OSCILLATOR**   **A and O Heil (Germany)**
(Early Magnetron)

Studies of the classical triode valve in which the anode current is controlled by the grid had shown that a fundamental difficulty for the highest frequencies was the excess grid control power needed due to electron inertia. In 1935 proposals were made by A Heil and O Heil for avoiding this limitation and also of avoiding the power dissipation limit of very high frequency circuits. These proposals were of particular importance since for the first time, a new mechanism specially suited for the generation of very high frequencies was suggested.

SOURCE: 'Microwave valves: A survey of evolution, principles of operation and basic characteristics' by C H Dix and W E Willshaw *J. Brit. IRE* p 580 (August 1960)

SEE ALSO: 'Eine neue methode zur erzeugung kurzer, ungedampfter, elektromagnetischer Wellen grosser Intensitat' by A A Heil and O Heil *Z. Phys.* vol 95, p 752 (1935)

1935   **SCANNING ELECTRON MICROSCOPE**   **M Knoll, M von Ardenne**
                                                (Germany) and D McMullan, C W
                                                Oatley (UK)

Postulated by Knoll in 1935, an early form of scanning electron microscscope was built by von Ardenne in 1938. However, the intensity of the electron beam at the specimen was very low (about 10–13 A) and it was therefore necessary to record the picture over a period of about 20 minutes in order to obtain an image of reasonable density on the photographic film.

Since the image was not visible until the film had been developed, focusing was a difficult proceeding, it being necessary to find the setting by trial and error.

The results with this microscope were inferior to conventional electron microscopes but von Ardenne pointed out that the scanning microscope should show advantages with thick specimens.

He also proposed that, instead of a photographic recording, the electron beam should be collected by an electrode, amplified and used to modulate a cathode-ray tube. The surfaces of opaque specimens could then be examined in terms of their secondary emitting properties

A scanning electron microscope designed especially for opaque speciments was made by Zworykin and others in 1942. The specimen was scanned by an electron spot as in von Ardenne's microscope, the main difference being that electrostatic lenses were used instead of magnetic ones. Some micrographs were published showing a resolution of about 500 A but the interpretation of them was inconclusive. It is well known that with primary voltages below a few thousand volts the secondary emission ratio is very dependent on the cleanness of the surface and in a demountable system with oil pumps it is practically impossible to prevent a thin layer of oil forming on the specimen, and this layer plays a significant part in determining the contrasts in the final micrograph. This difficulty has been overcome in the scanning electron microscope at Cambridge and, in addition, a number of other improvements have been incorporated including direct viewing of the picture before recording.

SOURCE: Letter from Dr D McMullan dated 16/10/77. Also 'The scanning electron microscope and the electron-optical examination of surfaces' by D McMullan *Electronic Engineering* p 46 (February 1953)

SEE ALSO: 'Aufladepotential und Sekundar-emission elektronbestrahlter Oberflachen' by M Knoll *Z. Tech. Phys.* vol 2, p 467 (1935)

'Das Elektron raster Mikroskop' by M von Ardenne *Z Tech. Phys.* vol 19, 407–16 (1938)

'An improved scanning electron microscope for opaque specimens' by D McMullan *Proc. IEE* vol 100, Part III, No 75, p 245 (June 1953)

K C A Smith and C W Oatley *Brit. J Appl. Phys.* vol 6, p 391 (1955)

'First international conference on Electron and Ion Beam Science and Technology' Edited by R Bakish (New York: John Wiley) (1965)

NOTE:

In 1957 a team of scientists at Cambridge University made a breakthrough in electron probe microanalysers which gave Britain a lead in this field that has so far been maintained.

Before the wholly-British development of scanning techniques, specimens had to be moved under static probes and the element distribution plotted laboriously and slowly. Scanning made it possible to display the information on a TV-type viewing system.

The following year, Tube Investments Research Laboratories found the value of scanning X-ray microanalysis, developed by the team at the Cavendish Laboratory was so great that they built an instrument of their own.

Early in 1959, the Cambridge Instrument Co. entered into an agreement with TI to manufacture such instruments. Production started later that year and the first Microscan was completed for the UKAEA, Aldermaston, and shown in the Cambridge Instrument Company's London office at the time of the Physical Society Exhibition in January 1960.

During the same period the company's efforts to improve the resolution of Microscan led to the merging of their work with that of the University's Engineering Department where scanning electron microscopes were being studied. Stereoscan was the result.

This instrument had a field of focus some 300 times greater than any previous microscope, optical or otherwise, and produced dramatic results of both rough and delicate surface alike. So revolutionary were the photographs taken on this instrument, that the company had to arrange a special demonstration before microscopists were convinced that they were true pictures of the surface.

SOURCE: 'From Microscan to Stereoscan.... Cambridge keeping Britain in front' by P Slater *Electronics Weekly* p 16 (10 January 1968)

1935   **CIRCUITRY (Constant Resistance Capacity Stand-Off   A D Blumlein (UK)
Circuit)**

Figures 11.9(a) and (b) show two versions of the basis of this invention, namely two-terminal arrangements of an inductor L, a capacitor C, and two equal resistors R having the property that the impedance measured between the two terminals is purely resistive, of value R at all frequencies, provided $L/C = R^2$. This property was known, but Blumlein adapted the circuit (particularly figure 11.9(b)) as a means of removing from critical points in a circuit (e.g. a wide-band amplifier) the stray capacity to earth of, for example, floating power supplies. The example shown in figure 11.9(c) is the application of the idea to the filament supply for the cathode follower output valve of the vision modulator for the original Alexandra Palace transmitter. The 'hardware' of this is preserved in the Science Museum.

SOURCE: 'The work of Alan Blumlein' *British Kinematography Sound and Television* vol 50, No 7, p 209 (July 1968)

SEE ALSO: British Patent Specification No 462 530 (1935)

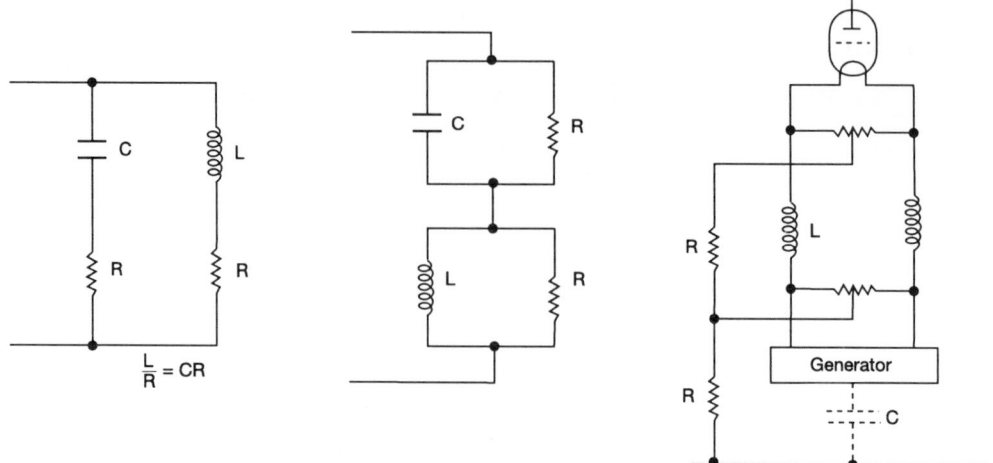

Figure 11.9. Constant resistance capacity stand-off circuit.

1935   **MULTIPLIER PHOTOTUBES**   **Zworykin, Morton & Malter (USA)**

One of the most important by-products of television research is the multiplier phototube. When electrons of one or several hundred electron volts energy impinge on a suitably prepared conducting surface, they eject 4 to 10 low-velocity electrons, multiplying the initial current by a corresponding factor of 4 to 10. Repetition of this process leads to current multiplication by an arbitrarily high factor, practically without the addition of amplification noise. If the initial current is derived from a photocathode, the tube output reflects the variation of the light incident on the photocathode with a precision which depends only on the quantum efficiency of the cathode; with proper design, the dispersion in the transit time of the electrons from the cathode to the final collector can be held to quantities of the order of 10^{-10} second.

In the earliest effective multipliers (Zworykin, Morton and Malter 1935) the electrons were guided from dynode to dynode along an approximately cycloidal path by crossed electric and magnetic fields. Purely electrostatic focusing and acceleration systems were developed subsequently by Zworykin and Rajchman and Rajchman and Synder as well as by Larson and Salinger. These may be regarded as the prototypes of present-day multiplier phototubes of RCA and DuMont. The venetian-blind design utilised in the image orthicon is also employed by EMI and RCA for multiplier phototubes. Finally, the early and very simple screen multiplier of Weiss (1936) does without focusing altogether, at the expense of materially lowered multiplication efficiency.

SOURCE: 'Beam-deflection and photo devices' by K Schlesinger and E C Ramberg *Proc. IRE* p 1001/2 (May 1962)

SEE ALSO: 'The secondary emission multiplier—a new electronic device' by V K Zworykin, G A Morton and L Malter *Proc. IRE* vol 24, p 351 (March 1936)

'The electrostatic electron multiplier' by V K Zvorykin and J A Rajchman *Proc. IRE* vol 27, p 558 (September 1939)

'Photocell multiplier tube' by C C Larson and H Salinger *Rev. Sci. Instrum.* vol 11, p 226 (July 1940)

'On secondary emission multipliers' by G Weiss *Z. Tech. Phys.* vol 17, p 623 (December 1936)

1935   **TRANSISTOR (Field Effect)**   **O Heil (Germany)**

In 1935, Oskar Heil of Berlin obtained a British patent on 'Improvements in or Relating to Electrical Amplifiers and Other Control Arrangements and Devices'. Figure 11.10 is the inventor's original illustration describing his device. The light area marked 3 is described as a thin layer of a semiconductor such as tellurium, iodine, cuprous oxide, or vanadium pentoxide; 1 and 2 designate ohmic contacts to the semiconductor. A thin metallic layer marked 6 immediately adjacent to but insulated from the semiconductorlayer serves as control electrode. Heil describes how a signal on the control electrode modulates the resistance of the semiconductor layer so that an amplified signal may be observed by means of the current meter 5. Using today's experience and language, one might describe this device as a unipolar field-effect transistor with insulated gate.

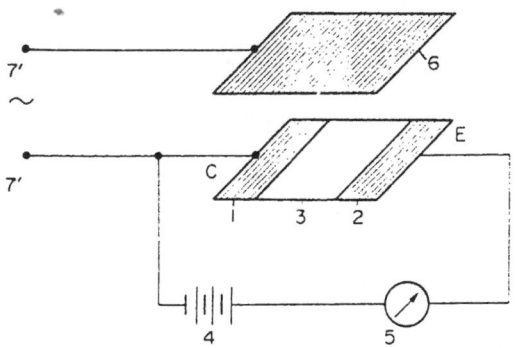

Figure 11.10. Field-effect transistor.

SOURCE: 'The field-effect transistor—an old device with new promise' by J T Wallmark *IEEE Spectrum* p 183 (March 1964)

1936 **CIRCUITRY (Long-Tailed Pair)**   **A D Blumlein (UK)**

This now familiar and much used circuit (figure 11.11(a) and (b)) was first needed in the amplifiers for the original video cable between points in Central London and Alexandra Palace. The cable was not the now familiar co-axial type but a shielded pair, and the problem was to obtain the 'push–pull' signal uncontaminated by 'push–push' interference pick-up. In telephone practice a transformer serves this purpose, but transformers to handle the video frequency range were not then available.

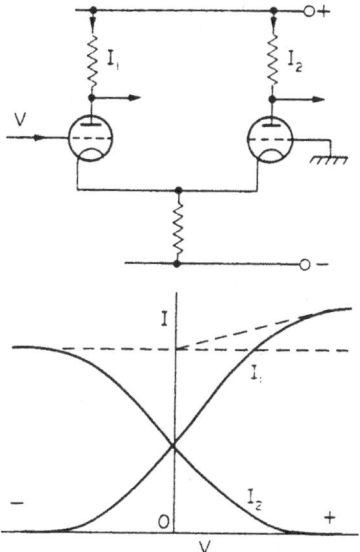

Figure 11.11. Long-tailed pair circuit.

The name of the circuit invented to do the job is the name given to it by Blumlein, and is so descriptive that it has stuck.

SOURCE: 'The work of Alan Blumlein' *British Kinematography Sound and Television* vol 50, No 7, p 209 (July 1968)

SEE ALSO: British Patent Specification No 482 740 (1936)

1936   COLD CATHODE TRIGGER TUBE   Bell Laboratories (USA)

Bell Telephone Laboratories announced the first cold cathode trigger tube in 1936. Using an activated cathode, the 313-A set the pattern for a number of trigger tubes made in Europe during the next two decades. The activated cathode led to anode maintaining and critical trigger voltages each of the order of 70 V.

SOURCE: 'A survey of cold cathode discharge tubes' by D M Neale *The Radio and Electronic Engineer* p 87 (February 1964)

SEE ALSO: 'The 313-A vacuum tube' by S B Ingram *Bell Lab. Rec.* p 114–6 (December 1936)

1936   WAVEGUIDES   J R Carson, S P Meade, S A Schelkunoff, G C Southworth (Bell Laboratories) (USA)

In 1936, from the Bell Telephone Laboratories, Carson, Meade and Schelkunoff published their mathematical theory on 'Hyper-Frequency Wave Guides' while G C Southworth published his experimental results. These papers provided the basis for the TECH mode cylindrical waveguide. In that same year W L Barrow of MIT published his work on the 'Transmission of Electromagnetic Waves in Hollow Tubes of Metal'. Before 1934 Southworth had transmitted telegraph and telephone signals at 15.cm wavelengths in a 5-in. diameter hollow metal pipe 875 ft. long with relatively small attenuation.

SOURCE: 'Microwave communications' by J H Vogelman *Proc. IRE* p 907 (May 1962)

SEE ALSO: (1) 'Hyperfrequency waveguides—mathematical theory' by J R Carson, S P Meade and S A Schelkunoff *Bell Sys. Tech. J.* vol 15, p 310–33 (April 1936)

1936   VOCODER   Bell Laboratories (USA)

In 1936, Bell Laboratories developed the voice coder, or vocoder, for analysing the pitch and energy content of speech waves. With later developments, vocoder output was digitised, encrypted and the digital signal transmitted within a voice channel. The vocoder has been used since World War II by the US Government for secure communications.

SOURCE: 'Mission Communications—the Story of Bell Laboratories' by Prescott C Mabon (Murray Hill, NJ: Bell Laboratories Inc.) p 171 (1975)

1937   XEROGRAPHY   C Carlson (USA)

An individual inventor, Chester Carlson, conceived the idea of Xerography. This is a new photographic process which in a relatively short time has found numerous industrial applications. It is completely dry and is based entirely upon principles of photoconductivity and electrostatics. The process:

'employs a plate which consists of a thin photoconductive coating on a metallic sheet. This coating can be electrically charged in the dark and will hold this charge until exposed to light. Thus an electrostatic image can be produced on the plate by exposing the plate to an optical image. When the plate is dusted with powder particles, the electrostatic image is transformed into a powder image which can be transferred to paper and fixed by fusing.'

The development of Xerography was turned over to Roland M Schaffert, a Battelle research physicist with some previous experience in printing. For a year he worked alone but after the war Battelle assigned a few assistants to help him. By the latter part of 1946 two important developments were completed: a high-vacuum technique for coating plates with selenium; and a corona discharge wire, both for applying the original electrostatic charge to the plate and for transferring powder from the plate to the paper. The most significant contribution was the discovery of a method to keep the image background from being filled with stray powder. Thus Battelle improved Xerography to the point where industry became interested.

SOURCE: *The Sources of Invention* by J Jewkes, D Sawers and R Stillerman (London: MacMillan & Co.) p 405, 408 (1958)

SEE ALSO: 'Printing with Powders' Fortune (June 1949)

'Xerography—From Fable to Fact' by W T Reid

'Developments in Xerography' by R M Schaffert (The Penrose Annual) (1954)

1937   **POLAR CO-ORDINATE OSCILLOGRAPH**   **M von Ardenne (Germany), J J**
                                                     Dowling and T G Bullen (Ireland)

A polar co-ordinate oscillograph usually employs a special form of cathode ray tube or other oscillographic device which has been specially designed for the purpose of depicting oscillograms in polar co-ordinates.

von Ardenne and Dowling and Bullen have independently developed polar co-ordinate cathode ray oscillograph tubes. von Ardenne's tube and the circuit employed with it are shown in figure 11.12.

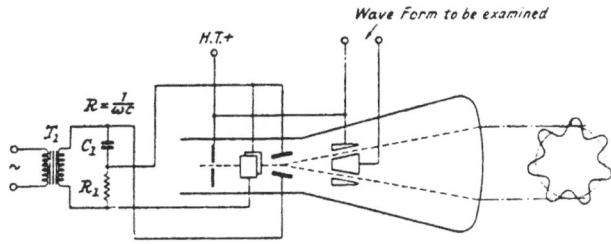

Figure 11.12. von Ardenae's polar co-ordinate oscillograph.

The tube contains two concentric cone-shaped deflectors across which the potential to be examined is connected. The resultant form of the image is shown in figure 11.13. This form of cathode ray tube has the advantage that the final anode potential remains fixed and it is, therefore, possible to obtain larger deflections without defocusing than is the case when the signal is applied as a modulation of the final anode potential.

von Ardenne has also employed electromagnetic deflection methods for this purpose.

The form of the image, as shown in figure 11.13, is not in true polar co-ordinates in the mathematical sense of the term, but it is difficult to find a name which truly describes the arrangement.

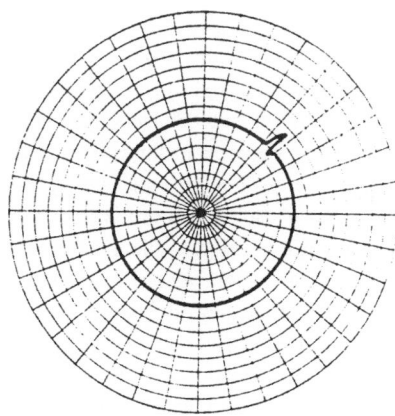

Figure 11.13. Polar co-ordinate oscillogram obtained with von Ardenne's polar co-ordinate oscillograph.

SOURCE: *Time Bases* by O S Puckle (London: Chapman & Hall) p 104 (1955)

SEE ALSO: 'A new polar co-ordinate cathode ray oscillograph with extremely linear time scale' by M von Ardenne *Wireless Engineer* vol 14, p 5 (1937)

'Precision measurements with a radial deflection cathode ray oscillograph' by J J Dowling and T G Bullen *Proc. R. Irish Acad.* A, vol 44, p 1 (1937)

1937   PULSE CODE MODULATION   A H Reeves (UK)

Pulse code modulation, or coded step modulation (which I think would have been an apter name), is a good example of an invention that came too early. I conceived the idea in 1937 while working at the Paris laboratories of International Telephone and Telegraph Corporation. When PCM was patented in 1938 and in 1942, I knew that no tools then existed that could make it economic for general civilian use. Only in the last few years, in this semiconductor age, has its commercial value been felt.

Pulse code modulation was invented mainly for line-of-sight microwave links or link sections, where in 1938 the needed extra bandwidth would have been cheap and easily obtainable, rather than for more limited frequency bands, as in cables, which are now in fact the main fields of application. It is this change of aim for PCM, for quite sound reasons, that has caused most of the technological difficulties so far in its application.

It was in the United States during World War II that the next step in PCM's progress was made, by Bell Telephone Laboratories. In this important stage, a team under Harold S Black designed a practical PCM system later produced in quantity for the US Army Signal Corps. Research was also done under Ralph Bown. It is appropriate that this early Bell work should be stressed, for it was the first time that the principles underlying the new system were translated into hardware.

SOURCE: 'The past, present and future of PCM' by A H Reeves *IEEE Spectrum* vol 3, No 5, p 58 (May 1965)

1937   RADAR AIMING ANTI-AIRCRAFT GUNS   P E Pollard (UK)

The first radar equipment for aiming anti-aircraft guns was devised by Mr P E Pollard in 1937. It was the basis of the first radar gun-laying equipment, GLI, brought into anti-aircraft service in 1939. It gave range up to 10 miles, with an accuracy of about 25 yards, but no angle of elevation, and was the only equipment of its kind available in the night attacks of 1940–41.

SOURCE: *Science at War* by J G Crowther and R Whiddington (London: HMSO) p 76 (1947)

1938   TELEVISION: SHADOW-MASK TUBE   W Flechsig (Germany)

The shadow-mask tube had its genesis in a 1938 conception of the German inventor Flechsig. However, at that time, there seemed to be no way to make Flechsig's device in which hundreds of fine wires had to be in exact alignment with an equal number of phosphor triads, each with a red-, green- and blue-emitting phosphor line. Modifications of the idea were proposed by Goldsmith and by Schroeder, at RCA Laboratories. Schroeder suggested a hexagonal array of circular holes in a metal mask, together with round phosphor dots and three closely spaced electron beams through a common deflection yoke. Prior to 1948, some experiments were done on methods of multicolour phosphor deposition, but the basic technology of aligning either holes or wires with phosphor dots or lines appeared well beyond reach. One of the experimenters on phosphor deposition, and a colleague of Schroeder's, was H B Law. When the RCA crash program to develop a colour tube started in 1949, Law elected to pursue Schroeder's idea. He then made a key invention, which he called the 'lighthouse'. This device permitted a photographic process to produce light shadows that were essentially the same as the electron-beam shadows. Application of any one of several photolithographic techniques then permitted deposition of phosphors in exactly the right place. Success came rapidly and in a few months Law photographically etched a metal mask with tiny holes, through which the three colour-emitting phosphors could be deposited, shifting the mask slightly for each deposition. Law's first tube displayed small but remarkably good colour pictures. Application to a larger screen (30-cm diagonal) was made by engineering teams at RCA's Lancaster and Harrison locations, and a single-gun version was developed by R R Law (unrelated to H B Law but also at RCA's Princeton Laboratories).

SOURCE: 'A history of colour television displays' by E W Herold *Proc. IRE* vol 64, No 9. p 1333 (September 1976)

SEE ALSO: W Flechsig: German Patent 736575 filed 1938

'Multi-colour television' A C Schroeder, US Patent No 2595 548 filed 1947, issued 1952

'Picture reproducing apparatus' A C Schroeder, US Patent No 2595 548 filed 1947, issued 1952

'A three-gun shadow-mask kinescope' by H B Law *Proc. IRE* vol 39, p 1186 (October 1951)

1938   **'GEE' NAVIGATION**   **R J Dippy (UK)**

Owing to the use of a chart covered with a network or grid of curves, the system was given the code name of GEE (see figure 11.14). It was invented by Mr R J Dippy and developed by his team at Telecommunications Research Establishment. In practice three stations are actually used. A master station A sends out a series of pulses. If we consider one of these, it will travel past station B and also to the aircraft P and beyond. It is arranged that when the pulse reaches B which is called a slave station, a transmitter is then activated which issues a second pulse, A second pulse from A also activates a transmitter in a second slave station at C. Then the cycle repeats.

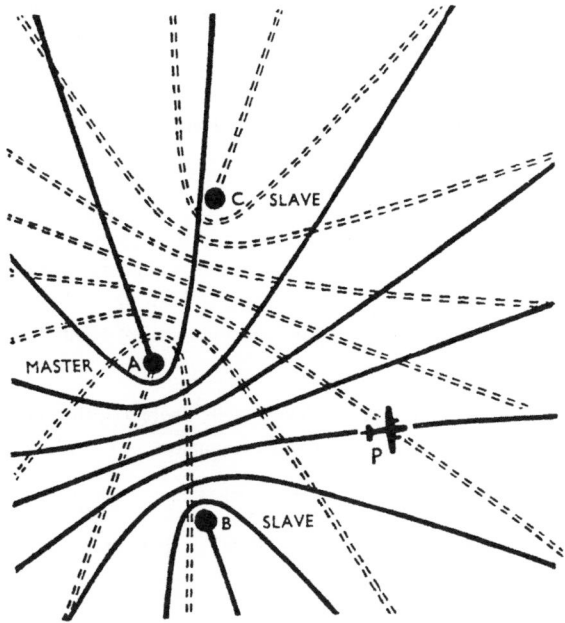

Figure 11.14. GEE navigational system.

The aircraft carries one cathode ray tube on which the arrival of the four pulses is recorded. The navigator is provided with a chart covered with two intersecting sets of curves, corresponding to stations A and C, and A and B respectively. This can be superimposed on a map of the country to be flown over. The target is marked on the map and hence its place in respect to the curves is seen.

SOURCE: *Science at War* by J G Crowther and R Whiddington (London: HMSO) p 54 (1947)

SEE ALSO: 'Gee—a radio navigational aid' by R J Dippy *Proc. IEE.* vol 93, pt. IIIA, p 468 (1946)

1938   **COMPUTERS (Information Theory)**   **C E Shannon (USA)**

Shows the analysis of complicated circuits for switching could be effected by the use of Boolean algebra.

SOURCE: 'A symbolic analysis of relay and switching circuits' by C E Shannon Trans. AIEE vol 57, p 713 (1938)

1938   **NUCLEAR FISSION**   **Fritsch and Meitner (Germany)**

Late in 1938, Hahn and F Strassman, in reviewing the chemical knowledge of the new substances, recognised that one of them was probably barium, whose atomic mass and number are only about half of that of uranium. This meant that they had previously been on a false trail. Frisch and Meitner, now in Scandinavia, immediately explained the significance of this discovery. Neutrons did not transmute uranium atoms into new atoms slightly heavier or more complicated, with a higher atomic number. They split these big uranium atoms into two roughly equal parts. The splitting could be done in a variety of ways. Uranium atom No 92 might be split into Barium No 56 + Krypton No 36; or into Strontium No 38 + Xenon No 54. Dr J R Dunning in America rapidly repeated the work on fission.

Here was the explanation of the chemical confusion: a wide variety of chemically different atoms was being produced by the disintegration. Frisch and Meitner named this new process of atom-splitting 'nuclear fission'. Nothing like this had been seen in heavy atoms before.

SOURCE: *Science at War* by J G Crowther and R Whiddington (London: HMSO) p 131 (1947)

1939   **RADIO ALTIMETER**   **Bell Laboratories (USA)**

The radio altimeter, by which a pilot can calculate his height above the ground, is developed in the USA at the Bell Laboratories. The altimeter bounces signals off the Earth and measures the time they take to return to the aircraft. The pilot then uses a calibrated indicator to translate this figure—the relative altitude—into a figure giving his absolute altitude.

SOURCE: *The Timetable of Technology* (London: Michael Joseph, Marshall Editions) p 117 (November 1982)

1939   **KLYSTRON**   **W C Hahn and Varian Bros (USA)**

Perhaps the first great step in understanding the phenomena in microwave tubes came with the invention of the klystron. Bruche and Recknagel discussed 'phase focusing' in 1938 and the work of the Varians, Webster's theoretical treatment of the klystron and the work of Hahn and Metcalf were published in 1939.

With the klystron came a well-thought-out theory of its operation, the concept of velocity modulation, and a full appreciation of the value of microwave resonators.

SOURCE: 'History of the microwave tube art' by J R Pierce *Proc. IRE* p 979 (May 1962)

SEE ALSO: 'High frequency oscillator and amplifier' by R H and S F Varian *J. Appl. Phys.* vol 10, p 321 (May 1939)

1939   **DOUBLE-BEAM OSCILLOGRAPH**   **B C Fleming-Williams (UK)**

A splitter plate is immersed in the beam and divides it into two separate sections. Two 'bucking' wires to which potentials are applied are used in order to cancel out mutual deflectional interference of the two beams. With this tube, the two beams are simultaneously deflected in the X axis but they are separately controlled in the Y direction. Since the Y deflecting potentials are necessarily unbalanced, (because only one plate is available for each beam) it becomes necessary to employ an anti-trapezium construction for the tube.

The production of a double beam is much better accomplished by means of Fleming-Williams' double-beam cathode ray tube than by means of an electronic switch since, in the former case, the two images appearing are coincident in time. This is not so with the electronic switch and, hence, it is possible for events which are not coincident in time to be assumed to be so.

SOURCE: *Time Bases* by O S Puckle (2nd edn) (London: Chapman & Hall) p 262 (1951)

SEE ALSO: 'The double-beam cathode ray oscillograph' by B C Fleming-Williams *Electronics & Short Wave World* vol 12, p 457 (1939)

1939   **COMPUTERS (Digital)**   **H H Aitken (USA) and IBM**

Utilizing twentieth-century advances in mechanical and electrical engineering, the Automatic Sequence Controlled Calculator, or Mark I, brought Babbage's ideas into being, giving concrete existence to much more at the same time. The Mark I, an electromechanical calculator 51 feet long and 8 feet high, was built by the International Business Machines Corporation between 1939 and 1944. It could perform any specified sequence of five fundamental operations, addition, subtraction, multiplication, division and reference to tables of previously computed results. The operation of the entire calculator was governed by an automatic sequence mechanism. The machine consisted of 60 registers for constants, 72 adding storage registers, a central multiplying and dividing unit, means of computing the elementary transcendental functions $\log_{10} x$, 10^x and $\sin x$, and three interpolators reading functions coded in perforated paper tapes. The input was in the form of punched cards and switch positions. The output was either punched into cards or printed by electric typewriters.

SOURCE: 'The evolution of computing machines and systems' by Serrell, Astrahan, Patterson and Pyrne *Proc. IRE* p 1043 (May 1962)

SEE ALSO: *The Computer from Pascal to von Neumann* by H H Goldstine (Princeton, NJ: Princeton University Press) p 118 (1972)

'Proposed automatic calculating machine' by H H Aitken *IEEE Spectrum* p 62 (August 1964)

1939   **BELL TELEPHONE LABS 'COMPLEX**   **G Stibitz *et al* (USA)**
         COMPUTER'

It is perhaps a little surprising that it was not until 1937 that Bell Telephone Laboratories investigated the design of calculating devices, although Andrews has stated that from about 1925 the possibility of using relay circuit techniques for such purposes was well accepted there. However, in 1937 George Stibitz started to experiment with relays and drew up circuit designs for addition, multiplication and division. At first he concentrated on binary arithmetic, together with automatic decimal–binary and binary–decimal conversion, but later turned his attention to a binary-coded decimal number representation. The project became an official one when, prompted by T C Fry, Stibitz started to design a calculator capable of multiplying and dividing complex numbers, which was intended to fill a very practical need, namely to facilitate the solution of problems in the design of filter networks, and so started the very important Bell Telephone Laboratories Series of Relay Computers.

In November 1938, S B Williams took over responsibility for the machine's development and together with Stibitz refined the design of the calculator, whose construction was started in April and completed in October of 1939. The calculator, which became known as the 'Complex Number Computer', often shortened to 'Complex Computer' and as other calculators were built, the 'Model 1' began routine operation in January 1940. Within a short time it was modified so as to provide facilities for the addition and subtraction of complex numbers, and was provided with a second, and then a third teletype control, situated in remote locations. It remained in daily use at Bell Laboratories until 1949.

SOURCE: *The Origins of Digital Computers* edited by B Randell (Berlin: Springer) p 238 (1973)

SEE ALSO: 'Computer' by G R Stibitz *The Origins of Digital Computers* edited by B Randell (Berlin: Springer) p 241 (1973)

1939   **MAGNETRON**   **J T Randall and H A H Boot (UK)**

In the autumn of 1939 the Admiralty asked Professor M L Oliphant and the physics department of the University of Birmingham to develop a high-power microwave transmitter. The majority of the scientists in the laboratory concentrated on the klystron, described by its inventors, R H and S F Varian of Stanford University, California, in 1939, which used for the first time closed resonators, described by W W Hansen, also of Stanford, in 1938, for the production of high-frequency power. J T Randall and H A H Boot, struck by the difficulty of getting enough power from the klystron, considered instead applying the resonator principle to the magnetron, which had been invented by A W Hull of the American General Electric firm in 1921 but which, in its conventional form, lacked the properties they were seeking. The result was the cavity magnetron, which proved to be the needed generator, producing high powers on centimetre wavelengths.

SOURCE: *The Sources of Invention* by J Jewkes, D Sawers and R Stillerman (London: MacMillan & Co.) p 348 (1958)

1939   **FREQUENCY STANDARDS (Caesium Beam)**   **I I Rabi (USA)**

The difficulties of bandwidth and low intensity are most easily overcome by using the atomic beam magnetic resonance method developed at Columbia University by Rabi and his co-workers. In this method which can be used with atoms possessing a magnetic dipole moment, a beam of atoms passes to a detector through a system of magnets and a region of field alternating at the Bohr frequency. The magnets have a nonuniform field and deflect the atoms in one direction or the other according to which of the two energy levels they are in. When the frequency of the RF field is exactly equal to the Bohr frequency and is of the right amplitude transitions are induced and the deflections in the second magnet B are the opposite from those in the first magnet A, and the atoms are thus focused on the detector.

SOURCE: 'Frequency and time standards' by L Essen *Proc. IRE* p 1162 (May 1962)

1939   **LARGE SCREEN TELEVISION PROJECTOR**   **Fischer (Switzerland)**

The first large-screen television projector was invented by Professor Fischer at the Swiss Federal Institute of Technology in 1939. At that time, Fischer thought that the growth of television would come from the development of networks of neighbourhood 'television theatres' and he invented the Eidophor with the capability of projecting TV pictures onto cinema-sized screens. The earliest Eidophors were cumbersome machines, which could project only black and white pictures in a darkened or semi-darkened room. They were not the most reliable of machines and for a number of years the Eidophor system was little known or used.

Later the American space programme called for a reliable, high performance, large-screen projection system capable of working for long periods of time, to provide data displays in NASA flight control centres. Gretag AG, Zurich, a subsidiary of Ciba Geigy and patent holders and manufacturers of the Eidophor, successfully developed the projector's capability to meet NASA specifications. The latest Eidophors are able to project full-colour television pictures onto screens 18 m wide.

SOURCE: 'Projection television—a review of current practice in large-screen projectors' by A Robertson *Wireless World* p 47 (September 1976)

1940   **CYBERNETICS**   **N Weiner (Germany)**

Cybernetics, as a science, was invented by Norbert Wiener in 1940, but the word was not coined until 1948 by Wiener and A Rosenblueth. It comes from the Greek word kybernêtés meaning a steersman or pilot.

Cybernetics is the study of automatic communication and control mechanisms in machines as well as in humans.

SOURCE: *The Book of Inventions and Discoveries* Associate Editor Valerie-Anne Giscard d'Estaing (UK: Queen Anne Press, Macdonald & Co.) (1990) p 116

1940   **PLAN POSITION INDICATOR**   **E G Bowen, W B Lewis, G W A Dummer and E Franklin (UK)**

The first radar PPI (Plan Position Indicator) to be used by the RAF, was designed by the authors (G W A Dummer and E Franklin) in 1940 and in view of the use since made of this device it may be of interest to recall early experimental work on radial time bases.

The possibility of the desirable PPI presentation of radar echoes had been realised by the pioneers in the early days of radar. It was not, however, until 1939–40 that the two developments of a radar station using a sufficiently narrow beam and the cathode ray screen with bright and lasting afterglow led to the development of a satisfactory PPI.

In 1939–40 work was proceeding on the design of a 'radio lighthouse' on a wavelength of 50 cm, It was envisaged that with the narrow beams then obtained on this wavelength it should be possible to rotate a time base in synchronism with the aerial rotation to give a radar 'map' of all surrounding aircraft. It was decided that the time base should take one of two forms:

1. An inductive or capacitive voltage split of X and Y vectors and recombination on an electrostatically deflected tube.

2. A mechanically rotated current time base on a magnetically deflected tube.

At that time 12-inch electrostatic tubes were in use on CH and CHL sets and magnetic afterglow tubes were not fully developed; it was therefore decided to adopt the first scheme.

Experimental work was also carried out on 'strobing' a portion of the time base, amplifying it and feeding it to another tube. By this means an enlarged PPI was developed. A dim 'square' (approximately 1 in side on a 12 in tube) was produced on the first tube by partial blackout to identify the area which was being enlarged and this area appeared as full size on the other tube. Owing to the 15° beamwidth of the polar diagram the system was not used, but it is interesting to record that a system of this type was operating in this country in 1941.

The PPI was next adapted for use in the first centimetre Al equipment and afterwards for H_2S and ASV. It has become one of the most widely used presentation systems for radar both in this country and the USA and it seems a far cry from the original 6 ft rack to the compact, efficient airborne PPIs in use today.

SOURCE: 'Radial time bases'—how they were developed for radar' by G W A Dummer and E Franklin *Wireless World* p 287 (August 1947)

SEE ALSO: *Three Steps to Victory* by Sir Robert Watson-Watt (London: Odhams Press) p 268 (1957)

1940   **'OBOE' NAVIGATIONAL SYSTEM**   **A H Reeves (UK)**

Yet another high-precision method of radar navigation removed the necessity for the pilot to find the target or even know where he was going, All of this could be managed from ground stations without the knowledge of the pilot who was relieved of much strain and decision, and could devote the whole of his attention to the controls of his aircraft.

In this system there are two fixed stations, A and B. Station A enables the aircraft P to fly along the circumference of a great circle, whose centre is at A, and whose circumference passes over the target C. This station, which pushes the aircraft this way and that, tracking it, as it were, and making it keep to the circle, is called the Cat station. The stations A and B send pulses which are picked up by the aircraft, magnified and returned. From these responses, the exact distance of P from A is recorded at A, and of P from B is recorded at B. The Cat station emits a signal as a result of its knowledge of the distance PA. If the aircraft has strayed to the right, so that AP is greater than AC, the pilot hears a series of Morse dashes in his earphones. If the aircraft strays to the left, too near to A, he hears a series of dots. But if he keeps exactly on the circle he hears a high-pitched continuous buzz. The Mouse station B watches the aircraft, ready to warn it when it reaches the target and should, as it were, dart down the hole. It gives the pilot a series of warning signals as he comes within range, and then a final signal at the right moment for releasing the bombs. It may even release the bombs without the pilot's intervention at all (see figure 11.15).

The inventor of this system, called Oboe, was Mr A H Reeves. Like Dippy and Lovell in their developments of Gee and H_2S, he worked in close collaboration with Group Capt (Now Air Vice-Marshal) D C T Bennett, the Pathfinder leader, who tried and adopted it. It was used by Pathfinder aircraft to mark special targets, which could then be attacked by following bombers.

SOURCE: *Science at War* by J G Crowther and R Whiddington (London: HMSO) (1947) p 58

SEE ALSO: 'Oboe—a precision ground controlled blind-bombing system' by F E Jones *Proc. IEE* vol 93, pt. IIIA, p 496 (1946)

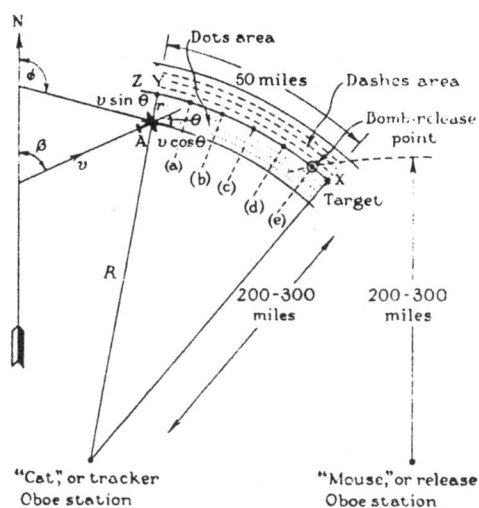

Figure 11.15. OBOE navigational system.

1940   SKIATRON CRO   A H Rosenthal (USA)

The Skiatron, or dark-trace tube, was developed to meet a radar requirement for a large-screen 'black on white' picture with a persistence of several seconds.

The Skiatron consists of a magnetically focused and deflected cathoderay tube with a screen consisting of a translucent micro-crystalline layer of potassium chloride. This screen is obtained by evaporation of the material in vacuum on to the tube face. An intensity-modulated scanning electron beam produces a picture by causing darkening in the areas bombarded. The picture is episcopially projected using external illumination from mercury-vapour lamps.

SOURCE: 'The Skiatron or dark-trace tube' by P G R King and J F Gittins *J. IEE* pt IIIA, vol 93, p 822 (1946)

SEE ALSO: 'A system of large-screen television based on certain electron phenonema in crystals' by A H Rosenthal *Proc. IRE* vol 28 p 203 (1940)

'Photography of cathode ray tube traces' by H F Roberts and F A Richards *RCA Review* vol 6 p 234 (1941)

1941   BETATRON   D W Kerst (USA)

A betatron is an electrical device in which electrons revolve in a vacuum enclosure, in a circular or a spiral orbit normal to a magnetic field, and have their energies continuously increased by the electric force resulting from the variation with time of the magnetic flux enclosed by their orbits. The betatron accelerates electrons to velocities approaching that of light. This magnetic induction accelerator was invented by D W Kerst of the University of Illinois in 1941. It is comparable to an ordinary transformer wherein the high voltage winding, or secondary, consists of an evacuated tube in which electrons moving at high velocity form the secondary circuit, The device has been designed to accelerate electrons to 340 million electron volts.

SOURCE: *Encyclopedic Dictionary of Electronics and Nuclear Engineering* by R I Sarbacher (London: Pitman) p 10 (1959)

1941   RADIO 'PROXIMITY' FUSE   W S Butement (UK)

One of the most brilliant innovations of Army radar was the V-T or self-acting radio fuse. This was proposed by Butement. It consists of a small radio transmitter and receiver fitted within a shell. When the shell is fired, the transmitter emits radio waves. These are reflected from the target. The time-interval between the emission of the waves and the return of their reflections is a measure of the distance between the shell and its target. By arranging that the shell explodes when the time-interval falls below a certain value, the shell is made to explode when it is within a certain distance of the target, and therefore virtually sure to inflict damage.

This very ingenious invention was based on an application of the Doppler principle, and on the use of very rugged radio valves which could be fired in a shell without being destroyed. Both of these original features were British inventions, the early successful rugged valves being made by the research team under Mr D I Lawson of Pye Ltd, while important contributions were made by the Research Laboratories of the General Electric Company Ltd.

The later development and manufacture of this fuse were taken over by American scientists and engineers. Many very difficult problems had to be solved before it could be produced reliably in quantity. This was done just in time to meet the flying-bomb menace to London, and at the end of that attack, nearly 100 per cent of all the flying-bombs approaching London were being shot down by anti-aircraft guns using V-T fuses invented in England, and developed and manufactured in the United States.

Our American allies were able to put 1500 persons on to the development of this fuse; at no time were we able to put more than 50 persons on to the same task. The Americans performed the prodigy of making 150 000 000 of the special valves for these fuses.

SOURCE: *Science at War* by J G Crowther and R Whiddington (London: HMSO) p 82 (1947)

1941 RADAR (H_2S) NAVIGATION SYSTEM   P I Dee, A C B Lovell, A D Blumlein *et al* (UK)

If you could see air targets with spiral scan AI equipment, why not try to see ground targets with them? Accordingly, in the autumn of 1941 Dee arranged that this experiment be tried. The aircraft flew from Christchurch Aerodrome, Hampshire. After four minutes a camp near Stonehenge and the City of Salisbury were identified on the cathode ray tube.

Thus it was proved that in the cathode ray tube picture of the general reflections from the ground of the waves from a 9 centimetre airborne equipment, certain areas of ground could be distinguished from others. The picture of the mass of echoes from the ground was not just a shimmering confusion, it had definite features which corresponded to different objects on the ground.

Specific equipment for scanning the ground was now made and fitted into a blister or dome on a heavy bomber, in the place of the under-turret. The blister was made of the synthetic plastic material perspex, which has the valuable property of being transparent to short radio waves, as well as visible light. The rotating scanning equipment is thus protected from the rush of the air. The bomber, Halifax V 9977, flew with the new equipment for the first time on 27 March 1942, and radar (magic eye) target-finding was born to live under the name of H_2S.

SOURCE: *Science at War* by J G Crowther and R Whiddington (London: HMSO) p 63 (1947)

1941 MICROELECTRONICS (Thick Film Circuits)   Centralab (USA)

During World War II, the Centralab Division of Globe-Union Inc., developed a ceramic-based circuit for the National Bureau of Standards. This 'printed circuit' used screen-deposited resistor inks and silver paste to support the miniature circuits in an Army proximity fuse (see figure 11.16). The PC board that followed stimulated manufacturers to develop components with radial leads and tubular shapes.

SOURCE: 'Solid state devices—packaging and materials' by R L Goldberg *Electronic Design* vol 24 (23 November 1972) p 127

1942 THE VELODYNE   F C Williams and A M Uttley (UK)

The Velodyne is an electromechanical system in which a speed of rotation is held closely proportional to an input voltage by feedback methods. In such a system the total number of revolutions of the output shaft is a measure of the time-integral of the input voltage.

 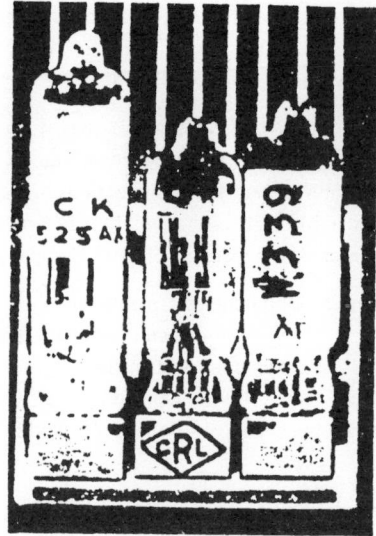

Figure 11.16. Proximity fuse circuits made in 1945.

The Velodyne has been applied to the solution of differential equations, and the TRE differential analyser uses it to solve simultaneous differential equations of importance in propagation theory; many simulators have been made which obey specific high-order differential equations and have been very useful, particularly in the design of complicated automatic flight-control equipment where computation of solutions would have involved prohibitive effort.

SOURCE: 'The Velodyne' by F C Williams and A M Uttley *Proc. IEE* vol 93, pt IIIA, p 1256 (1946)

1942   **SANATRON Linear Time Base Circuit**   **F C Williams and N F Moody (UK)**

THE SANATRON. The two valves are arranged in the form of a multivibrator having one stable and one semi-stable state, the former corresponding with the quiescent condition of the circuit and the latter having the duration of the linear discharge of a capacitor C_2. Operation is initiated by a negative trigger pulse and the semi-stable state is then maintained until the discharge of C_2 is completed when the circuit reverts to its quiescent condition.

As a time base generator the Sanatron leaves little to be desired, a sweep waveform being available, with amplitudes up to 250 V available from a 300-V supply line and rates from 1 V/s to 5×10^7 V/s are readily achieved. The upper limit of rate may be extended even further by using a modification of the circuit in which the control of the Miller integrator is applied to the screen instead of to the suppressor grid.

SOURCE: 'Ranging circuits, linear time-base generators and associated circuits' by F C Williams and N F Moody (Abstract of supporting paper on circuit techniques) *J. IEE* p 320

SEE ALSO: 'Linear time-base generators and associated circuits' *J. IEE* vol 93, Part IIIA, (1946)

1942   **PHANTRASTRON Linear Time Base Circuit**   **F C Williams and N F Moody (UK)**

THE PHANTRASTRON: The Phantrastron is a circuit which combines the Miller integrator and trigger properties of the Sanatron in a single valve. Its main use lies in delayed pulse generation and, like the Sanatron, it delivers a rectangular pulse during the linear sweep. Due to its simplicity it is widely used where a somewhat lower order of accuracy and linearity than that provided by the Sanatron is acceptable.

SOURCE: 'Linear time-base generators and associated circuits' by F C Williams and N F Moody *J. IEE* vol 93, part IIIA, p 1193 (1946)

1942 **LORAN**   **MIT (USA)**

LORAN (Long Range Air Navigation), which marks the world's air and sea lanes like streets, goes into operation at four stations between the Chesapeake Capes and Nova Scotia. LORAN, a natural successor to the GEE system developed in Britain and given to the USA in 1940 can, at long range, give an aircraft pilot or ship's captain his position to within a few hundred yards. The LORAN receiver, developed at MIT, picks up radio signals as pairs of 'pips' on a screen; when the pips intersect, the position is indicated.

SOURCE: *The Timetable of Technology* (London: Michael Joseph, Marshall Editions) p 124 (November 1982)

1943 **RELIABILITY (Sequential analysis)**   **A Wald (USA)**

A major technical breakthrough occurred in the spring of 1943, when the noted mathematical statistician, Abraham Wald, devised his now celebrated basic theory of sequential analysis for analysing US war problems. Initial application of the theory in analysing combat experience demonstrated its value in obtaining reliable conclusions from a minimum of information, swiftly and economically. This seemed like an ideal tool for use in quality control at this crucial period in our history when speed, precision, reliability and economy in production were of the essence. To give the new tool a thorough trial, the military released a multiple sampling plan based on Wald's sequential theory. This plan was first released to a limited number of strategic manufacturing firms for use in acceptance sampling. The early success of the plan and the subsequent widespread demands for its use resulted in removal of its 'restricted' classification in 1945.

SOURCE: 'The reliability and quality control field from its inception to the present' by C M Ryerson *Proc. IRE* p 1326 (May 1962)

1943 **TRAVELLING WAVE TUBE**   **R Kompfner, A W Haeff and J R Pierce (USA)**

It remains, however, for Kompfner to take the decisive step of reasoned approach and effective experiment which gave us the travelling-wave tube and led to a host of related devices.

Kompfner reasoned that an electromagnetic wave on a slow-wave structure, and especially a helix, should interact powerfully with a beam of electrons if the wave velocity and the electron velocity were nearly the same. He built a tube and found a gain of around 10 dB for a wave travelling in the same direction as the electrons. A mathematical analysis agreed with this performance.

The travelling-wave tube turned out to be a device which amplifies over unprecedently broad bands, and which can function over an astonishingly wide range of frequencies and powers. A travelling-wave tube with a helix circuit can amplify over a frequency range of more than an octave. More typically, bandwidth is limited to 500–1000 Mc by input and output couplers.

SOURCE: 'History of the microwave tube art' by J R Pierce *Proc. IRE* p 980 (May 1962)

SEE ALSO: 'The travelling wave valve' by R Kompfner *Wireless World* vol 52, p 369 (November 1946)

1943 **PRINTED WIRING**   **P Eisler (UK)**

This invention relates to the manufacture of electric circuits—such, for instance, as that of a telephone switchboard—and circuit components such as inductances, resistances, magnetic cores and their existing windings; and consists in producing them by the methods of the printing art or methods akin to them, methods that is to say, by which the conductor of the circuit is brought into existence in its final form, or a development of that form upon a plane or other surface, instead of being first produced as a linear conductor and afterwards given its three-dimensional form.

A typical instance of the invention comprises the steps of making a drawing of the electric or magnetic circuit, or of a development of it if it is of three dimensions: preparing from that drawing, by any of the well known methods of the printing art, a printing surface; making an imprint by the aid of the printing surface; and from that imprint producing the conductor.

SOURCE: UK Patent No 639 178 (2 February 1943)

SEE ALSO: *Technology of Printed Circuits* by P Eisler (London: Heywood) (1959)

1943   **ULTRASONIC RADAR NAVIGATION TRAINING   G W A Dummer and A W Smart**
       DEVICE   (UK)

One of the most important of the radar navigation devices designed during the war was H_2S This system was used by RAF Bomber Command for 'blind' navigation over Germany and made possible the accurate and heavy bombing of Berlin and other targets beyond the effective range of 'Gee' and 'Oboe'. To use this device, training was necessary and navigator/operators could be trained either by flying training aircraft equipped with H_2S or by using a synthetic training device on the ground.

The H_2S training device, originated in February 1943, was the first to use the new principle of a 'miniature' radar system using ultrasonic waves propagated through water in place of electromagnetic waves propagated through air. The long delay time of transmission of ultrasonic waves through a liquid was used to represent the radar delay times normally encountered in the operation of H_2S. The velocity of ultrasonic waves in water is 1.5×10^5 cm/s and as that of electro-magnetic waves in air is 3×10^{10} cm/s the scale on which the trainer operated was

$$\frac{1.5 \times 10^5}{3 \times 10^{10}} = \frac{1}{200\,000} \text{th}$$

of the radar scale. This meant that one 'ultrasonic' mile = 0.315 in, and a radar range of fifty miles could be simulated in a physical distance not exceeding 15 inches approximately.

If an X-cut quartz crystal is pulsed at its resonant frequency under water and a reflecting object is placed in the path of the transmitted wave, an echo will be re-radiated from the object in the same way as in radar and picked up on the crystal.

The crystal beam was projected on to a glass relief map so that the projection covered a range of 0.30 miles (0.10 in approximately), the crystal being used for both transmission and reception. The crystal was set back 1.25 in. from the axis of rotation to simulate 'ground returns' and picture as seen from an aircraft flying at 20 000 ft. On the glass map was reproduced (to a scale of 1:200 000) a simulation of the area in Germany over which training was to be effected. The amount of reflection of the radiated energy was proportional to the 'roughness' of the glass surface, and the map was sand-blasted or etched to represent land masses. Towns were built up of granules of Carborundum glued to the glass and the sea areas were left as plain glass. The map was placed at the bottom of a tank of water. The pulsed crystal was then rotated at 60 rpm. synchronously with the radial time base, and the signals received from the map were fed from the crystal to the input stage of the IF amplifier in the H_2S receiver and thence to the PPI.

The display produced was an excellent simulation of the actual H_2S picture seen in the air when flying over Germany.

SOURCE: 'H_2S trainer—use of ultrasonic reflections from submerged relief maps' by G W A Dummer *Wireless World* p 65 (February 1947)

SEE ALSO: 'Aids to training—the design of radar synthetic training devices for the RAF' by G W A Dummer *Proc. IEE* vol 96, pt. III, No 40, p 101 (March 1949)

1943   **COMPUTERS (ENIAC) (Electronic Numerator,   Moore School (USA)**
       Integrator and Computer)

The ENIAC was developed and built at the Moore School of Electrical Engineering of the University of Pennsylvania, beginning in 1942 and completed in 1946. Its principal object was the computation of firing and ballistic tables for the Aberdeen Proving Ground of the US Army Ordnance Corps. This computation required the integration of a simple system of ordinary differential equations involving arbitrary functions.

This equipment occupied a space 30 × 50 feet and contained 18 000 vacuum tubes, The computing elements consisted largely of decade rings, flip-flops and pentode gates. The input–output system consisted of modified IBM card readers and punches.

SOURCE: 'The evolution of computing machines and systems' by Serrell, Astrahan, Patterson and Pyrne *Proc. IRE* p 1044 (May 1962)

SEE ALSO: 'The Computer from Pascal to von Neumann' by H H Goldstine (Princeton, NJ: Princeton University Press) p 117 (1972)

1943 **MAGNETIC AMPLIFIER (TRANSDUCTOR)**　　　　**ASEA (Sweden)**

Pioneer work on transductors was carried out by the Swedish firm of ASEA who devised many techniques involving combinations of saturated reactors and metal rectifiers for supplying controllable d.c. power from single-, three- and six-phase networks in place of the conventional mercury arc rectifier systems. Much of this pioneer work is described in the classic work by Uno Lamm. Considerable development took place in Germany during the 1939–45 war on transductors as true magnetic amplifiers for various small power applications, mainly in association with servo systems. The V2 rocket incorporated a transductor for controlling the frequency of a 500 c/s, 150 VA alternator.

SOURCE: 'The Magnetic Amplifier' by J H Reynere (Rockcliff Pub. Corp.) p 17 (1950)

SEE ALSO: 'The Transductor' by A U Lamm (Stockholm: Essalte Aktiebolag) (1943)

'Some fundamentals of a theory of a transductor' by A U Lamm (AIEE Trans.) vol 66 (1947)

'Magnetic amplifier' by A G Milnes *J. IEE* vol 96, part I, p 89 (1949)

1943 **COLOSSUS—CRYPTANALYSIS MACHINE**　　　　**M Newman, A Turing, T H Flowers and A W M Coombs (UK)**

As far as building a British stored-program computer, the initial enthusiasm came largely from a group of people who had been involved in code-breaking activity at the government's Bletchley Park establishment. Much of the Bletchley work is still subject to the Official Secrets Act because the methods are (or were) in use by other countries after the war. As far as electronic digital techniques were concerned, the most interesting Bletchley cryptanalysis machine was the COLOSSUS, first operational in December 1943 and containing about 1500 thermionic valves. COLOSSUS was not, and had no need to be, a true stored-program computer, but it came quite close to being one. The extraordinary and vital contribution of COLOSSUS to the allied war effort has been told elsewhere. A view of the COLOSSUS machine is shown in figure 11.17.

SOURCE: 'The early days of British computers—1' by S H Lavington *Electronics & Power* p 827 (November/December 1978)

SEE ALSO: 'The Colossus' by B Randell *University of Newcastle Computer Science Technical Report 90* (1976). Reprinted in condensed form in *New Scientist* vol 73, pp 346–8 (1977)

The Ultra Secret by F W Winterbotham (Wiedenfelt and Nicholson) (1974)

'Colossus: godfather of the computer' by B Randell *New Scientist* vol 73, p 346 (10 February 1977)

1944 **RELIABILITY SAMPLING: INSPECTION TABLES**　　　　**H F Dodge and H G Romig (USA)**

After languishing in libraries for several years, the work of Dodge and Romig in acceptance sampling and the work of Shewhart on control charts finally was brought to light during World War II through the nationwide training programmes sponsored by the Office of Production Research and Development of the War Production Board.

Although the underlying concepts were developed by scientific investigators and statisticians in the preceding decades, the genius of Dodge, Romig and Shewhart lay in their recognition of basic principles

Figure 11.17. A view of the COLOSSUS machine (The Science Museum/ Science & Society Picture Library).

as an aid to solving practical problems, and their ability to recognize and formulate a systematic approach.

SOURCE: 'Treating real data with respect' by J A Henry *Quality Progress* p 18 (March 1976)

SEE ALSO: *Sampling Inspection Tables* 2nd edn by H F Dodge and H G Romig (New York: John Wiley & Sons) (1944)

1945   **COMPUTERS (Theory)**   **von Neumann (USA)**

Basic design of the electronic computer project of the Institute of Advanced Study incorporating ideas underlying essentially all modern machines.

REFERENCE: 'Memorandum on the program of the High Speed Computer' by von Neumann (8 November 1945)

SEE ALSO: 'The Computer from Pascal to von Neumann' by H H Goldstine (Princeton, NJ: Princeton University Press) p 255 (1972)

1945   **DECCA Navigation system**   **W O'Brien (UK) and H Schwartz (USA)**

DECCA, internationally regarded as the best navigation system, under goes crucial tests during the Allies' D-Day landings on the Normandy beaches. Developed by William O'Brien and Harvey Schwartz, in London and Hollywood, DECCA indicates on cockpit dials the position of an aircraft in three dimensions—latitude, longitude and altitude—and is accurate to within a few yards, Waves of radio signals, emitted from two transmitters, collide and are picked up in phase.

SOURCE: *The Timetable of Technology* (London: Michael Joseph, Marshall Editions) p 130 (November 1982)

1945   **COMPUTERS (Whirlwind)**   **MIT (USA)**

An assignment to build a real-time aircraft simulator was given in 1945 to the Digital Computer Laboratory of the Massachusetts Institute of Technology, at that time a part of the Servomechanisms Laboratory of MIT. Beginning in 1947, the major part of the effort was devoted to the design and construction of the electronic digital computer known as 'Whirlwind'. The project was sponsored by the Office of Naval Research and the United States Air Force. The machine was put in operation in March 1951.

'Whirlwind I' was a parallel, synchronous, fixed-point computer utilising a number length of 15 binary digits plus sign (16 binary digits in all). Physically, it was a large machine containing some 5000 vacuum tubes (mostly single pentodes) and some 11 000 semiconductor diodes. It consisted of an arithmetic 'element' including three registers; a control element including central control, storage control, arithmetic control and input–output control; a program counter—a source of synchronising pulses or master clock supplying Z megapulses per second to the arithmetic element and 1 megapulse per second to the other circuits; an internal storage element or memory, terminal equipment; and extensive test and marginal checking equipment.

SOURCE: 'The evolution of computing machines and systems' Serrell, Astrahan, Patterson and Pyne *Proc. IRE* p 1047 (May 1962)

SEE ALSO: *The Computer from Pascal to von Neumann* by H H Goldstine (Princeton, NJ: Princeton University Press) p 212 (1972)

1945   **COMMUNICATION (Satellite)**   **A C Clarke (UK)**

Early interest in space was concentrated upon the propulsion aspect, and the forthcoming marriage of space and electronics had to await the publication of A C Clarke's paper on communications satellites in 1945. The use of a satellite S above the radio horizons of both A and B permitted microwave transmission from A to S, and from S to B, thus bridging the oceans by microwave link. Clarke's paper further pointed out that at an orbital altitude intermediate between a 90-minute SPUTNIK and the 28-day orbit of the moon, there was an orbit taking one day, so that an easterly-launched satellite above the Equator would give the radio engineer an imaginary mast 22 300 miles high on which to place his aerials.

SOURCE: 'Electronics in space' by W F Hilton *The Radio and Electronic Engineer* vol 45, No 10, p 623 (October 1975)

SEE ALSO: 'Extra terrestial relays' by A C Clarke *Wireless World* vol 51, No 10, p 305 (October 1945)

1945–   **POTTED CIRCUITS**   **(UK) and (USA)**
1950

The potting of electrical apparatus in wax or bitumen compounds was carried out for many years, but it is only recently that plastics suitable for this purpose have become available, in the form of cold-polymerizing casting resins. There is no doubt that the small sub-unit is now as essential part of modern electronic equipment, and potting techniques lend themselves to this construction provided that means are available to dissipate the heat developed. The resins are relatively expensive, and mechanically and economically are not attractive for casting exceeding a few inches in major dimensions.

The casting resins are converted into rigid plastics by the addition of a catalyst and accelerator, without the application of the considerable pressures and temperatures normally associated with the polymerisation of thermosetting resins.

SOURCE: 'New constructional techniques' by G W A Dummer and D L Johnston *Electronic Engineering* p 456 (November 1953)

SEE ALSO: 'How plastics aid miniaturisation of electrical assemblies' by R J Bibbero and E B Chester *Mach. Des.* p 127 (October 1951)

'Potted Circuits—new development in miniaturisation of equipment' *Wireless World* vol 57, p 493 (1951)

'Cast resin embedments of circuit sub-units and components' *Elect. Mnfg.* vol 48, p 103 (1951)

1946   ACE (AUTOMATIC COMPUTING ENGINE)   A M Turing (UK)

Turing joined the new mathematics division at NPL, where he immediately set about designing a universal computer with characteristic energy. Turing had written an important theoretical paper on computers in 1936 and, although familiar through personal contact with von Neumann, he had no need or inclination to copy anyone else's design. On 19 February 1946, he presented to the Executive Committee of NPL probably the first complete design for an electronic stored-program computer, including a cost estimate of 11 200. It is likely that Sir Charles Darwin, the NPL Director, thought of Turing's proposal for an Automatic Computing Engine (ACE) in terms of a single national effort that would result in a computer housed at NPL and serving the needs of the whole country.

The Pilot ACE had a complicated 32-bit instruction format including provision for specifying one of 32 'sources', one of 32 'destinations' and the source of the next instruction. Instructions also specified the duration of a transfer, so that prolonging a transfer over several cycles could give the effect of shifting or multiplying operands by small integers. Many operations in the instruction repertoire were straightforward transfers, with the remaining instructions providing about a dozen conventional arithmetic or logical functions, including unsigned multiplication. Signed multiplication took just over 2 ms, being performed partly by subroutine. Other orders could be obeyed in as little as 64 μs (1024 μs in the worst case), depending on the position of the next instruction. Arithmetic was serial, with a 1μs digit period. The main store consisted initially of 128 32 bit words in mercury delay lines. This was extended to 352 words by the end of 1951, and a 4k drum was added in 1954. Since the NPL already had a large Hollerith punched card calculator, it was sensible to make cards the medium for both input to and output from the Pilot ACE.

The Pilot ACE contained 800 thermionic valves, the processor logic involving type ECC 81 double triodes. It first ran a program in May 1950. The computer is illustrated in figure 11.18.

SOURCE: 'The early days of British computers—1' by S H Lavington *Electronics & Power* p 828 (November/December 1978)

'The early days of British computers—2' by S H Lavington *Electronics & Power* p 40 (January 1979)

SEE ALSO: 'Proposals for the development in the Mathematics Division of an Automatic Computing Engine (ACE)' by A M Turing *Report E882, Executive Committee, INTPL* (Reprinted in April 1972 as NPL Report Com. Sci. 57.)

'The other Turing machine' by B E Carpenter and R W Doran *Computer J.* vol 20, pp 269–79 (1977)

1946   COMPUTERS (CRT Storage)   F C Williams (UK)

Storage of pulses on the face of a CRT as a memory device.

REFERENCE: 'A storage system for use with binary digital computers' by F C Williams and T Kilburn *Proc. IEE* vol 96, pt. 2, No 81, p 183 (1949)

SEE ALSO: *The Computer from Pascal to von Neumann* by H H Goldstine (Princeton, NJ: Princeton University Press) p 248 (1972)

1947   HIGH QUALITY AMPLIFIER CIRCUIT   D T N Williamson (UK)

The requirements of such an amplifier may be listed as:

(1) Negligible non-linear distortion up to the maximum rated output. (The term 'non-linear distortion' includes the production of undesired harmonic frequencies and the intermodulation of component frequencies of the sound wave.) This requires that the dynamic output–input characteristic be linear within close limits up to maximum output at all frequencies within the audible range.

(2) (a) Linear frequency response within the audible frequency spectrum of 10–20 000 c/s. (b) Constant power handling capacity for neglible non-linear distortion at any frequency within the audible frequency spectrum.

Figure 11.18. ACE stored programme computer (The Science Museum/ Science & Society Picture Library).

(3) Neglible phase-shift within the audible range.   Although the phase relationship between the component frequencies of a complex steady-state sound does not appear to affect the audible quality of the sound, the same is not true of sounds of a transient nature, the quality of which may be profoundly altered by disturbance of the phase relationship between component frequencies.

(4) Good transient response. In addition to low phase and frequency distortion, other factors which are essential for the accurate reproduction of transient wave-forms are the elimination of changes in effective gain due to current and voltage cut-off in any stages, the utmost care in the design of iron-cored components, and the reduction of the number of such components to a minimum.

(5) Low output resistance. This requirement is concerned with the attainment of good frequency and transient response from the loudspeaker system by ensuring that it has adequate electrical damping.

(6) Adequate power reserve. The realistic reproduction of orchestral music in an average room requires peak power capabilities of the order of 15–20 watts when the electro-acoustic transducer is a baffle-loaded moving-coil loudspeaker system of normal efficiency.

SOURCE: 'Design for a high-quality amplifier' part 1 by D T N Williamson *Wireless World* p 118 (April 1947)

'Design for a high-quality amplifier' part 2 by D T N Williamson *Wireless World* p 161 (April 1947)

1947   **CHIRP RADAR TECHNIQUES**   **Bell Laboratories (USA)**

The Chirp or pulse-compression technique for radar originated at Bell Laboratories in 1947. This technique, in which long, modulated pulses are transmitted and then compressed upon reception, permitted pulsed radar systems to have long range and high resolution while avoiding problems associated with generating and transmitting short pulses with high peak powers

SOURCE: *Mission Communications—the Story of Bell Laboratories* by P C Mabon (Murray Hill, NJ: Bell Laboratories Inc.) p 179 (1975)

1947   **ECME (Electronic Circuit Making Equipment)**   **J A Sargrove (UK)**

The process of John A Sargrove for the automatic assembly of electronic apparatus was the first modern approach to automatic operation in electronic manufacturing. In 1947 he built and operated a machine for the automatic production of two- and five-tube radio receivers.

The first operation of Sargrove's machine was to prepare the $\frac{1}{4}$-in molded-plastic plates by blasting with an abrasive grit to roughen both sides of the plates simultaneously. The plates were then triple-sprayed with zinc to form the conducting surface. The spraying machine consisted of eight nozzles arranged four to a side to allow simultaneous spraying of both sides of the plate once it was positioned. Materials to form resistance, capacitance and conductors were sprayed through stencils onto their proper positions on the plate. A typical plate is shown in figure 11.19.

Figure 11.19. A typical sprayed plate (Sargrove machine).

SOURCE: *Electronic Equipment Design and Construction* by G W A Dummer, C Brunetti and L K Lee (New York: McGraw-Hill) pp 192–3 (1961)

SEE ALSO: 'New methods of radio production' by J A Sargrove *J. Brit. IRE* vol 7(1), p 2 (January/February 1947)

'Automatic receiver production' *Wireless World* (April 1947)

1947   **COMPUTERS (EDVAC)**   **University of Pennsylvania (USA)**

The Electronic Discrete Variable Automatic Computer, or EDVAC, was built at the Moore School (University of Pennsylvania) between 1947 and 1950 for the Ballistic Research Laboratory at the Aberdeen Proving Ground. It is a serial, synchronous machine in which all pulses are timed by a master clock operating at 1 mega-pulse per second. It contains some 5900 vacuum tubes, about 1200 semiconductor diodes and utilizes the binary number system with a word length of 44 binary digits.

SOURCE: 'The evolution of computing machines and systems' by Serrell, Astrahan, Patterson and Pyne *Proc. IRE* p 1046 (May 1962)

SEE ALSO: *The Computer from Pascal to von Neumann* by H H Goldstine (Princeton, NJ: Princeton University Press) p 187 (1972)

1947 **COMPUTERS (UNIVAC) (Universal Automatic**   **P Eckert and J Mauchly (USA)**
      Computer)

The development of the Universal Automatic Computer, or UNIVAC was started about 1947 by Presper Eckert and John Mauchly who founded the Eckert-Mauchly Computer Corporation in December of that year. The first UMVAC I was built for the USA Bureau of the Census and was put in operation in the spring of 1951. (The Eckert-Mauchly Corporation later became a subsidiary of Remington Rand, forming the organisation which is now the Remington Rand UNIVAC Division of the Sperry Rand Corporation).

UNIVAC I was a direct descendent of the ENIAC and of the EDVAC in the development of which Eckert and Mauchly had both had an important part at the University of Pennsylvania. It was a serial, synchronous machine operating at a rate of 2.25 megapulses per second. It contained some 5000 tubes and several times as many semiconductor diodes in logic and clamp circuits. One hundred mercury delay lines provided 1000 twelve-decimal-digit words of internal storage. Twelve additional delay lines were used as input–output registers. Aside from console switches and an electric typewriter providing small amounts of information, the input–output medium was metal-base magnetic tape. Forty eight UNIVAC I machines were built.

SOURCE: 'The evolution of computing machines and systems' Serrell, Astrahan, Patterson and Pyne *Proc. IRE* pp 1048/9 (May 1962)

SEE ALSO: *The Computer from Pascal to von Neumann* by H H Goldstine (Princeton, NJ: Princeton University Press) p 246 (1972)

1947 **MOLECULAR BEAM EPITAXY**   **J Sosnowski, J Starkiewicz and O**
                                          Simpson (UK)

Molecular beam epitaxy (MBE) is a term used to denote the epitaxial growth of compound semiconductor films by a process involving the reaction of one or more thermal molecular beams with a crystalline surface under high vacuum conditions. MBE is related to vacuum evaporation, but offers much improved control over the incident atomic or molecular fluxes so that sticking coefficient differences may be taken into account, and allows rapid changing of beam speeds. For example, using MBE, it is possible to produce 'superlattice' structures consisting of many layers of GaAs and $Al_2Ga_{1-x}As$ with layer thickness as low as 10Å. Since electrically active impurities are added to the growing film with separate beams, the doping profile normal to the surface may be varied and controlled with a special resolution difficult to achieve by more conventional, faster growth techniques.

SOURCE: 'Molecular beam epitaxy' by A Y Cho and J R Arthur *J. R. Prog. in Solid State Chemistry* vol 16, part 3, p 157 (1975)

SEE ALSO: 'Lead sulphide photoconductive cells' by L Sosnowski, J Starkiewicz and O Simpson *Nature* vol 159, p 818 (14 June 1947)

'The structure and growth of Pbs deposits on rocksalt substrates' by J Elleman and H Wilman *Proc. Phys. Soc. (London)* vol 61, p 164 (1948)

1948 **COMPUTERS (SEAC)**   **National Bureau of Standards (USA)**

The Standards Electronics Automatic Computer, SEAC, was built by the staff of the Electronic Computer Laboratory of the National Bureau Standards. The design began in June 1948 and the machine was put in operation in May 1950. It was built under the sponsorship of the Office of the Air Controller, Department of the Air Force, principally to carry out mathematical investigations of techniques for solving large logistics programming problems.

SOURCE: 'The evolution of computing machines and systems' by Serrell, Astrahan, Patterson and Pyne *Proc. IRE* p 1046 (May 1962)

SEE ALSO: *The Computer from Pascal to von Neumann* by H H Goldstine (Princeton, NJ: Princeton University Press) p 315 (1972)

1948   **TRANSISTOR**   **Bardeen, Brattain and Shockley**
                                                              (USA)

Immediately hostilities ceased in 1945 Shockley organised a group for research on the physics of solids. On testing out experimentally Shockley's ideas, it was discovered that the projected amplifier did not function as Shockley had predicted; something prevented the electric field from penetrating into the interior of the semi-conductor. John Bardeen, a theoretical physicist, formulated a theory concerning the nature of the surface of a semi-conductor which accounted for this lack of penetration of field, and also led to other predictions concerning the electrical properties of semi-conductor surfaces. Experiments were carried out to test the predictions of the theory. In one of these, Walter H Brattain and R B Gibney observed that an electric field would penetrate into the interior if the field was applied through an electrolyte in contact with the surface. Bardeen proposed using an electrolyte in a modified form of Shockley's amplifier in which a suitably prepared small block of silicon was used. He believed that current flowing to a diode contact to the silicon block could be controlled by a voltage applied to an electrolyte surrounding the contact. In the earlier experiments testing Shockley's ideas, thin films with inferior electrical characteristics had been employed. Brattain tried Bardeen's suggested arrangement, and found the amplification as Bardeen had predicted, but the operation was limited to very low frequencies because of the electrolyte. Similar experiments involving germanium were successful, but the sign of the effect was opposite to that predicted. Brattain and Bardeen then conducted experiments in which a rectifying metal contact replaced the electrolyte and discovered that voltage applied to this contact could be used to control, to a small extent, the current flowing to the diode contact. Here again, however, the sign of the effect was opposite to the predicted one. Analysis of these unexpected results by the two scientists led them to the invention of the point contact transistor, which operates on a completely different principle from the one first proposed. Current flowing to one contact is controlled by current flowing from a second contact, rather than by an externally applied electric field. Brattain and Bardeen used extremely simple equipment, the most expensive piece of apparatus being an oscilloscope.

The Bell Telephone Laboratories announced the invention in June 1948, and since then, development work has proceeded rapidly. The first point-contact transistor had several limitations: it was noisy, it could not control high amounts of power and it had a limited applicability. Shockley had meanwhile conceived the idea of the junction transistor which was free of many of these defects and most of the transistors now made are of the junction type.

SOURCE: *The Sources of Invention* by J Jewkes, D Sawers and R Stillerman (London: MacMillan & Co.) p 400 (1958)

SEE ALSO: 'The First Five Years of the Transistor' by M Kelly *Bell Telephone Magazine* (Summer 1953)

1948   **HOLOGRAPHY**   **D Gabor (UK)**

With holography, one records not the optically formed image of an object but the object wave itself. This wave is recorded (usually on photo graphic film) in such a way that a subsequent illumination of this record called a 'hologram' reconstructs the original object wave. A visual observation of this reconstructed wavefront then yields a view of the object which is practically indiscernible from the original, including three dimensional parallax effects.

SOURCE: *The Encyclopaedia of Physics* (2nd edn) editor R M Besancon (New York: Van Nostrand-Reinhold & Litton Educational Pub. Inc.) p 426 (1974)

SEE ALSO: *Optical Holography* by R J Collier, C B Burzkhardt and L H Lin (New York: Academic Press) (1971)

1948   **EDSAC (Electronic Delay Storage Automatic**   **M V Wilkes (UK)**
        Calculator)

The EDSAC (electronic delay storage automatic calculator) is a serial electronic calculating machine working in the scale of two and using ultrasonic tanks for storage. The main store consists of 32 tanks, each of which is about 5 ft long and holds 32 numbers of 17 binary digits, one being a sign digit. This gives 1024 storage locations in all. It is possible to run two adjacent storage locations together so as to accommodate a number with 35 binary digits (including a sign digit); thus at any time the store may contain a mixture of long and short numbers. Short tanks which can hold one number only are used for accumulator and multiplier registers in the arithmetical united, and for control purposes in various parts of the machine.

A single address code is used in the EDSAC, orders being of the same length as short numbers.

SOURCE: *The Origins of Digital Computers* edited by B Randell (Berlin: Springer) p 389 (1973)

SEE ALSO: 'The design of a practical high-speed computing machine' by M V Wilkes, *Proc. R. Soc. London* A, vol 195, p 274 (1948)

1948   **TRANSISTORS (Single Crystal Fabrication   G K Teal & J B Little (USA)**
Germanium)

In the latter part of 1948 G K Teal and J B Little of BTL began experiments to grow germanium single crystals, selecting the pulling technique. They succeeded in growing large single crystals of germanium of high structural perfection. They also improved the impurity of the material by repeated recrystallization methods.

At BTL Teal, working with M Sparks, devised a unique method for preparing p-n junctions by modifying his crystal-pulling apparatus to allow controlled addition of impurities during crystal growth. Using ingots they prepared single crystals containing p-n junctions and soon afterwards n-p-n grown-junction transistors which had many of the properties predicted by Shockley.

SOURCE: 'Contributions of materials technology to semiconductor devices' by R L Petritz *Proc. ISE* p 1026 (May 1962)

'Growth of germanium single crystals' by G K Teal and J B Little *Phys. Rev.* vol 78, p 647 (June 1950)

'Growth of silicon single crystals and of single crystal silicon p-n junctions' *Phys. Rev.* vol 87, p 190 (July 1952)

1948   **COMMUNICATION (Information Theory)   C E Shannon (USA)**

The term 'Information Theory' is used in the current technical literature with many different senses. Historically it seems first to have been generally applied to describe the specific mathematical model of communication systems developed in 1948 by Shannon. In this pioneering paper, Shannon introduced a numerical measure, called by him and others entropy, of the randomness or uncertainty associated with a class of messages and showed that this quantity measures in a real sense the amount of communication facility needed to transmit with accuracy messages from the given class. He also showed (quite incidentally to his main argument) that this measure of uncertainty agreed in certain aspects with the common, vague intuitive notion of the 'information content of a message'. He accordingly used the words 'information content' as a synonym for the precisely defined notion of entropy. As a result, his work and its immediate extensions became known as information theory.

SOURCE: 'Information Theory' by B McMillan and D Slepian *Proc. IRE* pp 1151/2 (May 1962)

SEE ALSO: *The Mathematical Theory of Communication* by C E Shannon and W Weaver (Urbana, IL: University of Illinois Press) (1949)

1948   **FILM SOUND RECORDING (Magnetic Film)   RCA and others (USA)**

Although magnetic recording was one of the oldest methods known, it was not until World War II that this form of recording came into its own. During this period there was developed in Germany a fine grain, low-noise, magnetic oxide, and a process for uniformly coating it on a thin flexible base 4 inches in width. Use of this new tape in properly designed recorders and reproducers resulted in sound quality which was higher than had previously been obtained from either the film or the disk method.

Immediately after the war some of the German recorders were demonstrated in this country, and the potential impact of magnetic recording on the motion-picture industry was quickly recognised.

Early in 1948, oxide coated 35 mm film became available for use in motion picture sound recording. It was then possible to convert photographic sound recorders to combination units capable of recording either magnetic or photographic sound. Many recorders were converted as quickly as possible and were tried in motion-picture sound studios for original 'takes'. The tests of magnetic recording in the studios were immediately successful, not only because of its high-quality and large dynamic range, but also because magnetic film provided a more flexible and a more economical means of recording. Since re-recording of all original 'takes' to a composite photographic negative was already the accepted practice in the industry, little inconvenience was caused by the change.

SOURCE: 'Film recording and reproduction' by M C Batsel and G L Dimmick *Proc. IRE* p 749 (May 1962)

1949   MICROWIRE   Ulitovsky (USSR)

The microwire process for producing ultrafine wires was invented in Russia in 1949 by Professor Ulitovsky of the Baykov Institute. It is important to realise that up to then it was quite difficult to obtain fine insulated wire of reasonable quality and price.

Three processes have been developed as advances over the traditional die drawing techniques:

1. Wollaston process

2. Taylor process

3. Microwire process

In the Wollaston process wires are compound-drawn; that is, a platinum rod encased in a silver tube is drawn through wire dies, being finally subjected to a reduction of between 20 and 40 to one. The silver coated wire so produced is etched to remove the silver, leaving the fine platinum core behind. This procedure will give platinum wires down to 0.5μ.

In the Taylor process a composite rod is made consisting of a glass tube with a metal core cast into position by being aspirated, pipette-fashion, from a melting pot. This composite rod is attenuated in a muffle furnace, using the technique of pulling and stretching as practiced in the glass fibre industry. The metal usually has a lower melting point than the softening point of the glass, thus allowing the capillary to be filled with cast metal, Wires of lead, antimony, bismuth, gold, silver, tin, copper and others were successfully produced down to 0.25μ.

In the Microwire process this technique was improved. The basic procedure consists of melting the core metal by induction in a crucible formed by the walls of a glass tube extended upwards to a glass feeding device above the melt. The fact that the process takes place in a vertical line is an improvement immediately since there is less distortion of the melt and interference with the capillary by gravity.

SOURCE: *Materials for Conductive and Resistive Functions* by G W A Dummer (New York: Hayden Book Co.) p 62

SEE ALSO: 'Glass-coated microwire' by H Wagner *Wire and Wire Productions* (June 1964)

'Microwire. A new engineering material' by R G S Clarke *Electronic Components* (September 1963)

'The structure of copper microwire' by E M Nadgorny and B I Smirnov *Fizika Tverdogo Tele.* vol 2 (12), pp 3048–9 (1960)

1949   DIP-SOLDERING OF PRINTED CIRCUITS   S F Danko and Abramson (USA)

When Danko and Abramson of the Army Signal Corps invented dip soldering in 1949, a new era of automation came into being.

SOURCE: 'Packaging and materials' by R L Goldberg *Electronic Design* vol 24, p 126 (23 November 1972)

SEE ALSO: 'Autosembly of miniature military equipment' by S F Danko and S J Lanzalotti *Electronics* vol 24 (7), p 94 (July 1951)

'Printed circuits and microelectronics' by S F Danko *Proc. IRE* p 937 (May 1962)

Autosembly. US Patent No 2756485 assigned to US Army, 31 July 1956

1949   COLD CATHODE STEPPING TUBE   Remington Rand (USA)

The first published account of a multi-cathode stepping tube descibed a tube developed in America by Remington Rand. The first tubes to be widely used, however, were made in England by STC and by Ericsson Telephones. For some ten years following its introduction in 1949, the Ericsson 'Dekatron' became practically synonymous with the cold cathode stepping tube. The original double-pulse 'Dekatron' was followed in 1952 by the single-pulse tube operating up to 20 kc/s. Routing guides were added in 1955 to simplify the construction of reversible scalers. In 1962 auxiliary-anode tubes were introduced from which numerical indicators could be driven directly.

SOURCE: 'A survey of cold cathode discharge tubes' by D M Neale *The Radio and Electronic Engineer* p 87 (February 1964)

SEE ALSO: 'Poly-cathode glow tube for counters and calculators' by J J Lamb and J A Brustman *Electronics* vol 22, No 11, p 92–6 (November 1949)

1949–1950   ION IMPLANTATION IN SEMICONDUCTORS   R S Ohl, W Shockley (USA)

Ion implantation is a technique for modifying the properties of solids by injecting (implanting) charged atoms (ions) into them. The ions alter the electrical, optical, chemical, magnetic and mechanical properties of a solid by the interactions they have with the solid both as they slow down and by their presence after they have come to rest. The implantation idea is not a new one—work at Bell Laboratories during the late 1940s and early 1950s by Russel S Ohl and William Shockley pioneered the application of ion implantation to semiconductor device fabrication.

SOURCE: 'Ion implantation' by W C Brown and A U MacRae *Bell Laboratories Research* p 389 (November 1975)

SEE ALSO: 'Forming semiconductive devices by ionic bombardment' W Shockley: US Patent No 2787564 (28 October 1954)

'Ion implantation in semiconductor device technology' by J Stephen *The Radio Electronic Engineer* vol 42, No 6, p 265 (June 1972)

1950s   GLOBAL POSITIONING SYSTEM   Ivan Getting (USA)

GPS refers to a system that provides all users, by air, sea or land, an accurate position reading on a 24-hour basis in all weather conditions. The system uses a satellite-based radio positioning technique to provide three-dimensional position, velocity and time to all users equipped with a receiver that can intercept a certain minimum of visible satellite signals.

In the 1950s, Ivan Getting, an IEEE life fellow and the originator of GPS, realized that if a system of satellite transmitters was deployed such that a minimum of four were always in sight to any receiver on ground, it would be possible to know the location of that receiver in three dimensions, similar to the way a position is determined using the LORAN system, a ground navigation system effective in a limited geographical area.

The US Department of Defense approved funding for the development and deployment of a complete Navstar GPS at a cost of US$10 billion, and in 1978 launched the first of a series of a minimum 18 GPS satellites needed for the system to be fully operational. The program was to provide complete worldwide coverage by 1987, but due to the Challenger disaster and the resulting delays in satellite deployment, the system did not become fully operational until 1993.

The US Defense Department was doubtlessly unaware of the enthusiasm this system would arouse worldwide, far beyond the civil aviation community. New applications emerge on an almost daily basis, and military applications, which in the mid-1970s justified such a heavy investment in GPS, have been pushed to the sideline. The broad civil application base has made its case to the US DoD, and is now even challenging the exclusive control of military authorities over the system.

The GPS market is expected to grow by 45 percent this year, and GPS business is estimated to reach US$10 billion by the year 2000, with the military share being less than 10 percent.

SOURCE: 'GPS out of the shadow of the Defense Department' by R K Arora *The Institute* (IEE Inc.) p 3 (August 1996)

| 1950 | **FLOPPY DISCS** | **Y Nakamats (Japan)** |

Floppy discs, universally used on microcomputers, were invented in 1950 at the Imperial University in Tokyo by Doctor Yoshire Nakamats, an inventor who boasts of having 2360 patents which include golf clubs and loudspeakers. He granted the sales licence for the disc to IBM.

SOURCE: *The Book of Inventions and Discoveries* Associate Editor Valerie-Anne Giscard d'Estaing (UK: Queen Anne Press, Macdonald & Co.) p 124 (1990)

NOTE ON DEVELOPMENTS

| 1971 | **8 inch Floppy Disc** | **Schugzat/IBM (USA)** |

| 1978 | **5.25 inch Floppy Disc** | **Apple Computer, Radio Shack Tandy (USA)** |

| 1984 | **3.5 inch Floppy Disc** | **Sony (Japan)** |

SOURCE: Private communication from E Davies, London.

| 1950 | **COMPUTER GRAPHICS** | **Burnett (USA)** |

The art of computer graphics can be traced back to the graphics made for wallpaper by Burnett in California from 1937 onwards. These graphics were based on Lissajous figures. But it was Ben F Laposky who, in 1950, really founded the art of computer graphics. Computer graphics are pure products of computer technology, and a few years ago still represented something of a feat (both technically and financially). Today, still graphics are quite common and can be produced on microcomputers. As for animated graphics—which are frequently used in television advertisements, for example—these can be so perfect and real that they are sometimes quite disturbing.

The work of the companies Robert Abel Associates, Digital Equipment Corporation and Sogitec (France) has now become famous in this field.

SOURCE: *Inventions and Discoveries 1993* edited by Valerie-Anne Giscard d'Estaing and Mark Young (New York: Facts on File) p 219

| 1950s | **ULTRASOUND IMAGING** | **I Donald *et al* (UK)** |

The first published experimental ultrasound examinations of women, undertaken by Professor Ian Donald (1910–87) and his colleagues in Glasgow in the late 1950s, used an industrial flaw detector to show that echoes from within the patient's abdomen could be used to measure the size of ovarian cysts and other tumours. This is known as a one-dimensional scan because it shows how far apart things are without producing an image. In two-dimensional scans actual images are produced by the technique of moving the ultrasonic probe in a series of sweeping movements during scanning. Both techniques are now used when you go for a 'scan'.

Professor Stuart Campbell's (b 1936) work using the machine now in the Science Museum, made in about 1967, established many of the procedures which have made ultrasound an indispensable part of routing antenatal care. In a period of heightened awareness of the dangers of X-rays, ultrasound was seen to offer a safe way of examining all pregnant women for multiple foetuses and malformations. For example, in 1969, quintuplets, later successfully delivered, were diagnosed at nine weeks using the two-dimensional scan. The first abortion of a malformed foetus after an ultrasound examination occurred after this machine disclosed a case of amencephaly (the partial or total absence of a brain) in 1972. Diagnosis of spina bifida followed in 1975.

SOURCE: *Ultrasound Scanner: Making of the Modern World—Milestones of Science & Technology* edited by N Cossons *et al* (Published by John Murray, in association with the Science Museum)

1950   **VIDICON—TV Camera Tube**   **RCA (USA)**

The Vidicon, the first TV camera tube to use the principle of photoconductivity by which light of different intensities produces a change in electrical activity, is devised in the USA by RCA. After further developments, the Vidicon proves significantly more adaptable, more sensitive—and cheaper than previous TV cameras.

SOURCE: *The Timetable of Technology* (London: Michael Joseph, Marshall Editions) p 138 (November 1982)

1950   **HAMMING CODE**   **R W Hamming (USA)**

It was simple human frustration that led Dr Richard W Hamming, then a research mathematician at Bell Laboratories, to devise the first method for correcting machine-caused errors in digital computers.

Hamming's technique—which was a result of his research in pure mathematics—enabled computers to spot electrical errors in the data and instructions and tell where they occurred. And, for the first time, it enabled computers to correct those errors and go right on solving problems without interruption.

20 April 1980, marks the 30th anniversary of Hamming's pioneering work, which has evolved into a new field of research called error-correcting codes.

SOURCE: 'Bell Labs marks 30th anniversary of computer error-correcting codes' *Bell Laboratories Record* p 152 (May 1980)

SEE ALSO: 'Applying the Hamming code to microprocessor-based systems' by Ernst L Wall *Electronics* p 103 (22 November 1979)

1950   **COMPUTERS (IBM 650)**   **IBM (USA)**

The IBM 650, an intermediate-size, vacuum-tube computer was considered a workhorse of the industry during the late 1950s. Development began in 1949 and the first installation was made late in 1954. Over a thousand 650s have been in service since then.

The 650 operates serially by character on words of 10 decimal digits plus sign. A 2-out-of-5 decimal representation in storage is translated into a biquinary code in the operating registers, allowing a fixed-count check to detect the presence of more or less than 2 bits per character. The main store of the 650 is a 12 500 rpm 2000-word magnetic drum. A two-address instruction format accommodates, as part of each instruction, the location of the next programme step. This format allows the programmer to place instructions anywhere on the program drum, and makes it possible for him to minimise access times to successive instructions.

SOURCE: 'The evolution of computing machines and systems' by Serrell, Astrahan, Patterson and Pyne *Proc. IRE* p 1050 (May 1962)

SEE ALSO: *The Computer from Pascal to von Neumann* by H H Goldstine (Princeton, NJ: Princeton University Press) p 330 (1972)

1950   **PIN DIODE**   **J-I Nishizawa (Japan)**

The p-i-n sometimes called as the 'pin' diode in which an intrinsic or a high resistivity layer is sandwiched between a p layer and an n layer was invented by J Nishizawa in 1950. In the same year several manufacturing methods of the p-i-n diode, including the thermal diffusion, the chemical treatment, the anodic oxidation, the ion implantation and the nuclear transmutation by the bombardment of high energy particle were also invented by J Nishizaura. In 1958, his research group successfully realized the Si p-i-n diode, in which the characteristics were 2300 V reverse breakdown voltage and 1.5 V forward voltage drop at 100 A, by using the alloying method and the elaborate simple surface passivation technology prior to the Westinghouse Company. Also the 4000 V p-i-n diode was realized by his research group with the development of the original high purity epitaxial growth technology using $SiCl_4$ and the hydrogen system.

The p-i-n diode features a high reverse breakdown voltage up to ten thousands volts, low forward voltage drop and small junction capacitance. It is used all over the world for various application fields such as the industrial high power systems, which includes the famous Japanese bullet train, owing to its efficient high power handling capability and also in consumer electronics products such as TV.

The p-i-n diode is also used as a low loss switching device, a phase shifter and an attenuator in microwave circuits in radar systems and communication equipments.

The p-i-n photo diode also invented by J Nishizawa is an excellent optical detector and is widely used for the recent optical communication system. The fundamental patent also included the application of the high resistivity layer into a transistor as pnip or npin type.

SOURCE: 'Semiconductor device having the high resistivity region' by J Nishizawa and Y Watanabe; Japanese Patent No 205068 (Application date: 20 December 1950)

SEE ALSO: 'Chemical surface treatment method of semiconductor device' by J Nishizawa and Y Watanabe; Japanese Patent No 221722 (Application date: 11 September 1950)

ALSO: Patent Nos. 221695, 223246, 226589, 229685, 235980, 236731 and 226859 (1950)

1950   COMPUTERS (IBM 701)   IBM (USA)

The development of the IBM 701 Data Processing System began at the end of 1950. A model was operating late in 1951 and the first production machine was delivered at the end of 1952. The heart of the IBM 701 system was a 36-bit single, address, binary, parallel, synchronous processor employing vacuum tube flip-flops and diode logic at a rate of one megapulse per second. Multiple pluggable circuit packages were used. The arithmetic registers employed a recirculating-pulse bit-storage circuit, developed for the NORC in which a combination of diode gating and pulse delay made it possible to store, shift right, or shift left with one triode per bit. Computation was governed by a single address stored program of two 18-bit instructions per 36-bit word.

SOURCE: 'The evolution of computing machines and systems' by Serrell, Aastrahan, Patterson and Dyne *Proc. IRE* p 1050 (May 1962)

SEE ALSO: *The Computer from Pascal to von Neumann* by H H Goldstine (Princeton, NJ: Princeton University Press) (1972)

1950s   THERMO-COMPRESSION BONDING   O L Anderson, H Christensen and P Andreatch (USA)

In the 1950s, O L Anderson, Howard Christensen and Peter Andreatch of Bell Laboratories, discovered a new bonding technique particularly useful for connecting transistors to other elements in electronic circuits. The technique, pressing the connecting wire to the transistor mounting at low heat levels, provides a firm bond without introducing undesired electrical properties and has been widely used throughout the electronics industry (see figure 11.20). It is particularly advantageous in avoiding contamination, thus achieving long life and reliability.

SOURCE: 'Mission Communications—the Story of Bell Laboratories' by P C Mabon (Murray Hill, NJ: Bell Laboratories Inc.) p 173 (1975)

1950s   MODEM (MODulation & DEModulation)   MIT Bell Labs (USA)

The modem as we know it today is a product of research done for the defence department by MIT Lincoln Laboratory and Bell Telephone Laboratories in the 1950s. Encouraged by advances in data

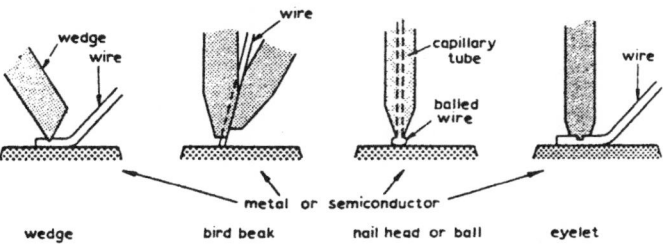

Figure 11.20. Thermo-compression bonding.

processing technology, scientists at these research centres concentrated on finding ways to improve the speed and the accuracy of data transmission using existing facilities of the nation's vast telephone network. A great deal of study went into understanding telephone line parameters and deriving new modulation techniques that could work efficiently with these line parameters. This early work paid off handsomely. In less than 20 years, reliable transmission speeds jumped 700 percent, from 1200 baud in the late 1950s to 9600 bps in the early 1970s.

SOURCE: 'Where are MODEM's going' by J L Holsinger *Telecommunications* p 12 (June 1977)

1950s **FERREED SWITCH**   **Bell Laboratories (USA)**

The ferreed switch, invented at Bell Laboratories in the late 1950s comprises two or four sealed-in-glass contacts and is controlled by magnetised wire coils. Ferreed switches, used to switch phone calls in most electronic switching systems, are smaller, faster operating and require less power than older switching devices.

SOURCE: 'Mission Communications—the Story of Bell Laboratories' by P C Mabon (Murray Hill, NJ: Bell Laboratories Inc.) p 177 (1975)

1950s **APL (A Programming Language)**   **K Iverson (USA)**

APL, which stands for A Programming Language, was developed in the late 1950s by Kenneth Iverson, a professor at Harvard University. He invented this simple, elegant notational system to fill a need for a pithy way to represent mathematical expressions, describe and analyse various topics in data processing, and teach his classes.

In 1960, Iverson joined IBM Corp. There, with the help of Adin Falkoff and other interested researchers, an interpretive version of the language was adapted for the System/360. In 1973, IBM released APLSV. The appended SV stands for Shared Variables—a means whereby a number of users may communicate information. More recently, the language has surfaced on less expensive machines: IBM's 5100 series of business computers, Digital Equipment Corp.'s DEC system 2020, Hewlett-Packard Co.'s 3000, and the newly introduced Interactive Computer Systems Inc.'s System 900, to name a few of the computers in question.

APL's primitive functions, of which there are about 60, fall into two categories, scalar and mixed. Scalar functions can be used with scalar arguments and arrays on an item-by-item basis. Mixed functions apply to arrays with various ranks and may produce results that vary from the original arguments in rank and shape. The scalar functions can be subclassified as monadic and dyadic, which are defined for one and two arguments, respectively. The primitive operators, which currently number five, modify the action of scalar dyadic functions and some mixed functions, resulting in a great number of new functions.

APL uses alphanumerics, Greek letters, and some uncommon mathematical symbols to represent the functions and operators. These make APL programs appear cryptic to the beginner; in fact, the language is easy to learn. With a little practice, powerful routines can be generated with a few simple key strokes.

1951   **QUALITY CONTROL**   **J M Juran (USA)**

First edition of 'Quality Control Handbook' written by J M Juran, published by McGraw-Hill, New York.

An authorative treatise on all aspects of quality control.

1951   **AUTOMATIC CIRCUIT ASSEMBLY ('TINKER**   **National Bureau of Standards (USA)**
       TOY' System)

In 1950 the Navy Bureau of Aeronautics asked the National Bureau of Standards to study further automation of circuit assembly. The process that followed in 1951—developed by Robert Henry of the Bureau of Standards was dubbed Project Tinkertoy. It provided for the automatic assembly and inspection of circuit components, and it led to the first modular package (see figure 11.21).

The system started with individual components mounted on steatite ceramic wafers 7/8-inch square by 1/16-inch thick. The components were machine-printed or mounted over printed wiring. Four to six wafers were then automatically selected, stacked and mechanically and electrically joined by machine-soldered riser wires, which were attached at notches along the sides of each wafer. The resulting module generally had a tube socket on the top wafer.

Figure 11.21. A Tinkertoy assembly model.

Though this modular approach to packaging was used for production items, it faded in the late 1950s as the transistor began to replace the vacuum tube.

SOURCE: 'Solid state devices—packaging and materials' by R L Goldberg *Electronic Design* vol 24, p 126 (23 November 1972)

1951   **MICROPROGRAMMING COMPUTERS**   **M V Wilkes (UK)**

Credit for the original microprogramming concept is generally given to Britisher M V Wilkes of Cambridge University's Mathematical Laboratory. In a paper he presented at Manchester University's Computer Conference in July 1951, Wilkes discussed 'The Best Way to Design an Automatic Calculating Machine'. Wilkes' intention, ironically enough, was to simplify the design of a hardwires machine. Today microprogramming is used to replace hardwired logic altogether.

As Samir Husson notes in his book 'Microprogramming Principles and Practices', the technique attracted little attention before the 1960s because it was too expensive to implement. The first commercial microprogrammed computer was IBM's 7950, introduced in 1961. Other early appearances of microprogramming were in such machines as the IBM System/360, the RCA Spectra/70, and the Honeywell H4200.

A variety of memory technologies were used for the read-only memories needed for the microinstruction store. Among these were traditional ferrite cores, cores cut into an E-shape to create a transformer memory, and arrays of diodes or capacitors. A novel approach was taken by IBM, which devised a card capacitor ROM for several of its System/360 models.

Microprogramming came to minicomputers early in the 1970s in units such as Microdata's Micro 800, which used diode arrays as ROMs, and the later 3200, Hewlett-Packard's 2100 family, and Digital Equipment's PDP-11 line. The increasing availability of low-cost and fast semiconductor ROMs, however, has recently caused the technique to boom in popularity.

Virtually all mainframes and minicomputers today are microprogrammed. Many of the machines now use random-access memory to store the microinstruction making it easier for the user or the manufacturer to change the machine. Also, the ease with which microprogramming allows one machine to emulate another has resulted in its widespread use in the IBM-compatible computer area.

SOURCE: 'How microprogramming started' by A Durniak *Electronics* p 126 (9 November 1978)

1951   **IMAGE ANIMATION**   **MIT (USA)**

Computerised image animation was first experimented on at the Massachusetts Institute of Technology in 1951. But it was not until the early 1960s that the potential of the technique was fully understood. Today it is used in the fields of medicine, architecture (with models in three dimensions), space exploration and chemistry.

SOURCE: 'Inventions and Discoveries 1993' edited by Valerie-Anne Giscard d'Estaing and Mark Young *Facts on File* New York p 219

1952   **ALLOYED TRANSISTOR**   **RCA (USA)**

In 1952 an announcement was made by the Radio Corporation of America that they had successfully made transistors by heating small quantities of impurities, on the surface of germanium chips, so as to make alloy regions with the germanium. The resulting penetration of the chips was made from both sides.

SOURCE: *Semiconductors for Engineers* by D F Dunster (London: Business Books Limited) p 20

1952   **MICROELECTRONICS (Integrated Circuit Concept)   G W A Dummer (UK)**

In a paper read at the IRE Symposium in Washington DC on 5 May 1952, entitled: 'Electronic components in Great Britain' G W A Dummer stated:

'At this stage, I would like to take a peep into the future. With the advent of the transistor and the work in semiconductors generally, it seems now possible to envisage electronic equipment in a solid block with no connecting wires. The block may consist of layers of insulating, conducting, rectifying and amplifying materials, the electrical functions being connected directly by cutting out areas of the various layers'.

SOURCE: *Proc. IRE Symposium on 'Progress in Quality Electronic Components* Washington, DC, p 19 (May 1952)

SEE ALSO: 'A history of microelectronics development at the Royal Radar Establishment' by G W A Dummer *Microelectronics and Reliability* vol 4, p 193 (1965)

'Solid circuits' *Wireless World* (November 1957)

'The semiconductor story—3; solid circuits—a new concept' by K J Dean and S White *Wireless World* p 137 (March 1973)

'The genesis of the integrated circuit' by M F Wolff *IEEE Spectrum* p 45 (August 1976)

1952   ZONE MELTING OF GERMANIUM AND SILICON   W G Pfann (USA)

W G Pfann discovered a simple method for repeating the action of normal melting and freezing, which avoided handling the material between each operation. This resulted in material of extremely high purity which was then grown into single crystals bv the pulling technique. Pfann also developed the zone levelling technique, which distributes impurities uniformly through a rod. He grew single crystals in his zone levelling apparatus using seeding techniques. The combination of zone levelling and horizontal growth of single crystals has become the standard technique used in today's transistor manufacturing operations.

SOURCE: 'Contributions of materials technology to semiconductor devices' by R L Petriz *Proc. IRE* p 1027 (May 1962)

SEE ALSO: 'Segregation of two solutes, with particular reference to semiconductors' by W G Pfann *J. Metals* vol 4, p 861 (August 1952)

'Techniques of zone melting and crystal growing' *Solid State Physics* (New York: Academic Press) vol 4, p 423 (1957)

1952   DARLINGTON PAIRS, DIRECT-CONNECTED   S Darlington (USA)
TRANSISTOR CIRCUIT

In one illustrative embodiment of this invention, a translating device comprises a pair of similar junction transistors, the collector zones of which are electrically integral and the base zone of one of which is tied directly to the emitter zone of the other. Individual connections are provided to the other emitter and base zones. The device constitutes an equivalent single transistor having emitter and collector resistances substantially equal to those of one of the component transistors, but having a current multiplication factor substantially greater than that of either of the components.

SOURCE: 'Semiconductor Signal Translating Device' US Patent No 2663806, Bell Telephone Laboratories (dated 9 May 1952)

1952   DIGITAL VOLTMETER   A Kay (USA)

The digital revolution started in 1952, when Andy Kay unveiled the first digital voltmeter. The model 419 was crude, compared with today's DVMs. But both the idea and Non Linear Systems—the company formed around the idea—took off like a rocket, Today, almost every instrument from signal generators to multimeters to scopes is digitised, thanks to Andy Kay and to the commercial digital readout tube, introduced by Burroughs (then Haydu) just one year before. (Burroughs' familiar Nixie tube actually had a rival in its early days—the Inditron, which was developed by National Union Radio Corporation and which did not survive.)

SOURCE: 'Solid state devices—instruments' by S Runyon *Electronic Design* vol 24, p 102 (23 November 1972)

1952   NEGATIVE FEEDBACK TONE CONTROL   P J Baxandall (UK)
CIRCUIT

The circuit to be described is the outcome of a prolonged investigation of tone-control circuits of the continuously-adjustable type, and provides in dependent control of bass and treble response by means of two potentiometers, without the need for switches to change over from 'lift' to 'cut'. Unusual features are the wide range of control available, and the fact that a level response is obtained with both potentiometers at mid-setting. The treble-response curves are of almost constant shape, being shifted along the frequency axis when the control is operated, and there is practically no tendency for the curves to 'flatten off' towards the upper limit of the audio range. The shape of the bass-response curves, though not constant, varies less than with most continuously-adjustable circuits.

SOURCE: 'Negative-feedback tone control' by P J Baxandall *Wireless World* p 402 (October 1952)

1952   **COMPUTERS (SAGE)**   **IBM, MIT (Lincoln Labs) (USA)**

Descendents of MIT's 'Whirlwind I' and of the IBM 701, the AN/FSQ-7 air defence computers for the SAGE system began, in 1952, as a co-operative IBM–MIT Lincoln Laboratories effort based upon previous studies and specifications by Lincoln Laboratories SAGE, a real-time communication-based digital computer control system, accepts radar data over phone lines, processes, displays information for operator decisions and guides interception weapons. The first engineering model of the computer was delivered by IBM in 1955, and production deliveries began in June 1956.

SOURCE: 'The evolution of computing machines and systems' by Serrell, Astrahan, Patterson and Pyne *Proc. IRE* p 1051 (May 1962)

1952   **TRANSISTORS (Single Crystal Fabrication—Silicon)   G K Teal and E Buehler (USA)**

Large single crystals of silicon and silicon p-n junctions were prepared by Teal and Buehler by an extension of the pulling technique developed for germanium. These crystals were used by G L Pearson to prepare p-n junction diodes by the alloy method.

SOURCE: 'Contributions of materials technology to semiconductor devices' by R L Petritz *Proc. IRE* p 1028 (May 1962)

SEE ALSO: 'Growth of silicon single crystals and of single crystal p-n junctions' by G K Teal and E Buehler *Phys. Rev.* vol 8, p 190 (July 1952)

1953   **TRANSISTOR (Unijunction)**   **UJT GEC (USA)**

Engineers at General Electrics electronics advanced semiconductor laboratory in Syracuse, NY, experimented in the early 1950s with germanium alloy tetrode devices in search of a semiconductor that could be used at frequencies higher than the 5-megahertz operating level of existing conventional bipolar transitory The tetrode structures, it was found, produced a transverse electric field that boosted the device's cutoff frequency and lessened the semiconductor's input impedance—the limiting factor regarding frequency.

But in 1953, while examining the waveforms on the structure's terminals, I A Lesk of the laboratory noticed that an oscillatory signal was present on the tetrode's emitter. And when the collector supply was removed, the oscillations persisted for a while. The researchers realised that they had stumbled on a new switching-type device. The GE engineers added to the development tetrode models, a double-based diode structure that was being studied because of a negative-resistance property.

SOURCE: 'Solid State—a switch in time' by W R Spofford Jr and R A Stasior *Electronics* p 118 (19 February 1968)

1953   **TRANSISTOR (Surface Barrier)**   **Philco (USA)**

Advancing the early trend toward higher frequencies, Philco developed the jet-etching technique in 1953. Here electrochemical machining was used to fabricate the necessary thin base layers. A major product of this process was the surface-barrier transistor, which boosted the upper frequency limit of transistors into the megahertz region.

SOURCE: 'Solid state devices—the processes' by E A Torrero *Electronic Design* vol 24, p 73 (23 November 1972)

1953   **COMPUTERS (IBM 704, 709 and 7090)**   **IBM (USA)**

The development of the IBM 704, descended from the 701, began in November 1953 and the first system was delivered in January 1956. The 704 featured higher speed, magnetic-core memory, floating point and indexing. It was followed in 1958 by the 709, featuring simultaneous read, write and compute by means of Data Synchronizer units that allowed the input–output channels to operate independently, as well as several special operations, including a table look–up instruction and indirect addressing. A 32 768-word core memory was installed on a 704 in April 1957. The IBM 7090, the first units of which were delivered in 1959, is a transistorised system compatible with the 709. About five times faster than the 709 on typical problems, the 7090 incorporates a 2.18-μs core memory and improved magnetic-tape units.

SOURCE: 'The evolution of computing machines and systems' by Serrell, Astrahan, Patterson and Pyne *Proc. IRE* p 1052 (May 1962)

SEE ALSO: *The Computer from Pascal to von Neumann* by H H Goldstine (Princeton, NJ: Princeton University Press) (1972)

1953	**AUTOMATIC ASSEMBLY SYSTEMS**	**Autofab (General Mills), IBM Machine (IBM), United Shoe Machinery Co., GE/Signal Corps. and Mini-Mech (Melpar) (USA)**

In attempting to develop standards which would allow unrestricted electronics design and production processes sufficiently flexible to permit rapid transition from design to manufacturing, the following conclusions are reached:

1. The use of a printed-circuit board is basic to practically all automatic approaches.

2. A generic grid pattern to govern circuit layout must be used in order to prevent obsolescence of tooling (0.1 and 0.025 in are commonly accepted).

3. In conjunction with the standard grid pattern, standard mounting dimensions for all components and parts must be used.

The use of single-head component-insertion machines will satisfy these fundamental requirements. The flexibility of such a machine permits the insertion of a variety of components by recycling the printed-circuit modules through the unit for each component. When higher volumes are obtained, several machines can be utilized. By this evolutionary method one is able to create any degree of automation which the manufacturer wishes to attain. It is possible to stop with the production ot the printed circuit and to use one or several component-attaching or inserting machines. Dip-soldering can be mechanised or performed manually. It would be impractical to consider taking the giant step into completely automatic production at the beginning of the factory's use of automation principles. Even for mass producers, it is necessary to develop specific procedures for particular requirements documented by many years of experience in the step-by-step evolution toward a mechanised operation (see figure 11.22).

SOURCE: *Electronic Equipment Design and Construction* by G W A Dummer, C Brunetti and L K Lee (New York: McGraw-Hill) p 185–6 (1961)

SEE ALSO: Proceedings of Symposium on Automatic Production of Electronic Equipment, sponsored by Stanford Research Institute and the US Air Force (April 1954)

1953	**MASER (Microwave Amplification by Stimulated Emission of Radiation)**	**C H Townes and J Weber (USA), N G Basov and A M Prokhorov (USSR)**

The first clear recognition of the possibility of amplification of electromagnetic radiation by stimulated emission seems to have been by a Russian, Fabrikant, who filed a patent in 1951 (although it was not published until 1959) and who had discussed various aspects of his thesis of 1940. However, his attempts to produce optical amplification in caesium were unsuccessful.

The first statement in the open literature about amplification was by Weber in 1953, followed by the detailed proposals of Basov and Prokhorov for a beam-type maser in 1954. However, the real excitement was caused by the short article of Gordon, Zeiger and Townes, in the same year, announcing the operation of the first maser using ammonia. Townes had conceived the required experimental arrangement three years earlier, based on his experience in microwave spectroscopy. In the years immediately following many other techniques were studied, but the only one to give any degree of practical success was the three-level maser of Bloembergen which resulted in the ruby maser amplifier.

Figure 11.22. Sylvania in-line assembly system (courtesy Sylvania Corp.).

SOURCE: 'Lasers and optical electronics' by W A Gambling *The Radio & Electronic Engineer* vol 45, No 10, p 538 (October 1975)

SEE ALSO: 'Mekhanizm izlucheniya gazovogo razryada' by V A Fabrikant in *Elektronnye i ionnye pribory* (Electron and ion devices); Trudy Vsesoyuznogo Elektrotekhnicheskogo Instituta (Proceedings of the All-Union Electrotechnical Institute) vol 41, p 236 (1940)

'Evolution of masers and lasers' by B A Lengyel and V A Fabrikant *Am. J. Phys.* vol 34, p 903 (1966)

'Amplification of microwave radiation by substances not in thermal equilibrium' by J Weber *IRE Trans. on Electron Devices* vol PGED-3, pp 1–4 (June 1953)

'Application of molecular beams to radio spectroscopic studies of rotation spectra of molecules' by N G Basov and A M Prokhorov *J. Exp. Theor Phys. (USSR)* vol 27, pp 431–8 (1954)

'Molecular microwave oscillator and new hyperfine structure in the microwave spectrum of NH_3' by J P Gordon, H J Zeiger and C H Townes *Phys. Rev.* vol 95, pp 282–4 (1 July 1954)

'Proposals for a new type solid-state maser' by N Bloembergen *Phys. Rev.* vol 104, pp 324–7 (15 October 1956)

'Forgotten inventor emerges from epic patent battle with claim to laser' *Science* vol 198, p 379 (28 October 1977)

1953   **CONNECTION TECHNIQUES (Wire Wrapped**   **R F Mallina *et al* (USA)**
        Joints)

The mechanical basis of the wire wrapped joint (see figure 11.23) was investigated very extensively by workers at Bell Telephone Laboratories 20 years ago. Their very full analysis of the joining system is still considered to be essentially correct. The work includes photoelastic observations on a wrapped joint model to investigate strain patterns produced by wrapping, and the study of stress relaxation as a function of time and temperature.

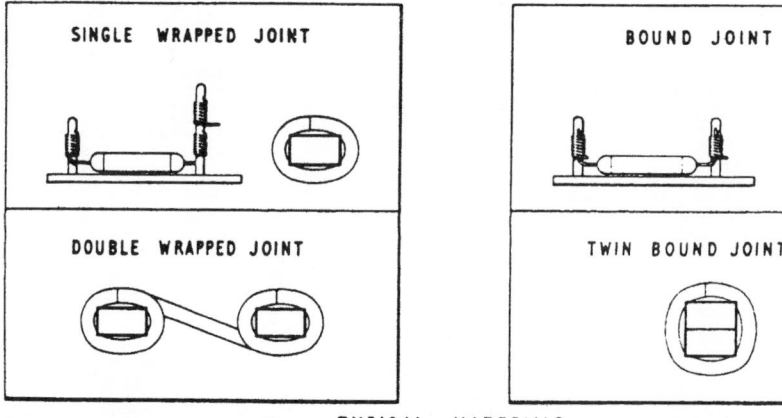

Figure 11.23. Wire-wrapped joints.

The wire wrapped joint consists of a wire which is tightly wrapped around a sharp cornered terminal. Sufficient deformation is engendered in the many notches created by the terminal in the wrapping wire to create metal to metal interfaces with a high level of integrity. The wrapping wire, which is bent several times during wrapping before final positioning in the wrap, is under a high level of tensile stress during wrapping. The tensile strain which is caused remains in the wire after wrapping because the stretched wire is locked by the notches formed in it.

SOURCE: 'Wire wrapped joints—a review' by P M A Sollars *Electrocomponent Science and Technology* vol 1, p 17 (1974)

SEE ALSO: 'Solderless wrapped connections' *The Bell System Technical Journal* (May 1953)

Introduction: J W McRae

Part 1 Structure & Tools by R F Mallina

Part 2 Necessary conditions for obtaining a permanent connection by W P Mason and T F Osmer

Part 3 Evaluation & Performance tests by R H Van Horn

1953   **TRANSISTORS (Floating Zone Refining of Silicon)**   **P H Keck, R Emeis and H C Theurer (USA)**

An improved silicon purification technique was developed which produced material of sufficient quality that alloy silicon transistors could be fabricated with good yields. This was a novel variation of zone refining called 'floating zone refining' developed by P H Keck and independently by R Emeis and H C Theurer. This operation employs a vertical system and uses surface tension to support a stable liquid zone formed by induction heating. Hence, the crucible is completely eliminated. Silicon with thousands of ohm-cm resistivity and minority carrier lifetimes of greater than 100 μs can be produced by this method.

SOURCE: 'Contributions of materials technology to semiconductor devices' by R L Petritz *Proc. IRE* p 1028 (May 1962)

SEE ALSO: 'Crystallisation of silicon from a floating liquid zone' by P H Keck and M J E Golay *Phys Rev.* vol 89, p 1297 (March 1953)

'Growing single crystals without a crucible' *Z. Naturforsch.* vol 9A p 67 (January 1954)

'Removal of boron from silicon by hydrogen water vapour treatment' *J. Metals* vol 8, p 1316 (October 1956)

1954   **TRANSISTOR (Interdigitated)**   **N H Fletcher (USA)**

The father of the interdigitated transistor is N H Fletcher, an engineer with Transistor Products Inc. When, in 1954, he hit upon the idea of elongated emitter areas, Fletcher was seeking a means to increase the power handling capability of devices, not a way to boost their cut-off frequency levels.

His discoveries were applied by other firms to most transistor types over the next decade, but his own company realised few benefits from his work.

SOURCE: 'Solid state—fingers in the die' by J E Tatum *Electronics* p 94 (19 February 1968)

SEE ALSO: 'Some aspects of the design of power transistors' by N H Fletcher *Proc. IRE* p 551 (May 1955)

1954   **TRANSISTOR RADIO SET**   **Regency (USA)**

In 1954 the first transistor radio, the Regency, appeared on the market. Although not a commercial success, it introduced the transistor into the consumer market and gave transistor makers the impetus they needed to develop mass production techniques. That, coupled with an awakening of interest by the military, increased transistor sales meteorically in the mid-fifties.

SOURCE: 'Silicon, germanium & silver—the transistor's 25th anniversary' by C P Kocher *The Electronic Engineer* p 30 (November 1972)

1954   **SOLAR BATTERY**   **D M Chapin, C S Fuller and G L Pearson (USA)**

As an outgrowth of work on transistors, Bell Laboratories scientists D M Chapin, C S Fuller and G L Pearson in 1954 invented the silicon solar battery—an efficient device for converting sunlight directly into electricity. Arrays of these devices are used to power satellites and as energy sources for other uses.

SOURCE: *Mission Communications—the Story of Bell Laboratories* by P C Mabon (Murray Hill, NJ: Bell Laboratories Inc.) p 172 (1975)

1955   **CRYOTRON**   **D Buck (USA)**

A super-conducting switching element was first examined by de Haas and Casimir-Jonker in 1935; however, it was not until 1955 that Buck demonstrated a practical device which he called the cryotron. The basic principle of the cryotron depends on the existence of a critical magnetic field above which the superconducting metal becomes a normal conductor. The original cryotron utilized a small tantalum rod wound with a niobium wire At liquid helium temperatures the tantalum wire has a critical field of the order of several hundred gauss, whereas that of niobium is of the order of 2000 gauss. Consqently, when the niobium wire is pulsed with a suitable current, the magnetic field that it creates is sufficient to destroy the superconductivity in the tantalum but not in itself. The current in the niobium wire can be smaller than that in the tantalum wire so that a small current can control a larger one, thus producing a current gain in the device.

SOURCE: 'Solid state devices other than semiconductors' by B Lax and J G Mavroides *Proc. IRE* p 1016 (May 1962)

SEE ALSO: 'The cryotron—a superconductive computer component' by D Buck *Proc. IRE* vol 44, p 482 (April 1956)

1955   **INFRA-RED EMISSION FR OM GALLIUM**   R Braunstein (USA)
       ARSENIDE SEMICONDUCTORS

Radiation produced by carrier injection has been observed from GaSb, GaAs, InP and the Ge;Si alloys at room temperature and 77 K. The spectral distributions of the radiation are maximum at energies close to the best estimates of the band gaps of these materials; consequently, the evidence is that the radiation is due to the direct recombination of electron–hole pairs.

SOURCE: 'Radiative transitions in semiconductors' by R Braunstein *Phys. Rev.* vol 99, p 1892 (1955)

1956   **DIFFUSION PROCESS**   C S Fuller and H Reis (USA)

The next major advance in device technology was the diffusion process. Research on the diffusion of III–V impurities into germanium and silicon by Fuller at the BTL, and by Dunlap at GE laid the foundation for transistor fabrication using diffusion as a key process step. The BTL was the first to fully integrate these results into germanium and silicon transistors.

Diffusion techniques have proved to be one of the best controlled methods for preparing p-n junctions. Because the common doping impurities diffuse very slowly in semiconductors at rates which can be varied by adjusting temperatures, close control and reproducibility of the impurity distributions can be achieved. Hence, control over the electrical parameters of the resulting devices may be maintained. The ability to form base regions only a fraction of a micron thick allows very high-frequency transistors to be fabricated.

SOURCE: 'Contributions of materials technology to semiconductor devices' by R L Petritz *Proc. IRE* p 1029 (May 1962)

SEE ALSO: 'Diffusion processes in germanium and silicon' by H Reis and C S Fuller. Chaper 6 of *Control of composition in semiconductors by freezing methods* edited by N B Hannay. (New York: Reinhold Pub. Corp.) (1959)

1956   **SOLID ELECTROLYTE CAPACITOR**   D A McLean and F S Power (USA)

From the time transistors began to be produced commercially, the need for a solid electrolytic capacitor as a coupling capacitor or a bypass capacitor for electronic equipment has increased. Since McLean and Power announced the development of the tantalum solid electrolytic capacitor in 1956, studies of this capacitor began to appear with great frequency in literature through the world and applications of the solid capacitor gradually began to spread.

SOURCE: 'Miniaturised aluminium solid electrolyte capacitors using a highly effective enlargement technique' by K Hirata and T Yamasaki *IEEE Trans. on Parts, Hybrids & Packaging* vol PHP-12 No 3, p 217 (September 1976)

SEE ALSO: 'Tantalum solid electrolyte capacitor' by D A McLean and F S Power *Proc. IRE* vol 44, p 872 (July 1956)

1956   **VALVES—VAPOUR COOLING**   C Beutheret (France)

The first vapour-cooled tubes were made by Beutheret who used an anode with teeth approximately 10 mm square tapering to 5 mm square over 20 mm protruding from the surface. The object of the teeth was to stabilise the anode temperature and prevent a sudden catastrophic increase in anode temperature known as calefraction.

SOURCE: *Electronic Engineer's Reference Book* (London: Newnes-Butterworth) Chapter 7, pp 7–47 (1976)

SEE ALSO: 'The Vaportron Technique' by C Beutheret *Rev. Tech. Thomson-CSF* vol 24 (1956)

1956   **'FLOWSOLDERING' OF PRINTED CIRCUITS**   Fry's Metal Foundries Ltd (UK)

In the Flowsolder dipping unit, developed by Fry's Metal Foundries Ltd which avoids some of the difficulties inherent in the conventional flat dipsoldering of printed circuits, a stationary wave of molten solder is created by pumping the metal upwards through a rectangular nozzle and the pre-fluxed circuit panels are passed through the crest of the wave. It is claimed that this unit, which is used with specially developed fluxes, facilitates the soldering of printed circuits, free from faulty joints or bridging (see figure 11.24).

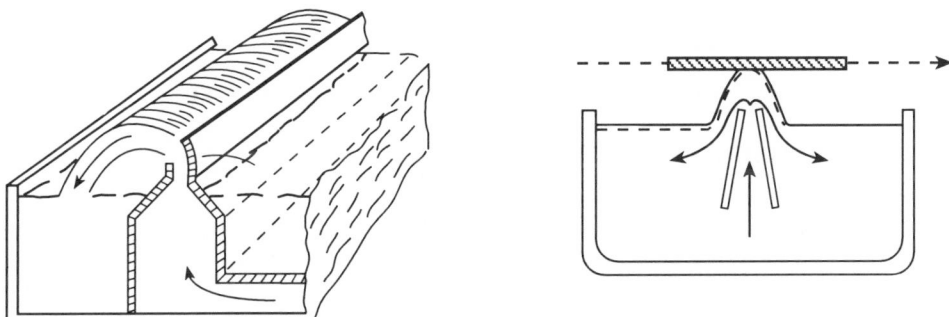

Figure 11.24. Two systems of wave-soldering.

SOURCE: 'Flowsolder Method of Soldering Printed Circuits' R Strauss & A F C Barnes *Electronic Engineering* vol 28, No 345, pp 494–6 (November 1956)

1956	**SEMICONDUCTOR DIODE FUNCTION CAPACITOR**	**L J Giacoletto and J O'Connell (USA)**

A semiconductor junction when biased in the reverse (non-conducting) direction is a capacitance which can be varied by the bias voltage. When biased in this direction the mobile charge carriers are moved away from the junction, leaving uncompensated fixed charges in a region near the junction. The width, and hence the electrical charge of this space-charge layer, depends on the applied voltage, thus giving rise to a junction transition capacitance.

SOURCE: 'History, present status and future developments of electronic components' by P S Darnell *IRE: Transactions on Component Parts* p 128 (September 1958)

SEE ALSO: 'A variable capacitor germanium junction diode for UHF' by L J Giacoletto and J O'Connell *RCA Review* vol 17, p 68 (March 1956)

1956	**TRANSATLANTIC TELEPHONE CABLE (TAT-1)**	**(UK/USA)**

25 September 1956 was an auspicious date for international telecommunications being the day that the first transatlantic telephone cable (TAT-1) entered service.

The history of the Atlantic cables started in 1858 with the laying of the first telegraph cable which did not have a very long life but did prove the feasibility of the operation. By 1956 there were 28 transatlantic telegraph cables.

As far as a telephone service was concerned the main problem lay with the fact that repeaters were necessary to amplify the already weak signals on their long journey, By 1920 repeatered telephone links were in the course of preliminary experiment. In 1946 a repeatered link was laid between the United Kingdom and mainland Europe. With this impetus development work was brought to the point where it was felt that the repeaters would exhibit the 20 year life required of them for deep water service.

Consequently, TAT-1 runs from Oban in Scotland under the Atlantic to Clarenville in Newfoundland, using 51 American made repeaters. Then the line goes overland to Terenceville where again it takes to the water for its journey to Sydney Mines in Nova Scotia. In the latter run the cable uses 16 British made repeaters, the first two of which are buried in Newfoundland.

For the cable's 36 telephone channels two cables are used, one for each direction, in the transatlantic section while from Clarenville the cable becomes a 60-circuit single cable system, connecting Newfoundland to Canada.

In the main section one telephone channel had been reserved as a United Kingdom–Canada telegraph link. From Sydney Mines, the circuit takes a short radio hop to Spruce Lake where it is divided to allow circuits to be directed to Montreal and New York.

Although in 1956 the 36-channel cable was hailed as an historical achievement only 20 years later the latest in the family of TAT cables, TAT-6, sports over 100 times the capacity while TAT-7 and TAT-8 are already being considered.

SOURCE: "TAT-1' 20 years old' *ITU Telecommunication Journal* vol 43, XII, p 734 (1976)

1956   MAGNETIC MATERIAL—YTTRIUM IRON   F Bertaut and F Forrat (France)
GARNET

The crystallographic structure of yttrium iron garnet (YIG) was discovered by Bertaut and Forrat in 1956 and very soon afterwards large bulk crystals (several centimetres in one dimension) began to be grown by the molten flux technique. Yttrium iron garnet was found to be cubic, to be essentially an insulator, to have a saturation magnetisation of 1750 gauss, and, unlike the previously available ferrite materials, to have a ferrimagnetic resonance linewidth of the order of 1 oersted at 10 GHz.

SOURCE: 'Epitaxial magnetic garnets' by J H Collins and A B Smith *The Radio and Electronic Engineer* vol 45, No 12 p 707 (December 1975)

SEE ALSO: 'Structure of ferrimagnetic ferrites of rare earths' by F Bertaut and F Forrat *C. R. Acad. Sci. Paris.* vol 242, p 382 (1956)

1956   TRANSISTORIZED COMPUTER   Bell Laboratories (USA)

Bell Telephone Laboratories in New York, the place where the transistor was invented in 1947, builds the Leprechaun, the first experimental transistorized computer. The on–off switching transistor fathers a new breed of more reliable, more economical machines. IBM, Philco and General Electric quickly follow suit with 'second generation' computers.

SOURCE: *The Timetable of Technology* (London: Michael Joseph, Marshall Editions) p 153 (November 1982)

1956   RADIO PAGING   Multitone (UK)

Concurrently with the development of two-way radio communication, there has been a remarkable development of radio-paging equipment and services. The first paging systems were established in the mid-1950s using a magnetic loop around the building to be served and operated on very low frequencies around 70 kHz. One of the first of these systems was installed by the Multitone Company at St Thomas's Hospital, London in 1956. Later the technique changed to v.h.f. radiating system using frequencies in the 27, 150 and 450 MHz bands. Development is now very widespread with over 2000 systems and 100 000 paging receivers in use in Britain alone. The receivers involved are very small, weighing only a few ounces.

SOURCE: 'Fifty years of mobile radio' by J R Brinkley *The Radio and Electronic Engineer* vol 45, No 10, p 556 (October 1975)

1957   FULL FREQUENCY-RANGE ELECTROSTATIC   P J Walker (UK)
LOUDSPEAKER

Much work was done on electrostatic loudspeakers in the 1920s and 30s, but all early workers seem to have made the mistake of assuming the basic principle to be necessarily that the force between two capacitor plates is proportional to the square of the voltage between them. Inventors sought to overcome the non-linearity distortion inherent in this square-law relationship by employing push-pull arrangements, in which the diaphragm, at a high and constant DC voltage, was placed between a pair of fixed perforated electrodes across which the audio programme voltage was applied. But linearity was only achieved if the vibration amplitude was kept small compared with the plate spacing, and this in practice confined all these early electrostatic loudspeakers to high audio frequencies only.

Professor F V Hunt of Harvard University seems to have been the first to appreciate that the above non-linearity could theoretically be totally removed, even for very large diaphragm amplitudes, if the electrically conductive diaphragm had a constant electric charge on it instead of being held at a constant voltage. The principle is then simply that the force on a constant charge is proportional to the product of the charge and the strength of the electric field in which it is situated, There is a tacit assumption, however, that all parts of the diaphragm move equally. This does not happen in practice, causing charge to move about on the diaphragm surface, thus preventing each part of the diaphragm from operating under the desired constant-charge condition that gives low distortion.

P J Walker and D T N Williamson discovered that this latter major cause of distortion could be removed by making the diaphragm of very thin plastic film, treated to be sufficiently conductive to allow it to be charged up in a reasonable time, but not sufficiently to allow any significant moving about of charge during a low-frequency audio cycle. It then became possible to achieve large acoustic outputs at bass frequencies with very low distortion. The plastic diaphragm also greatly reduced the danger of spark damage occurring at high signal levels, and led ultimately, after many other engineering problems had been solved, to the marketing of the first successful full-frequency-range electrostatic loudspeaker in 1957.

Recent improvements by P J Walker have been largely concerned with making the polar radiation characteristic of the loudspeaker vary less with frequency than in the earlier designs, and in a smooth and controlled manner. The electrode area is divided into a number of annular sections, the signals being fed to the outer sections via a delay line. The wavefront radiated is then as if from a point source located behind the loudspeaker, but since the actual radiating area is quite large, high volume levels can be produced.

SOURCE: Communication from P J Baxandall, Malvern (22 July 1982)

SEE ALSO: Walker P J 'Wide Range Electrostatic Loudspeakers' *Wireless World* (May, June, August 1955)

Walker P J 'New Developments in Electrostatic Loudspeakers' *J. Audio Eng. Soc.* (November 1980)

Walker P J and Williamson D T N 'Improvements Relating to Electrostatic Loudspeakers' British Patent No 815, 978. (Application dated 20 July 1954. Complete filed 19 October 1955. Complete published 8 July 1959)

Hunt F V *Electroacoustics* Chapter 6 (Cambridge, MA: Harvard University Press/John Wiley) (1954)

1957   PLUMBICON TV CAMERA TUBE   Philips (Holland)

Philips introduce the Plumbicon TV camera tube, a greatly improved version of its predecessor, the Vidicon. In Britain the BBC bases all its plans for the development of colour TV on the new tube, which soon becomes universal in colour TV design.

SOURCE: *The Timetable of Technology* (London: Michael Joseph, Marshall Editions) p 157 (November 1982)

1957   PLATED-WIRE MEMORIES   U F Gianole (USA)

The plated-wire memory uses the principle of the direction of magnetization in a material to store digital information. The original concept of the wire memory was invented in 1957 by U F Gianole of Bell Laboratories. Plated-wire memories require no standby power, are non-volatile, inexpensive to manufacture and will work in a high electrical noise environment.

SOURCE: *Mission Communications—the Story of Bell Laboratories* by P C Mabon (Murray Hill, NJ: Bell Laboratories) p 179 (1975)

1957   RESISTORS (Nickel-Chromium Thin Film)   R H Alderton and F Ashworth (UK)

It is well known that nichrome is the material most used today in thin-film resistors. Alderton and Ashworth stressed the importance of the following parameters: the source temperature, the degree of vacuum maintaining in the system, and the temperature of the receiving surface during deposition from the vapour phase They state that stable films can only be made if the substrate temperature is greater than 350°C and the vacuum in the system better than 10^{-4} torr. They also stated that the maximum surface resistivity that produced stable films was $300\Omega/\square$. These results are still used today as a guide in the production of nichrome films. Alderton and Ashworth measured the reistivity as a function of thickness and obtained a temperature coefficient of resistance from 100–200ppm/°C.

SOURCE: 'Resistive thin films and thin film resistors—history, science and technology' by J A Bennett *Electronic Components* p 748 (September 1964)

SEE ALSO: 'Vacuum deposited films of a nickel-chromium alloy' by R H Alderton and F Ashworth *Brit. J. Applied Physics* vol 8, p 205 (1957)

1957   SPUTNIK 1 Satellite   (USSR)

Launched 4 October 1957. First artificial satellite. Study of ionosphere, radio wave propagation. Batteries. Transmitted for 21 days. Decayed on 4 January, 1958.

SOURCE: Table of Artificial Satellites Launched Between 1957 and 1976 (Geneva: International Telecommunication Union) (1977)

1957   SPUTNIK 2 Satellite   (USSR)

Launched 3 November 1957. Study of ultraviolet rays and X-rays from the sun. Study of cosmic rays. Medico-biological study of the dog Laika. Transmitted for seven days. Decayed on 14 April 1958.

SOURCE: Table of Artificial Satellites Launched Between 1957 and 1976 (Geneva: International Telecommunication Union) (1977)

1957   TRANSISTORS (Oxide Masking Process)   C J Frosch (USA)

Another important technological advance in this period was the development of oxide masking for silicon by C J Frosch of BTL. He observed that a thermally grown oxide on silicon impeded the diffusion of certain impurities, including boron and phosphorus. This technique, coupled with photograshic masking against etching, provides a powerful tool for silicon processing.

SOURCE: 'Contributions of materials technology to semiconductor devices' by R L Petritz *Proc. IRE* p 1030 (May 1962)

SEE ALSO: 'Surface protection and selective masking during diffusion in silicon' by C J Frosch and L Derick *J. Electrochem. Soc.* vol 104, p 547 (September 1957)

1958   TECHNETRON Field Effect (FET)   S Teszner (France)

The first commercial FET was produced in France in 1958, by Stanislas Teszner, a Polish scientist employed by CFTH, a General Electric Company affiliate. Called the Technetron, Teszner's device was a germanium alloy semiconductor. It had a transonductance of 80 micro-ohms, a pinchoff of 35 volts, a gate leakage current of 4 microamps, and a low gate capacitance of 0.9 picofarads. The low trans-conductance and high leakage severely limited its applications. But the high pinchoff voltage was closer to the operating levels of some tubes, and its gate capacitance permitted it to be operated at a few megahertz.

SOURCE: 'Solid state—an old-timer comes of age' by J M Cohen *Electronics* p 123 (19 February 1968)

1958   PEDESTAL PULLING OF SILICON   W C Dash (USA)

The 'pedestal' method was devised to avoid oxygen contamination and at the same time achieve the high perfection attainable with the Czochralski technique. In this method the melt is an inductively heated mound held on top of a solid silicon support by surface tension and electromagnetic levitation. The support is sectored to inhibit electromagnetic coupling to the pedestal. A seed is inserted and the growing crystal is withdrawn at a rate which may vary from 3 cm per minute at the start to 3 or 4 mm per minute during the major part of its growth.

SOURCE: 'Growth of silicon crystals free from dislocations' by W C Dash. *J. Appl. Phys.* vol 30, No 4, p 459 (April 1959)

SEE ALSO: W C Dash *J. Appl. Phys.* vol 29, p 736 (1958)

1958   TUNNEL DIODE   **L Esaki (Japan) now (USA)**

First in chronological sequence came the tunnel diode, first described by Esaki in 1958. Again with hindsight, what a beautifully simple idea—to form a p-n junction between two such highly-doped regions that, in equilibrium, the continuity of the Fermi level across the junction would result in an energy barrier to the flow of carriers in the 'forward' direction. The device thus presents a high impedance at low forward bias, progresses through a region of negative impedance and then into a fairly normal 'forward' region of positive resistance.

SOURCE: 'Semiconductor devices—portrait of a technological explosion' by I M Mackintosh *The Radio and Electronic Engineer* vol 45, No 10, p 517 (October 1975)

SEE ALSO: 'New phenomenon in narrow germanium p-n Junctions' by L Esaki *Phys. Rev.* vol 109, p 603 (1958)

1958   FIELD-EFFECT VARISTOR   **Bell Laboratories (USA)**

This device, closely related in principle to the field-effect transistor, has a constant-current feature which makes it ideally suited for a current regulator in circuits where either the load or supply voltage varies over wide limits. It can also be used as a current limiter or pulse shaper. Its a.c. impedance is very high, making it useful as a coupling choke or as an ac switch. This device is based on the field effect principles developed by Shockley, Dacey and Ross.

SOURCE: 'History, present status and future development of electronic components' by P S Darnell *IRE Transactions on Components Parts* p 128 (September 1958)

SEE ALSO: 'A Field Effect Varistor' *Bell Labs. Record* vol 36, p 150 (April 1958)

1958   VIDEO TAPE RECORDER   **Ampex (USA)**

The first battery of 'video' tape recorders, a system called Ampex, was installed in the largest American television studios early in 1958. This system used tape moving at a speed of 200 inches per second but only half an inch wide; the recording being done on three tracks, two for storing the video signals and one for sound. A special machine for the cutting and editing of the tape had to be devised as it could be edited visually like cine film. Today, the majority of television programmes are recorded on videotape before transmission, in black-and-white as well as in colour; cine-film material shot for television can also easily be transferred on to videotape. Whether live, film or video, the viewer cannot detect the difference; the quality is equally high.

An important development of video recording is the 'canned' television programme for homes and schools, There are three rival systems. Two work with 'cassettes', which are inserted into a special replay unit plugged into the television set; one system uses Ampex tape, the other 8 mm film with two parallel tracks for sound and vision, electronically recorded. The third system a British–German venture, works with a fast-rotating (1500 rpm) disc and a pickup, providing a monochrome or colour programme of up to 12 minutes.

SOURCE: *A History of Inventions* by E Larsen (London: J M Dent & Sons/New York: Roy Publishers) p 330 (1971)

1958   EXPLORER 1 Satellite   **(USA)**

Launched 1 February 1958. Measurement of cosmic radiation and micrometeorites. Discovery of the than Allen radiation belt. Batteries. Transmitted up to 23 May 1958. Decayed on 31 March 1970.

SOURCE: Table of Artificial Satellites Launched Between 1957 and 1976 (Geneva: International Telecommunication Union) (1977)

1958   **VANGUARD-1 Satellite**   **(USA)**

Launched 7 March 1958. Part of the International Geophysical Year Programme. Permitted the discovery of the 'pear-shaped' earth. Studied the earth and measured the 'far out' density of the atmosphere. Batteries and solar cells. Transmitted until 12 February 1965.

SOURCE: Table of Artificial Satellites Launched Between 1957 and 1976 (Geneva: International Telecommunication Union) (1977)

1958   **AUTOMATIC CIRCUIT ASSEMBLY**   **US Army Signal Corps (USA)**
       ('Micro-Module' System)

By 1957 the goal for packaging had shifted from automation to miniaturisation. Working with the Army Signal Corps, RCA suggested an approach that was similar to Tinkertoy's but with smaller wafers. Using wafers 310 mils square, spaced 10 mils apart, RCA encapsulated the assembled module with an epoxy resin to increase mechanical strength and provide environmental protection.

With RCA as the prime contractor for an $18-million contract, the Signal Corps promoted micromodule as a standard package. A Signal Corps team headed by Daniel Elders, Stan Danko and Weldon Lane, established a continuing development programme for the micromodule (see figure 11.25).

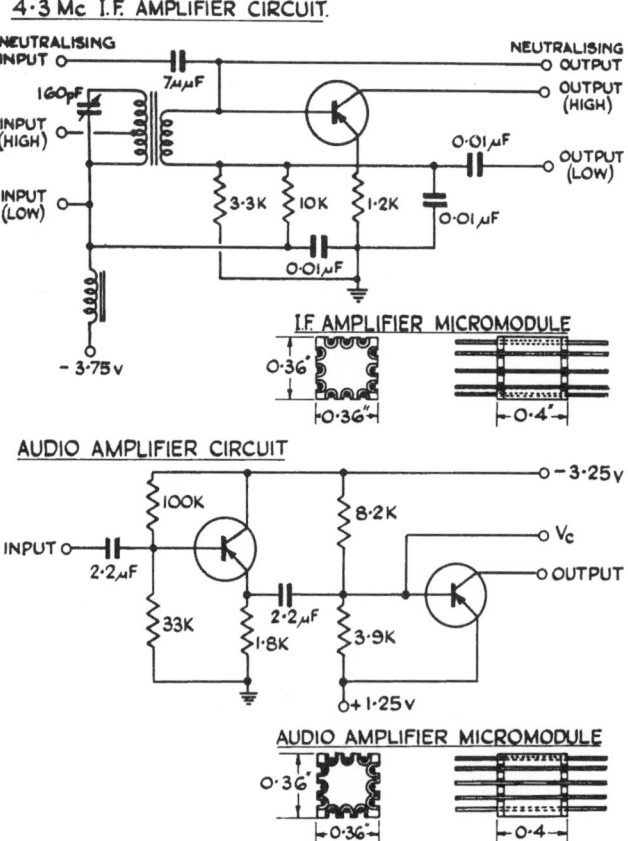

Figure 11.25. Typical micromodule circuit assemblies.

The micromodule approach combined high packaging density, machine assembly and modular design. It was the first attempt at functional modular replacement, where the entire module was treated as a single component. The programme established a compact universal packaging system using standard-shaped parts. But just as micromodule was gaining popularity in the early 1960s, the IC deflated its chance of achieving sufficient volume for a competitive price.

SOURCE: 'Solid state devices—packaging and materials' by R L Goldberg *Electronic Design* vol 24, p 126/7 (23 November 1972)

SEE ALSO: Signal Corps. Contract DA-36-039-SC-76968 RCA Camden, NJ

1958   **PIONEER-1 Satellite**   **(USA)**

Launched 11 October 1958. Moon probe. Failed to reach the moon. Sent data for 43 hours. Batteries.

Decayed on 12 October 1958 coming back to earth and burning in the atmosphere.

SOURCE: Table of Artificial Satellites Launched Between 1957 and 1976 (Geneva: International Telecommunication Union) (1977)

1958   **SCORE Satellite**   **(USA)**

Launched 18 December 1958. Signal Communication by Orbiting Relay Equipment. First communication satellite. Transmitted taped messages for 13 days. Decayed on 21 January, 1959.

SOURCE: Table of Artificial Satellites Launched Between 1957 and 1976 (Geneva: International Telecommunication Union) (1977)

1958   **MÖSSBAUER EFFECT**   **R L Mössbauer (Germany)**

The Mössbauer effect is the phenomenon of recoiless resonance fluorescence of gamma rays from nuclei bound in solids. The extreme sharpness of the recoiless gamma transitions and the relative ease and accuracy in observing small energy differences made the Mössbauer effect an important tool in nuclear physics, solid state physics etc.

SOURCE: *The Encyclopaedia of Physics* (2nd edn) edited by R M Besancon (New York: Van Nostrand, Reinhold & Litton Educational Pub. Inc.)

1958   **LASER (Light Amplification by Stimulated Emission**   **A L Schalow and C H Townes (USA)**
       of Radiation)

Because of the great interest aroused by masers it was not until 1957 that further serious attention was given to the idea of producing an optical version of the maser.

In their classic article of 1958 on the principles of laser action Schalow and Townes suggested potassium vapour as a possible medium and much effort was devoted to it but with no success. The reasons for this failure were rather puzzling, especially as other works later found caesium vapour to behave as predicted. Another medium under consideration was ruby (Cr^{3+} in Al_2O_3) although an internal report at Bell Telephone Laboratories concluded that the existing material was much too poor to give any hope of success and the experts of the time expected that the first laser would be based on a gas. Great was the surprise and general jubilation, therefore, when Maiman, who had persevered with ruby, achieved laser action in 1960. Maiman's own jubilation was short-lived as the manuscript which he prepared announcing his remarkable result was rejected by *Physical Review Letters* and an historic scoop of scientific journalism was achieved by the journals *Nature* and *British Communications and Electronics* which carried the first announcement in the established scientific literature.

Some months later the helium/neon laser was successfully operated and there followed over the next few years a tremendous explosion of publications on laser transitions in hundreds of different materials and on the properties of laser devices.

SOURCE: 'Lasers and optical electronics' by W A Gambling *The Radio and Electronic Engineer* vol 45, No 10, p 539 (October 1975)

SEE ALSO: 'Infrared and optical masers' by A L Schalow and C H Townes *Phys. Rev.* vol 112, p 1940 (15 December 1958)

'Infrared and optical masers' *Quantum Electronics* edited by C H Townes (New York: Columbia University Press) (1960)

'Stimulated optical production in ruby' by T H Maiman *Nature* vol 187, p 493 (6 August 1960)

1958   PACEMAKER   A Senning (Sweden)

The cardiac pacemaker was invented in 1958 by Doctor Ake Senning of Sweden. The first implants took place in the early 1960s. The pacemaker is capable of stimulating other organs as well as the heart.

In October 1986 a baby was born in Manchester suffering from a congenital malformation of the heart. It was given a cardiac pacemaker when it was only three days old. This was the first operation to be carried out in Europe on such a young child.

SOURCE: *The Book of Inventions and Discoveries* Associate Editor Valerie-Anne Giscard d'Estaing (UK: Queen Anne Press, Macdonald & Co.) p 96 (1990)

NOTE: The first self-contained pacemaker was by Chandack, Gage and Greatbach in 1960.

SOURCE: 'Making of the Modern World' Science Museum/John Murray (1992)

1958   VIDEO RECORDER IN COLOUR   Ampex (USA)

The first colour video recorder must also be credited to Ampex. It was presented in 1958, two years after the first video recorder, under the name VR 1000 B. It was followed in 1963 by a transistor version, the VR 110.

Meanwhile the Japanese had been steadily working on video recorder technology:

in 1958 Toshiba announced the first singlehead video recorder;

in 19S9 JVC developed the first two-head video recorder, the KVI;

in 1962 Shiba Electric (now Hitachi), in cooperation with Asahi Broadcasting, presented a professional transistorised video recorder;

in 1964 Sony marketed the first video recorder for the general public;

in 1965 Shiba Electric marketed a small portable video recorder;

in Europe Philips launched their VR 650 in 1964.

SOURCE: *The Book of Inventions and Discoveries* Associate Editor Valerie-Anne Giscard d'Estaing (UK: Queen Anne Press, Macdonald & Co.) p 240 (1990)

1959   INTRINSIC 10μ PHOTOCONDUCTORS (Cadmium   W D Lawson, S Nielsen, E H Putley, Telluride and Mercury Telluride)   A S Young (UK)

The elements mercury, cadmium and tellurium have been purified and crystals of the compounds CdTe and HgTe, and of the mixed compounds CdTeHgTe have been prepared. X-ray and cooling-curve data have established that coefficient and conductivity measurements show that HgTe is a semiconductor with a very low activation energy (0.01 eV) and a high mobility ratio ($\sim$ 100). HgTe is opaque to infrared radiation out to a wavelength of 38μ, but the mixed crystals show absorption edges which vary in position with composition from 0.8μ in pure CdTe to 13μ in crystals containing 90 per cent HgTe. Photoconductivity has been observed in filamentary detectors made from the mixed crystals.

SOURCE: 'Preparation and properties of HgTe and mixed crystals of HgTe-CdTe' by W D Lawson, S Nielsen, E H Putley and A S Young *J. Phys. Chem. Solids.* vol 9, p 325 (1959)

1959   THIN FILM CIRCUITS—TANTALUM   Bell Laboratories (USA)

Tantalum film circuitry is a single material technology in that capacitors, resistors and elementary interconnections are all derived from tantalum. Use of tantalum for this purpose is based on its chemical and structural stability, and on its capability of being anodized to form dielectrics for capacitors, and to protect and adjust resistors. In addition to the general value of tantalum film circuitry, tantalum film resistors and resistance networks, especially when sputtered in nitrogen, have independent interest as exceptional circuit elements.

SOURCE: 'Developments in tantalum nitride resistors' by D A McLean N Schwartz and E D Tidd *IEEE International Convention* (26 March 1964)

SEE ALSO: 'Microcircuity with Refractory Metals' by D A McLean *IRE Wescon Convention Record* vol 3, part 6, pp 87–91 (1959)

1959 **MICROELECTRONICS (INTEGRATED**   **J S Kilby (USA)**
      CIRCUIT-PATENT)

'It is, therefore, a principal object of this invention to provide a novel miniaturised electronic circuit fabricated from a body of semiconductor material containing a diffused p-n junction wherein all components of the electronic circuit are completely integrated into the body of semiconductor material.'

SOURCE: US Patent No 3 138 743 filed 6 February 1959 (J S Kilby)

SEE ALSO: US Patent No 3 261 081 patented 19 July 1966 (J S Kilby and Texas Instruments).

1959 **LUNIK-1 (Mechta) Satellite**   **(USSR)**

Launched 2 January 1959. In solar orbit. Moon probe passed within 600 km of moon. Equipment for studying circumterrestrial and circumlunar space. No magnetic field detected when passing close to the moon. Emission of a sodium vapour cloud. Batteries.

SOURCE: Table of Artificial Satellites Launched Between 1957 and 1976 (Geneva: International Telecommunication Union) (1977)

1959 **DISCOVERER-1 Satellite**   **(USA)**

Launched 28 February 1959. Stabilization defective: mission not fully accomplished. Decayed early March 1959.

SOURCE: Table of Artificial Satellites Launched Between 1957 and 1976 (Geneva: International Telecommunication Union) (1977)

1959 **PLANAR PROCESS**   **J A Hoerni (USA)**

At Fairchild, Dr Jean A Hoerni, a physicist, was trying to develop a family of double-diffused silicon mesa transistors. But instead of mounting the base layer on top of the collector, the traditional mesa approach, Hoerni diffused it down into the collector and protected the base-collector junction on the top surface with a layer of boron-and-phosphorous diffused silicon oxide. This first planar transistor was less brittle than the mesa and far more reliable—dust or other foreign matter could not contaminate the protected p-n junction. In 1959 Fairchild started marketing planar transistors and shortly thereafter applied the planar technique to the new integrated circuits.

SOURCE: 'Silicon, germanium & silver—the transistor's 25 anniversary' by C P Kocher *The Electrical Engineer* p 30 (November 1972)

1960 **COMPUTER-AIDED DESIGN**   **Military (USA)**

Computer-aided design (CAD) began in the 1960s in the context of US military aeronautics design programs.

The term refers to a set of techniques that can be used to create data that describe an object to be designed, to manipulate that data in a conversational mode and to arrive at a finished form of the design.

After its adoption by military, CAD penetrated civil aeronautics and the auto and computing industries. It enables an object (for example, a car) to be drawn in three dimensions and to be examined in a great number of theoretical circumstances, even before the building has begun. Today CAD plays an essential role in almost all fields of industry.

SOURCE: *Inventions and Discoveries 1993* edited by Valerie-Anne Giscard d'Estaing and Mark Young (New York: Facts on File) p 219

1960   **NEURISTOR**   **H D Crane (USA)**

A neuristor may be visualised as a channel having energy available to it everywhere along its length. The line or channel includes certain triggerable (active) processes arranged so that when any section of line is triggered the locally available energy is itself converted into 'trigger form'. Thus, successive neighbouring sections of line are activated. A signal propagating in this manner is generally referred to as a discharge, since it is continually regenerated by discharging available energy into the line as it propagates.

SOURCE: 'Neuristor—a novel design and system concept' by H D Crane *Proc. IRE* vol 50, p 2048 (October 1962)

SEE ALSO: 'Neuristor studies' by H D Crane *Stanford Electronics Lab. Tech. Rept.* No 1056-2 AD240306 (11 July 1960)

'An integrable MOS neuristor line' by C Kulkarni-Kohli and R W Newcombe *Proc. IEEE* p 1630 (November 1976)

1960   **FEMITRON (Field Emission Microwave Amplifier)**   **W P Dyke (USA)**

Microwave devices incorporating field-emission cathodes have been described by Charbonnier *et al*, and Dyke. The term 'femitron' was, in fact, first used by Dyke to describe a microwave amplifier resembling a 2-cavity klystron but incorporating a field emission cathode in the input-cavity gap. The femitron and derivatives of it, in particular frequency-multiplying devices, have also been investigated by Fontana and Shaw.

SOURCE: 'Field emission microwave amplifier: a reappraisal' by A J Sangster *IEE Solid-State & Electron Devices* vol 1, No 5, p 151 (September 1977)

SEE ALSO: 'Basic and applied studies of field emission at microwave frequencies' by F M Charbonnier, J P Barbour, L E Garrett and W P Dyke *Proc. IEEE* vol 51, pp 991–1004 (1963)

'Field emission, a newly practical electron source' by W P Dyke *IRE Trans.* vol 4, pp 38, 45 (1960)

'Microwave devices with field emission cathodes' by J R Fontana and H J Shaw *Trans. Amer. Inst. Engrs.* vol 81, pp 43–8 (1962)

'The carbon fibre field emitter' by F S Baker *et al J. Appl.Phys.* vol 7, pp 2105–15 (1974)

1960   **SUB-MILLIMETRE PHOTOCONDUCTIVE**   **E H Putley (UK)**
       DETECTOR (n-Type InSb)

Photoconductivity has been observed using a cryostat fitted with a light pipe so that a specimen[2] could be illuminated with radiation of wavelength between 0.1 and 4.0 mm. The sources of radiation were a mercury lamp and grating spectrometer covering the range 0.1 to 1.4 mm. and a Philips DX151 klystron and harmonic generator operating at 2 mm and 4 mm. The light-pipe was fitted with a black paper filter at room temperature and a black polythene filter in the helium to remove short-wave radiation.

The sample dimensions were 0.5 cm × 0.5 cm × 1.0 cm and indium electrodes were applied to opposite 0.5 cm × 1.0 cm faces. The magnetic field was applied at right angles to the direction of current flow and of the incident radiation. The radiation was directed along the long direction of the sample. For the majority of these experiments the radiation was modulated at 800 c/s and detected using a tuned amplifier and phase sensitive detector.

---

[2] n-type indium antimonide.

When the temperature was reduced to below 1.5 K and a magnetic induction of 6–8000 gauss applied, the sample resistance was about 10–30Ω. The sample was able to detect the applied radiation the minimum detectable energy per unit bandwidth being approximately 5×10^{-10} W at 0.5 mm, 5×10^{-11} W at 2 mm and 10^{-10} W at 4 mm.

SOURCE: 'Impurity photoconductivity in n-type InSb' *Proc. Phys. Soc.* vol 76, part 5, No 491 (1 November 1960) p 802

1960 **COMPUTERS (CD 1604)**   **Control-Data Corporation (USA)**

The CDC 1604 is a general-purpose data-processing system manufactured by the Control-Data Corporation. The first installation was made in January 1960. The entire system includes some 100 000 diodes and 25 000 transistors. The internal number system is the binary, with a word length of 48 bits. There are 62 24-bit one-address instructions (2 per word) each including a six-bit operation code, a three-bit index and 15 bits for the address. Indirect addressing is built-in and six index registers are provided. Arithmetic is performed with fixed or floating point in the parallel synchronous mode concurrently with other operations. Addition requires 4.8 to 9.6 μs, multiplication 25.2 to 63.6 μs including storage access. 32 768 words of magnetic-core storage are provided. Input–output equipment includes paper-tape typewriter, punched cards, magnetic tape (up to 24 units) and a 667–1000 lines-per-minute printer.

SOURCE: 'The evolution of computing machines and systems' by Serrell, Astrahan, Patterson and Dyne *Proc. IRE* p 1054 (May 1962)

1960 **PRINTED WIRING MULTILAYER BOARDS**   **Photocircuits Corp. (USA)**

Miniaturised replacement for back-panel wiring in computers may be accomplished by printed circuit sandwiches produced in many layers and laminated together under heat and pressure. The components are being made by Photocircuits Corporation, 31 Sea Cliff Ave., Glen Cove, New York. The manufacturer believes the new development will have uses in circuits having multiple crossovers and complicated interconnections among closely spaced component leads. Connections between different levels in a multilayered circuit are made through use of Tuf-Plate plated-through-holes. A typical six-layered printed circuit sandwich measures only 0.026 inch in thickness compared with a thickness of 0.062 inch for a conventional single circuit board.

SOURCE: 'On the market—PC sandwich six-layered unit' *Electronics* p 90 (8 April 1960)

1960 **COMPUTERS (HONEYWELL 800)**   **Honeywell (USA)**

The Honeywell 800 is a general purpose data-processing system capable of running as many as 8 distinct programmes simultaneously without special instructions. The first installation was made in December 1960.

The system includes 30 000 diodes and 6000 transistors, excluding peripheral equipment. The internal number structure is binary and binarycoded decimal with a word length of 48 bits, or 12 decimal digits. These 48 bits are assignable to numerical, alphanumerical or pure binary information. There are 59 basic instructions, each consisting of a twelve-bit operation core and 3 twelve-bit addresses. Eight index registers are available for each of the 8 programmes which can be run concurrently. Other special-purpose registers are available. Arithmetic is performed in a synchronous parallel-serial-parallel mode, concurrently with other operations. Addition requires 24 μs, multiplication 162 ps including storage access. Up to 32 000 words of magnetic-core storage can be used. Input-output equipment includes punched cards, paper tape and a 900 lines-per-minute printer. Up to 64 magnetic tape units can be connected to the system.

SOURCE: 'The evolution of computing machines and systems' by Serrell, Astrahan, Patterson and Dyne *Proc. IRE* p 1054 (May 1962)

1960   **CIRCUITRY (Linear Integrated Circuits)**   **Various (USA)**

Linear ICs came into their own in the 1960s. Starting with op amps, linear monolithics grew steadily in complexity and functions.

Monolithic op amps were first introduced in the early 1960s. At least two manufacturers—Texas Instruments and Westinghouse—were selling models. Then Fairchild, in 1964, came out with the 702, the result of the first collaboration between Bob Widlar and Dave Talbert. The 702 found limited acceptance—more significantly, its development led to the 709, one of the biggest success stories in an industry accustomed to them. The 709 was a revolution of sorts. Rather than translate a discrete design into a monolithic form, the standard approach, Widlar played the linear microcircuit game by a different set of rules: use transistors and diodes, even matched transistors and diodes—with impunity, but use resistors and capacitors—particularly those of large value— only where necessary. Even where use of a big resistor seemed inevitable, Widlar put a dc-biased transistor in its place. He exploited the monolith's natural ability to produce matched resistors and only assumed loose absolute values.

SOURCE: 'Integrated circuits' by E A Torrero *Electronic Design* vol 24, p 77 (23 November 1972)

1960   **LIGHT EMITTING DIODE (LED)**   **J W Allen and P E Gibbons (UK)**

It has been known for some time that rectifying contacts to GaP emit light when current is passed through them (Wolff *et al* 1955). Experiments at this laboratory and elsewhere suggest that the electronic transitions involved in this electroluminescence are different for the two directions of current flow. We consider alloyed or point-contact junctions on n-type GaP. Then the light emitted with forward bias has a spectrum which is a comparatively narrow band, the position of the band depending on the impurities present in the GaP. If the junction is biased in the reverse direction the current flowing is small until a certain voltage is reached. Beyond this voltage the current increases rapidly and orange light is emitted which has a very broad spectrum extending from the infra-red down to the absorption edge or beyond (Loebner and Poor 1959). It would seem that electroluminescence in the forward direction is due to radiative recombination of injected carriers via impurity levels, while that in the reverse direction is due to emission by 'hot' carriers produced by an avalanche break down (Chynoweth and McKay 1956).

SOURCE: 'Breakdown and light emission in gallium phosphide diodes' by J W Allen and P E Gibbons *Journal of Electronics* vol VII, No 6, p 518 (December 1959)

1960   **COMPUTERS (UNIVAC Solid State 80/90)**   **Sperry Corporation (USA)**

The UNIVAC Solid State 80/90 was designed as a medium-sized dataprocessing system. The term 'solid-state' refers to the use of 'Ferractor' magnetic amplifiers and transistors. The system consists of a central processor, a read-punch unit, a 450 card-per-minute card reader and a 600 lines-per-minute printer. The card equipment can be obtained for either the 80-column or the 90-column punched-card system. The first installation was made in January 1960.

SOURCE: 'The evolution of computing machines and systems' by Serrell, Astrahan, Patterson and Dyne *Proc. IRE* p 1053 (May 1962)

1960   **TIROS-1 Satellite**   **(USA)**

Launched 1 April 1960. First meteorological satellite. Sent 22 952 photos up to 17 June 1960. 9000 solar cells. Batteries.

SOURCE: Table of Artificial Satellites Launched Between 1957 and 1976 (Geneva: International Telecommunication Union) (1977)

1960   **ECHO-1 Satellite**   **(USA)**

Launched 12 August 1960. Passive telecommunication satellite. Relayed voice and television signals. 70 solar cells and batteries.

SOURCE: Table of Artificial Satellites Launched Between 1957 and 1976 (Geneva: International Telecommunication Union) (1977)

1960   **COURIER-1B Satellite**   **(USA)**

Launched 4 October 1960. First active repeater communication satellite. Operated for 17 days. 19 152 solar cells. Batteries.

SOURCE: Table of Artificial Satellites Launched Between 1957 and 1976 (Geneva: International Telecommunication Union) (1977)

1960   **TRANSIT-1B Satellite**   **(USA)**

Launched 13 April 1960. First navigation satellite. Transmitted until 12 July 1960. Solar cells. Batteries. Decayed on 5 October 1967.

SOURCE: Table of Artificial Satellites Launched Between 1957 and 1976 (Geneva: International Telecommunication Union) (1977)

1960   **EPITAXIAL CRYSTAL GROWTH**   **H H Loor, H Christensen, J J Kleimock & H C Theurer (USA)**

Until 1960 the semiconductor industry followed a pattern of starting with a crystal as pure as needed in the initial stage, and each step added impurities in a controlled manner. In June 1960, the Bell Telephone Laboratories announced a new method of fabricating transistors using epitaxial single crystals grown from the gas phase with controlled impurity levels. This technique had been studied at a number of laboratories, but it was not until the 1960 announcement that the potential was fully grasped by the semiconductor industry. Its unique advantage is the ability to grow very thin regions of controlled purity.

SOURCE: 'Contributions of materials technology to semiconductor devices' by R L Petritx *Proc. IRE* p 1030 (May 1962)

SEE ALSO: 'New advances in diffused devices' by H H Loor, H Christensen, J J Kleimock and H C Theurer. Presented at the IRE/AIEE Solid State Device Research Corp. Pittsburgh, PA (June 1960)

'Epitaxy—a fresh approach to semiconductor circuit design' Materials Dept. Motorola S/C Products Division *International Electronics* p 24 (March 1964)

1960   **TELEPHONE ELECTRONIC SWITCHING SYSTEM**   **Bell Labs (USA)**

Historically, electronic switching systems for telephone communications began in the US and saw their pioneering development undertaken there. The world's first electronic switching field experiment took place in Morris, Ill (1960), and the first production system was placed in service in the United States (No 101 ESS in 1963).

While the development of electronic switching is proceeding rapidly in many countries, the bulk of installed electronic equipment is now found in the US. More than 82 percent of the world's telephone lines that are switched electronically, as well as more than 48 percent of the world's electronic central office systems, are located in the United States. One of the most significant developments made possible by electronic switching has been stored program control flexibility, and here the US accounts for over 80 percent of the world's switching entities.

SOURCE: 'ESS: 'Minimonster' by A E Joel Jnr. *IEEE Spectrum* p 33 (August 1976)

SEE ALSO: 'Morris electronic telephone exchange' by W Keister, R W Ketchledee and C A Lovell *Proc. IEEE* 107 Suppl, No 20, p 257 (November 1960)

'Electronic PBX telephone switching systems (ESS101)' by W A Depp and M A Townsend *IEEE Trans. Communications.* vol 83, p 329 (July 1964)

1960–   **CIRCUITRY (Logic Circuits)**   **Various (USA)**
1964

Much of the early activity was involved with digital logic families. Almost from the beginning, a host of semiconductor manufacturers were attempting to establish the dominance of one logic family over the other or were second-sourcing the strong suit of a competitor.

At the start resistor-transistor logic (RTL) seemed to be the way to go Fairchild and Texas Instruments were strongly promoting it. Then diode-transistor logic (DTL) appeared in 1962 from the recently formed Signetics, and it took off. Transistor–transistor logic (TTL) emerged in Sylvania's Universal High Level Logic (SUHL) in 1963 and again, more permanently, in Texas Instruments' 5400 series in 1964.

SOURCE: 'Integrated Circuits' by E A Torrero *Electronic Design* vol 24, p 76 (23 November 1972)

1961   TAPE CASSETTE   Philips (Holland)

It was in 1961 that the Dutch company Philips developed the first mini tape cassette, which was 3.9 in long and designed for stereo and mono recordings. This cassette, along with the first cassette recorder, was unveiled in Berlin, Germany in 1963. Philips decided to allow manufactures to use its patent free of charge so as to encourage the spread of the system throughout the world.

SOURCE: *Inventions and Discoveries 1993* edited by Valerie-Anne Giscard d'Estaing and Mark Young (New York: Facts on File) p 138.

1961   TRANSFERRED ELECTRON EFFECT   B K Ridley and T B Watkins (UK)

The possibility of obtaining negative resistance effects in a new way in semiconductors is discussed. The principle of the method is to heat carriers in a high mobility sub-band with an electric field so that they transfer when they have a high enough 'temperature' to a higher energy low mobility sub-band.

SOURCE: 'The possibility of negative resistance effects in semiconductors' by B K Ridley and T B Watkins *Proc. Physical Soc.* vol LXXVIII, p 293 (1961)

SEE ALSO: H Kromer *Phys. Rev.* vol 109, p 1856 (1958)

1961   TRANSFERRED ELECTRONIC DEVICE   C Hilsum (UK)

In some semiconductors the conduction band system has two minima separated by only a small energy, and the lower minimum has associated with it a smaller electron effective mass than the upper minimum. At high electric fields it should be possible to transfer electrons to the upper minimum where they will have a power mobility. The conductivity of a homogeneous crystal bar can therefore decrease as the field is increased and it is conceivable that a differential negative resistance could be obtained. The conditions needed for obtaining negative resistance are examined, and calculations made for GaSb and semi-insulating GaAs. It appears that negative resistances should be observable in both these materials.

SOURCE: 'Transferred electron amplifiers and oscillators' by C Hilsum *Proc. IRE* vol 50, No 2, p 185 (February 1962)

SEE ALSO: 'Proposed negative mass microwave amplifier' by H Kromer *Phys. Rev.* vol 109, p 1856 (March 1958)

'Indium phosphide: a semiconductor for microwave devices' by H D Rees and K W Gray *IEE Solid State & Electronic Devices* vol 1, No 1, p 1 (September 1976)

'Three-level oscillator: a new form of transferred-electron device' by C Hilsum and H D Rees *Electron. Lett.* vol 6, p 277 (1960)

1961   LIQUID PHASE EPITAXY   H Nelson (USA)

An apparatus and procedures have been developed for the epitaxial growth of GaAs and Ge from the liquid state. The resulting technology has been found to posess advantages over vapour-phase epitaxy in some applications demanding highly doped epitaxial films and high-quality p-n junctions at the substrate-film interface. In this connection, it is an important feature of the liquid phase process that chemical impurities and mechanical damage of the substrate are removed when material is initially dissolved from the substrate surface prior to epitaxial growth. A clean interface p-n junction is thus obtained. Since liquid-phase epitaxy also favours the achievement of high doping and a steet concentration gradient at the p-n junctions, the process has proved itself eminently suitable for application in the manufacture of Ge tunnel diodes. In its application to the fabrication of the GaAs laser diode, it is an additional advantage of the liquid-phase process that the interface p-n junction is formed on a (100) crystal plane. As a consequence, this p-n junction is perfectly planar and also perpendicular to the (110) cleavage planes of the wafer. An optimum geometry (plane-parallel ends perpendicular to a perfectly flat p-n junction) is thus insured for diodes cleaved from (100) oriented GaAs wafers whose p-n junction has been formed by liquid-phase epitaxy.

SOURCE: 'Epitaxial growth from the liquid state and its application to the fabrication of tunnel and laser diodes' by H Nelson *RCA Review* p 603 (December 1963)

SEE ALSO: 'Epitaxial growth from the liquid phase' by H Nelson *Solid State Device Conference* Stanford University (26 June 1961)

'Properties and applications III-V compound films deposited by liquid phase epitaxy' by H Kressel and H Nelson *Physics of Thin Films* vol 7 (New York: Academic Press) (1973)

1961	**VENUS-1 Satellite**	**(USSR)**

Launched 12 February 1961. Automatic interplanetary station. Reached Venus in the second half of May 1961. Minimum distance to Venus 100 000 km. Investigation of interplanetary ionised gas and of solar corpuscular radiation. Investigation of the radiation belts and of space radiation. Magnetic measurements. Investigation of meteoristic dust. Solar cells, batteries.

SOURCE: Table of Artificial Satellites Launched Between 1957 and 1976 (Geneva: International Telecommunication Union) (1977)

1961	**VOSTOK-1 Satellite**	**(USSR)**

Launched 12 April 1961. First manned satellite. Pilot Yuri Gagarin. Returned to earth in the USSR after one orbit and 1.8 hours in space on 12 April 1961 near Smelovka, 800 km south-east of Moscow.

SOURCE: Table of Artificial Satellites Launched Between 1957 and 1976 (Geneva: International Telecommunication Union) (1977)

1961	**ELECTRONIC CLOCK**	**P Vogel & Cie (Switzerland)**

According to the present invention there is provided an electronic clock comprising no macroscopic moving parts which comprises an oscillator for delivering electrical pulses at a given frequency, distributing means arranged to be controlled by said oscillator for delivering at outputs thereof the pulses delivered by the oscillator, a counting device arranged to be controlled by the oscillator for delivering signals of a frequency of n cycles per hour, where n is an integral factor of 60, and 1 cycle per minute, an electronic switch arranged to be controlled by the distributing means for delivering signals corresponding to the state of the counting device said signals being associated with hours and minutes successively, to a distribution matrix for controlling a display device.

SOURCE: British Patent Specification No 995 546 'Improvements in or relating to electronic clocks' Application made in the USA (No 94832) on 10 March 1961.

1961	**MERCURY-ATLAS-4 Satellite**	**(USA)**

Launched 13 September 1961. Test of a cabin with a dummy on board. Checking of ground equipment performance (tracking stations). Cabin recovered in the Atlantic 260 km east of the Bermudas after 1st orbit on 13 September 1961. Batteries.

SOURCE: Table of Artificial Satellites Launched Between 1957 and 1976 (Geneva: International Telecommunication Union) (1977)

1961	**OSCAR-1 Satellite**	**(USA)**

Launched 12 December, 1961. Orbital Satellite Carrying Amateur Radio. Transmitted for 18 days. Decayed on 31 January 1962.

SOURCE: Table of Artificial Satellites Launched Between 1957 and 1976 (Geneva: International Telecommunication Union) (1977)

1961   MINICOMPUTER   Digital Equipment Inc. (USA)

It is generally accepted that the first minicomputer was designed by Digital Equipment in 1961—a 12-bit 4K word memory machine selling for approximately 15 000. Many applications were found for a machine of this type, and the market blossomed rapidly with a number of manufacturers designing specialised machines with 8, 12 or 16 bits and varying memory sizes. The direct descendants of such machines, with increased power are still available at prices around 1500, i.e. a reduction of 10:1 over ten years. This price reduction, and/or increase in performance, was made possible by the introduction of integrated circuits, MSI and SI logic, which allowed computers to become physically smaller while at the same time increasing performance. Since those early days when machines had very little software and peripheral support great strides have been made with the addition of extras, such as disks and magnetic tapes which allow the provision of operating systems running high level languages.

SOURCE: 'How minicomputers can produce an integrated solution to the running of a business' by I Evans. Paper read at Seminex, London (25 March 1976)

1962   SATELLITE (Telstar I)   Various (USA)

The first earth satellite was launched by the USSR on 4 October 1957. Telstar I, the first communication satellite, successfully transmitted high-definition television pictures across the Atlantic on 10 July 1962, and its successors promise a new form of global communication by sound and vision. Telstar I, now silent, has orbited the Earth about 17 000 times and is expected to remain in orbit for some 200 years; the 170 lb satellite was powered by nickel–cadmium batteries, recharged by 3600 solar cells, and contained 1064 transistors and a single electron tube (a travelling-wave tube for amplifying signals). By 23 July 1962, 16 European countries were exchanging live television with the United States and Telstar 2, launched on 7 May 1963, paved the way for the world's first commercial communication satellite, Early Bird.

SOURCE: 'The scope of modern electronics' by F A Benson *Electronics & Power* p 13 (January 1969)

1962   MERCURY-ATLAS-6 'FRIENDSHIP-7' Satellite   (USA)

Launched 20 February 1962. Investigation of man's capability in space. First United States manned spacecraft: astronaut John H Glenn Jr. Capsule recovered on 20 February 1962 after three orbits and 4.6-hour lifetime. Batteries.

SOURCE: Table of Artificial Satellites Launched Between 1957 and 1976 (Geneva: International Telecommunication Union) (1977)

1962   OSO-1 Satellite   (USA)

Launched 7 March 1962. Orbiting Solar Observatory. Measurement of solar electromagnetic radiation in the ultraviolet, X-ray and gamma-ray regions. Investigation of dust particles in space. Transmitted data on 75 solar flares until 6 August 1963. 1860 solar cells.

SOURCE: Table of Artificial Satellites Launched Between 1957 and 1976 (Geneva: International Telecommunication Union) (1977)

1962   RELAY-1 Satellite   (USA)

Launched 13 December 1962. Active telecommunication satellite to test microwave communications. Measurement of energy levels of space radiation. Study of radiation effect on solar cells and electronic components. Transmission of one television broadcast. 12 simultaneous 2-way phone calls or 144 teleprinter circuits 8215 solar cells, batteries. Experiments conducted until February 1965.

SOURCE: Table of Artificial Satellites Launched Between 1957 and 1976 (Geneva: International Telecommunication Union) (1977)

1962   **MOS (Metal-Oxide-Semiconductor) INTEGRATED   S R Hofstein and F P Heiman (USA)
        CIRCUIT**

Who made the first MOS integrated circuit? Undoubtedly, Drs Steven R Hofstein and Frederick P Heiman, who working under the direction of Thomas O Stanley, head of the Integrated Electronics Group at the RCA Electronic Research Laboratory in Princeton, NJ, were the first to succeed in late 1962 in integrating a multipurpose logic block of 16 MOS transistors into a silicon chip, 50 × 50 mils. They reported their success at the 1962 Electron Devices Meeting in October 1962.

SOURCE: 'The first MOS' by A Socolovsky *The Electronic Engineer* p 56 (February 1970)

1962   **'DUANE' RELIABILITY GROWTH THEORY   J T Duane (USA)**

Historically, the subject of reliability growth theory of electronic systems has received an abundance of attention and concern. Beginning with J T Duane in 1962, the literature on this subject has proliferated.

Duane and other investigators developed reliability growth analysis techniques based on actual data and used these data to test their models. However, these efforts had no statistically developed theory of inherent or analytical design reliability to fix their initial or starting points for their growth curves. However, in the structural reliability analysis, data relative to inherent or analytical reliability has been available and evolving since 1955. This data became available with the publication of Jablecki of data obtained from first time static structural tests of major aircraft structural subsystems under relatively controlled conditions at Wright-Patterson Air Force Base in the years 1940 to 1949.

SOURCE: 'A theory of reliability growth in structural systems' by H B Chenoweth *Proceedings Annual Reliability and Maintainability Symposium* p 106 (1980)

SEE ALSO: 'Learning curve approach to reliability monitoring' by J T Duane *IEEE Transactions on Aerospace* vol 2, p 563–6 (1964)

'Analysis of premature structural failures in static tested aircraft' by L S Jablecki *Dissertation* Die Eidgenossichen Technesche Hochschule, Zurich, Switzerland (1955)

1962   **JOSEPHSON EFFECT   B D Josephson (UK)**

In spite of the fact that the history of the Josephson effect is quite long, it is attributed to Josephson (1962) for its theoretical prediction. Prior experimental results published by Hahn and Meissner (1932) by Dietrich (1952) and by Giaver and Megerle (1961) have been given without definite conclusions or in doubt about effects, so that they could not be decisive for discovery. The experimental confirmation of the Josephson absorption effect (known as the a.c. Josephson effect) is attributed to Shapiro (1962, 1963) for first published results. The current discontinuities in the current-voltage characteristic of the Josephson junction, introduced by the macroscopic quantum absorption effect, thus became the generally accepted fact in physics. In addition Janson *et al* (1965) described the Josephson emission effect (frequently termed as the d.c. Josephson effect). Further, first-order discoveries based on the Josephson effect have been: the Mercerau effect or macroscopis quantum interference, frequency multiplication (Mercerau *et al* 1964), frequency mixing (Grimes and Shapiro 1966), e/h ratio measurements (Langenberg *et al* 1966), followed by a series of other applications. The crucial paper in Josephson voltage introduction is published by Finnegan *et al* (1971). This paper presents clear and firm experimental evidence that the Josephson voltage stability exceeds the best Weston-cell batteries used as national primary voltage standards.

SOURCE: 'An analysis of the inflexion point structure of Josephson absorption effect current steps' by Ranko Mutabzija *Int. J. Electronics* vol 42, No 3, p 241 (1977)

SEE ALSO: 'New superconducting devices' by B D Josephson. *Wireless World* p 484 (October 1966)

I Dietrich *Z. Phys.* vol 133, p 499 (1952)

T F Finnegan, A Denenstein and D N Langenberg *Phys. Rev.* B vol 4, p 1487 (1971)

I Giaver and K Megerle *Phys. Rev.* vol 122, p 1101 (1961)

C C Grimes and S Shapiro *Phys. Rev.* vol 169, p 186 (1966)

R Hahn and W Meissner *Z. Phys.* vol 74, p 715 (1932)

B D Josephson *Physics Lett.* vol 1, p 251 (1962)

D N Langenberg, W H Parker and B N Taylor *Phys. Rev.* vol 150, p 186 (1966)

J E Mercerau, R C Jaklevic, J J Lambe and A H Silver *Phys. Rev. Lett.* vol 12, p 274 (1964)

S Shapiro *Phys. Rev. Lett.* vol 11, p 80 (1963); *Phys. Rev.* vol 169, p 186 (1967)

1962   ELECTRONIC WATCH   P Vogel & Cie (Switzerland)

An electronic timepiece, comprising an oscillator unit; a frequency divider unit and a time display unit, in which each unit comprises or consists of a layer of semi-conductive material, the layers being sandwiched together and having their interfaces insulated from each other except at selected points at which the units are electrically connected together.

SOURCE: British Patent Specification No:1 057 453 'Electronic Timepieces'. Application made in Switzerland (No 13423) on 16 November 1962

1962   MICROELECTRONICS (Flat-Pack)   Y Tao (USA)

With the emergence of the IC as the modern circuit element of the early 1960s, transistor packages were found to lack sufficient heat sinking and adequate interconnections. To dissipate heat and provide a standard package size, Yung Tao created the flatpack in 1962 while at Texas Instruments. It was 1/4 by 1/8 inch and originally had 10 leads.

SOURCE: 'Solid state devices—packaging and materials' R L Goldberg *Electronic Design* vol 24, p 127 (23 November 1972)

1962   SEMICONDUCTOR LASER   R N Hall, J D Kingsley, G E Fenner, T J Soltys and R O Carlson (USA)

Coherent infrared radiation has been observed from forward biased GaAs p-n junctions. Evidence for this behaviour is based upon the sharply beamed radiation pattern of the emitted light, upon the observation of a threshold current beyond which the intensity of the beam increases abruptly, and upon the pronounced narrowing of the spectral distribution of this beam beyond threshold. The stimulated emission is believed to occur as the result of transitions between states of equal wave number in the condition and valence bands.

SOURCE: 'Coherent light emission from GaAs junctions' by R N Hall, G E Fenner, J D Kingsley, T J Soltys and R O Carlson *Phys. Rev. Lett.* vol 9, No 9, p 366 (1 November 1962)

SEE ALSO: M I Nathan and G Lasher (USA)

A characteristic effect of stimulated emission of radiation in a fluorescing material is the narrowing of the emission line as the excitation is increased. We have observed such narrowing of an emission line from a forward-biased GaAs p-n junction. As the injection current is increased, the emission line at 77 K narrows by a factor of more than 20 to a width of less than $kT/5$. We believe that this narrowing is direct evidence for the occurrence of stimulated emission.

SOURCE: 'Stimulated emission of radiation from GaAs p-n junctions' by M I Nathan, W P Dumke, G Burns, F H Dill Jr and G Lasher. *App. Phys. Lett.* vol 1, No 1, p 62 (1 November 1962)

1962   MARINER-2 Satellite   (USA)

Launched 27 August 1962. Data on interplanetary space during trip to Venus, survey of the planet; magnetic fields, charged particle distribution and intensity flux and momentum of cosmic dust, flow and density of solar plasma and energy of its particles. Flew by the planet and scanned it on 14 December 1962. Contact lost on 3 January 1963 at 87 390 000 km from earth. 9800 solar cells (222 W) batteries.

SOURCE: Table of Artificial Satellites Launched Between 1957 and 1976 (Geneva: International Telecommunication Union) (1977)

1962   **MARS-1 Satellite**   **(USSR)**

Launched 1 November 1962. Long-term space exploration. Establishment of interplanetary space radiocommunications. Lost earth lock at 106 Mkm. Passed within 193 000 km of planet. Solar cells.

SOURCE: Table of Artificial Satellites Launched Between 1957 and 1976 (Geneva: International Telecommunication Union) (1977)

1962   **ARIEL-1 Satellite**   **(UK)**

Launched 26 April 1962. Ionspheric satellite launched by United States rocket. Transmitted ionospheric, X-ray and cosmic data until November 1964. Solar cells.

Decayed on 24 May 1976.

SOURCE: Table of Artificial Satellites Launched Between 1957 and 1976 (Geneva: International Telecommunication Union) (1977)

1963   **INK JET PRINTING PROCESS**   **R G Sweet (USA)**

In the early 1960s, Sweet developed a method of forming, charging and electrostatically deflecting a high-speed stream of small ink drops to produce high frequency oscillograph traces in a direct-writing, signal-recording system. Each drop is given an electrostatic charge that is a function of the instantaneous value of the electrical input signal to be recorded. The drop is then deflected from its normal path by an amount that depends on the magnitude of its charge and in a direction that is a function of the polarity of the charge. As deflected drops are deposited on a strip of moving chart paper, a trace is formed that is representative of the input signal.

Lewis and Brown extended Sweet's technique to permit the printing of characters. Character images are stored in binary form in a character generator. An encoded signal addresses the character generator to select a desired character. The binary image of that character is then used to generate the drop charging signals necessary to deflect drops to the appropriate character matrix positions.

SOURCE: 'Application of ink jet technology to a word processing output printer' by W L Buehler, J D Hill, T H Williams and J W Woods *IBM Journal of Research & Development* p 2 (January 1977)

SEE ALSO: 'High frequency recording with electrostatically deflected ink jets' by R G Sweet *Stanford Electronics Laboratory Technical Report* No 1772-1. Stanford University, CA (1964)

'High frequency recording with electrostatically deflected ink jets' by R G Sweet *Rev. Sci. Inst.* vol 36, 131 (1965)

'Fluid Droplet Recorder' by R G Sweet *US Patent* 3 576 275 (1971)

'Electrically operated character printer' by A M Lewis and A D Brown *US Patent* 3 298 030 (1967)

1963   **GUNN DIODE OSCILLATOR**   **J B Gunn (USA)**

The observation is described of a new phenomenon in the electrical conductivity of certain III–V semiconductors. When the applied electric field exceeds a critical value, oscillations of extremely high frequency appear in the specimen current.

SOURCE: 'Microwave oscillations of current in III–V semiconductors' by J B Gunn *Solid State Communications* vol 1 p 88 (1963)

1963   **SYNCOM-1 Satellite**   **(USA)**

Launched 14 February 1963, active telecommunication satellite. Successfully injected into a near synchronous orbit but then lost contact with ground station. 3840 solar cells (135 W) battery. Syncom-3 is shown in figure 11.26.

Figure 11.26. Syncom-3 (courtesy Mark Williams, Space Technology Consultant, and Hughes Aircraft).

SOURCE: Table of Artificial Satellites Launched Between 1957 and 1976 (Geneva: International Telecommunication Union) (1977)

1963   **ELECTRONIC CALCULATOR**   **Bell Punch Co (UK)**

The first electronic calculators, containing discrete semiconductor components wired to printed circuit boards were produced in 1963 by a British firm, the Bell Punch Company. The machine was produced under licence in America and in Japan, where the advantage of cheaper Japanese labour for the hundreds of connections required led to a Japanese domination in the manufacture of calculators throughout the 1960s. Integrated circuitry was, of course, the perfect technology for the calculator, and MOS—slower but more compact and cheaper than bipolar integration—the most appropriate of the integrated circuit technologies. By the second half of the sixties, calculators using MOS integrated circuits were available.

The first American company to make calculators was a firm called Universal Data Machines, operating from a warehouse in Chicago. The company bought chips from Texas Instruments and, using cheap immigrant labour from Vietnam and South America, assembled five or six thousand calculators a week for sale through a local department store. Probably the second company to enter what was to become a particularly vicious race was the Canadian firm, Commodore, newly moved from Toronto to Silicon Valley. Commodore also used a Texas Instruments MOS chip, but adopted a technology developed by a component supplier, Bowmar, for making a particularly compact calculator. Bowmar had chosen not to make calculators itself and had found no interest in its technology among the established manufacturers of electromechanical calculators. Although these first mass-produced calculators dropped rapidly in price from about $100 in 1971 to $40 or $50 the following year, the profits of these small entrepreneurial companies remained high.

The situation had to change as it became staringly obvious where the profits in the exploding new market lay. By 1972 Bowmar was struggling to get back into the business it had earlier farmed out and was joined by other semiconductor manufacturers, including Texas Instruments. The calculator provides perhaps the best example of rapid vertical integration in the semiconductor industry, but if small firms had not demonstrated the viability of the new product, it is doubtful whether such integration would ever have taken place or, indeed, whether the calculator would ever have gained the acceptance it has.

SOURCE: *Revolution in Miniature—The History & Impact of Semiconductor Electronics* by E Braun and S Macdonald. To be published by Cambridge University Press.

SEE ALSO: 'Coming of age in the Calculator business' by N Valery *New Scientist Calculator Supplement* vol 68, 975, pp ii–iv (1975)

ALSO: 'Electronic Calculator' *Wireless World* vol 78, 1442, p 357 (1972)

1963   **ION PLATING OF PLASTIC & METALS**   **D H Mattox (USA)**

World-wide interest in ion plating stems from the new characteristics or colours the ion plated coats have when compared with films produced by other coating techniques. Among these properties are:

(1) excellent adhesion of incompatible substrates such as metals on plastics;

(2) irregular surfaces of many types such as screw threads, bearings and tubes can be coated in one operation because of the good 'throwing power' obtained in the process;

(3) compact pin-hole-free structures are formed with outstanding friction and wear characteristics and soldering to the metal coat is no problem;

(4) oxides and ceramic coating is possible with reactive ion plating;

(5) high corrosion resistant coating can be produced;

(6) high rate production of coatings is possible because the soft vacuum required means that long pump down times are not necessary. However, when good adhesion to metals is required, the metal substrate must be ion etched for periods up to 30 minutes to remove the oxide before deposition commences.

Although the technique was invented in 1963 in the United States by D M Mattox of Sandia Laboratories, New Mexico, the true potential of ion plating had not been appreciated until quite recently. In the beginning, the process is just like a conventional evaporation one, in that the bell jar is evacuated first with a rotary pump and then a diffusion pump. After the pressure reaches about 10^{-5} torr, argon is admitted through a needle valve until the pressure rises to about 2×10^{-2} torr. This soft vacuum is kept constant by controlling the needle valve and partially opening the baffle valve to the diffusion pump.

In the case of a metal, ion etching of the substrate is carried out first by striking a discharge between the substrate and top and base plate. A negative voltage of from 1 kV to 6 kV is applied to the substrate, and the argon ion discharge remains so long as the cathode voltage or the argon pressure is not allowed to fall too low. The bombardment of the substrate with neutral and ionised argon atoms sputters off metal oxide and metal and etches the surface. When cleaning and etching is complete, the metal source filament is switched on and the metal is evaporated into the argon discharge.

SOURCE: 'Ion plating—coat of many colours' *New Scientist* p 588 (9 June 1977)

1963   **SURFACE ACOUSTIC WAVE DEVICES**   **J H Rowen and E K Sittig (USA)**

'I filed a patent application describing a number of Surface Acoustic Wave devices in December 1963 and described these structures in a post deadline paper presented at the 1964 Symposium on Sonics and Ultrasonics in Santa Monica, California, October 14–16 1964. Dr Ehrhardt Sittig, who worked in my department at that time, constructed working models of these devices and subsequently filed an application describing an interdigital electrode structure for balanced (vs unbalanced) excitation of Rayleigh surface waves on such devices. I believe these efforts, which predate Professor R M White's work by at least three years, constitute the original invention of surface acoustic wave devices.'

SOURCE: Letter from Bell Laboratories, Murray Hill, NJ dated 20 July 1977

SEE ALSO: 'Tapped ultrasonic delay line and uses therefore' J H Rowen *USA Patent* No 3 289 114 dated 29 November 1966

'Elastic wave delay device' E K Sittig *USA Patent* No 3 360 749 dated 26 December 1967

NOTE: Lord Rayleigh first described the equations governing the propagation of surface elastic plane waves along the stress-free boundary of a semi-infinite, isotropic and perfectly elastic solid. Professor White's paper, in 1967, is summarised as follows.

Surface elastic-wave propagation, transduction and amplification (in a piezoelectric semiconductor) are discussed with emphasis on characteristics useful in electronic devices. Computed curves show the dependendence on distance from the surface of the elastic and the electric fields associated with surface elastic-wave propagation in cadmium sulfide. The interaction impedance, relating the external electric field to power flow, is computed for propagation on the basal plane of CdS and found to be low in comparison with values characteristic of electromagnetic slow-wave circuits. Amplification with a continuous drift field in cadmium sulfide is reported and differences between surface and bulk-wave amplifiers are discussed. Some operating characteristics and fabrication techniques for making electrode transducers on piezoelectric crystals are given, together with experimental results on several passive surface-wave devices.

SOURCE: 'Surface elastic wave propagation and amplification' by R M White *IEEE Trans. on Electron Devices* vol ED-14, No 4, p 181 (April 1967)

SEE ALSO: 'On waves propagated along the plane surface of an elastic solid' by Lord Rayleigh *Proc. London Math. Soc.* vol 17, p 4 (November 1885)

'Surface waves in anisotropic media' by J L Synge *Proc. Royal Irish Acad.* (Dublin) vol A58, p 13 (November 1956)

'Surface waves in anisotropic elastic media' by V T Buchwold *Nature* vol 191, p 899 (August 26 1961)

'Design of surface wave delay lines with interdigital transducers' by W R Smith, H M Gerard, J H Gollins, T M Reader and H J Shaw *IEEE Trans. on Microwave Theory & Techniques* vol MTT17, No 11, p 865 (November 1969)

'Passive interdigital devices using surface acoustic waves' *IEE Reprint Series 2* edited by D P Morgan (Peter Peregrinus) (May 1976)

1963   SILICON ON SAPPHIRE TECHNOLOGY   Various (USA)

The technology of silicon-on-insulating substrates, specifically silicon on sapphire, dates back to the beginning of practical MOS technology in 1963. The technology is known by different abbreviations, such as: SIS, ESFI (epitaxial silicon films on insulators), SOS, SOSL (silicon on spinel), etc.

The principal advantage of SOS circuitry is the inherent dielectric isolation, both dc and ac. The absence of silicon, except in the active device areas, significantly reduces parasitic capacitance between lines and essentially eliminates the parasitic capacitance to the substrate. Diffusion of device electrodes through the silicon film to the sapphire reduces electrode capacitance by several orders of magnitude because of the reduction in junction area. This significant reduction in electrode and interelectrode capacitance enables many devices to achieve their maximum band-width and frequency response; it allows for very high speeds, minimum speed-power products (below 0.5 pJ) on SOS CMOS and for very high frequency linear elements ($f_r > 2 \times 10^9$ Hz) such as dual-gate MOSFETs (tetrodes).

SOURCE: 'Recent SOS technology advances and applications' by R S Ronen and F B Micheletti *Solid State Technology* p 39 (August 1975)

SEE ALSO: Early publications on SOS Technology e.g. Material and Devices, include: H M Manasevit and W I Simpson, 'Single crystal silicon-on-Sapphire substrate' *J. Appl. Phys.* vol 35, 1349 (1964); C W Mueller and P H Robinson 'Grown-film silicon transistors on sapphire' *Proc. IEEE* vol 52, p 1487 (1964)

1964   **NIMBUS-1 Satellite**   **(USA)**

Launched 28 August 1964. Meteorological satellite to achieve a precise continuously earth-pointing orientation to evaluate the advanced vidicon camera system (AVCS) to provide improved pictures of local clouds by means of automatic picture transmission (APT) and to evaluate the high resolution infrared radiometer (HRIR) system for mapping global night-time cloud cover. 10 500 solar cells (450 W) batteries, 27 000 cloud cover photos returned until 23 September 1964.

SOURCE: Table of Artificial Satellites Launched Between 1957 and 1975 (Geneva: International Telecommunication Union) (1977)

1964   **VOSKHOD-1 Satellite**   **(USSR)**

Launched 12 October 1964. Manned spacecraft. First three-man crew: V Komarov, K Feokistov, B Yegorov. Landed after 16 orbits (24.3 hours) 305 km northwest of Kustanay, Kazakhstan.

SOURCE: Table of Artificial Satellites Launched Between 1957 and 1976 (Geneva: International Telecommunication Union) (1977)

1964   **PACKET-SWITCHNG-COMMUNICATIONS**   **P Baran (USA)**

Packet-switching demands a different kind of communications network from the normal telecommunications patterned. The channels, whether wire or radio, may be the same; but the switching points and exchanges have 'intelligence'—some form of computing device—which can accept a packet, look at it, and send it on its way according to the address and instructions it carries.

There is, however, also another major difference. The traditional communications network is essentially serial. To make a connection between two subscribers to such a system requires that the connection be established for the duration of the call. This means that the right switches have to be opened/closed and kept so for that duration.

Such a requirement is not necessary with a packet-switched system or network. In this latter case you transmit your data to the network, which then takes over, either sending it on or holding it until the addressee's receiver facilities are free and able to take it. The speed of transmission thus becomes a function of the weight of loading within the network. At the conceptual level, this is quite a radical approach to telecommunications.

To have a fail-safe network, in the terms that Baran proposed there should be 'over connection'. In other words, there should be not just one path in or out for a packet, but several. What would then determine the routing is the availability of a channel at a particular time. With this sort of network, the reliability can be far lower than would be necessary for a 'normal' linear communications system,

People have been trying to build such networks now for around 10 years. Today, though no-one can be certain how many are being planned or built, most of the communications networks of which we have high expectations are packetswitched—among them the experimental Arpanet in the United States, the commercial Telenet network (also American), Europe's interbank, Swift networks, Euronet, and the European Informatics Network.

SOURCE: 'Packet-switching's unsung hero' by R Malik *New Scientist* p 606 (8 September, 1977)

1964   **GEMINI-1 Satellite**   **(USA)**

Launched 8 April, 1964. Testing of the GEMINI launch vehicle compatibility and the structural integrity of the GEMINI spacecraft. The satellite re-entered the atmosphere and disintegrated on 12 April 1964. Batteries.

SOURCE: Table of Artificial Satellites Launched Between 1957 and 1976 (Geneva: International Telecommunication Union) (1977)

1964   **'IMPATT' DIODE**   **R L Johnston and B C deLoach**
                                                                (USA)

In 1964, Bell Laboratories scientists R L Johnston and B C deLoach discovered the IMPATT (IMPact Avalanche Transit Time) diode, subsequently shown to operate by an effect proposed earlier by W T Read Jr, also of the Laboratories. IMPATT diodes— semiconductor devices that generate microwaves directly when a DC voltage is applied to them—are becoming increasingly important in the design of microwave systems because of their high reliability and low cost.

SOURCE: *Mission Communications—the Story of Bell Laboratories* by P C Mabon (Murray Hill, NJ: Bell Laboratories Inc.) p 173 (1975)

SEE ALSO: 'A proposed high frequency, negative resistance diode' by W T Read *Bell Syst. Tech. J.* vol 37, p 401 (1958)

1964   **TRANSISTOR MODELLING**   **H K Gummel (USA)**

Since the original paper by Gummel in 1964, a great deal of literature has appeared on the subject of fundamental, or exact, transistor modelling. Gummel was the first to solve the semiconductor partial difference equations with no basic simplifications in their one-dimensional steady-state form. His integral formulation appeared in an improved form in the work of De Mari, who then went on to tackle the time-dependent 1-dimensional system. This required the use of a finite-difference formulation from which he could obtain a current driven transient solution for a diode. A simple spatial discretization was used and solutions were obtained for two implicit time integration methods, a Crank Nicolson scheme and a pure implicit first-order scheme. This was followed by the analysis of a Read diode by Gummel and Scharfetter, also using a 1-dimensional implicit scheme, but introducing a new and important spatial finite difference formulation.

SOURCE: 'Fundamental one-dimensional analysis of transistors' by A M Stark *Philips Research Reports Supplements* No 4, p 1 (1976)

SEE ALSO: 'A self-consistent iterative scheme for one-dimensional steady state transistor calaculations' by H K Gummel *IEEE Trans.* vol ED-11, pp 455–65 (1964)

'An accurate numerical steady state one-dimensional solution of the p-n junction' by A de Mari *Solid State Electronics* 11, pp 33–58 (1968)

'An accurate numerical one-dimensional solution of the p-n junction under arbitrary transient conditions' by A de Mari *Solid State Electronics* vol 11, pp 1021–53 (1968)

Finite Difference Methods for Partial Difference Equations by G E Forsythe and W R Wasow (New York: J Wiley & Sons Inc.) pp 101 *et seq* (1970)

'Large-signal analysis of a silicon Read diode oscillator' by D L Scharfetter and H K Gummel *IEEE Trans.* vol ED-16, 64–77 (1969)

1964   **TRANSISTOR (Overlay)**   **RCA (USA)**

The overlay transistors, first introduced in 1964, was developed at RCA under a contract from the Army Electronics Command, Ft Monmouth, New Jersey, as a direct replacement for the vacuum tube output stages then used in military transmitting equipment. The first commercial overlay, the 2N3375, produced 10 watts of output power at 100 MHz and could handle 4 watts at 400 MHz. Comparable interdigitated structures of that day were capable of 5 watts at 100 MHz and 0.5 MHz and 0.5 watts at 400 MHz.

SOURCE: 'Solid state—a worthy challenger for RF power honors' by D R Carley *Electronics* p 100 (19 February 1968)

SEE ALSO: 'The overlay—a new UHF power transistor' by D R Carley, P L McGeough and J F O'Brian *Electronics* p 70 (23 August 1965)

1964   **MICROELECTRONICS (Beam Lead)**   **M Lepselter (USA)**

In 1964, Martin Lepselter of Bell Telephone Laboratories invented the beam lead as a mechanical and electrical interconnection between the IC and its case.

SOURCE: 'Solid state devices—packaging and materials' R L Goldberg *Electronic Design* vol 24, p 127 (23 November 1972)

SEE ALSO: 'Beam lead technology' by M P Lepselter *The Bell System Tech. Journal* vol XVL, No 2, p 233 (February 1966)

1964   **TELEMEDICINE**   **Various (USA *et al*)**

The development of telemedicine in the United States can be divided into three stages: 1964–69, 1969–73 and 1973–present.

The first stage involved experimentation by medical practitioners on the clinical applications of telecommunications technology. The primary concern was the feasibility of two-way transmission of diagnostic information and clinical encounters via microwave links and video equipment.

Starting in 1964, the first interactive TV telemedicine project for the delivery of health care was carried out—a closed-circuit TV link between Nebraska Psychiatric Institute, Omaha, Neb., and Norfolk State Hospital, 112 miles away—under financing by the National Institute of Mental Health.

In 1967, an interactive TV link was installed between Massachusetts General Hospital and Logan International Airport, Boston, Mass., with financial support from the United States Public Health Service (later expanded to a Massachusetts General–Bedford VA Hospital link with Veterans Administration funds).

While the Nebraska telemedicine system was used primarily for psychiatric consultation and administrative purposes, the Massachusetts General Hospital–Logan Airport system was the first programme to use telemedicine in physical diagnosis and general patient care. The medical procedures used in physical diagnosis that were found to provide effective treatment over the interactive TV link were teleradiology, telestethoscopy and teleauscultation, speech therapy, teledermatology and telepsychiatry. The successful demonstration of physical diagnosis procedures provided additional incentives for Federal agencies to encourage further developments in the field.

The second telemedicine stage was characterised by a trend toward the exchange of knowledge and experience among the participants, and by Government support and sponsorship of research and demonstration programmes. The major supporter was the Health Care Technology Division in the Department of Health, Education and Welfare, which funded seven research and demonstration projects during 1972: Illinois Mental Health Institute, Chicago, Ill.; Case Western Reserve University, Cleveland, Ohio; Cambridge Hospital, Cambridge, Mass; Bethany/Garfield, Chicago, Ill.; Lakeview Clinic, Waconia, Minn; Dartmouth Medical School's INTERACT, Janover NH; and Mount Sinai School of Medicine New York, NY. In addition, the National Science Foundation funded two telemedicine projects in 1973: the Boston Nursing Home project for geriatric patients in nursing home that usually refer patients to Boston City Hospital, and the Miami-Dade project between Dade County and Jackson Memorial Hospital, Miami Fla.

During this stage, issues other than technical ones received some attention. These included consideration of the appropriate organisational and environmental settings for telemedicine implementation, manpower mixes and the role of non-MD, providers, and rudimentary approaches to evaluation of telemedicinels impact on healthcare delivery. The contributions of telemedicine to society as a whole were variously presented, but although some evaluation projects were started during this period, there were no significant efforts to investigate or document those benefits.

The initial evaluation efforts did not reveal conclusive results, but a comparison between the telephone and interactive TV encounters showed the former to be of shorter duration and more efficient for some aspects of patient care.

The third, and present, stage started in 1973 and its characteristic feature is the idea of telemedicine as an innovative mode of medical-care delivery. Two factors must be dealt with during this stage; sooner or later, telemedicine has to become self-supporting, or at least economically viable on its own; and the evaluation of telemedicine has to follow the concepts and method of evaluation in the medical-care field—i.e. evaluation in terms of structure, process and outcome variables The major new challenge for telemedicine has become its economic viability—how to make it pay for itself. To date, for telemedicine programmes to survive, they have had to be heavily subsidised.

It has been recognised that various problems in medical care may be redressed by telemedicine, but these depend on the vantage point of the user. For those persons where time and distance barriers make it difficult to obtain access to medical care, telemedicine is obviously very useful. The providers recognise potential benefits, including greater opportunities to interact with other physicians, to consult with specialists without worrying about the possibility of closings their patients, to have more free time, and to supervise the work of a nurse practioner or physician assistant in a remote clinic, The benefits to the system of medical care lie in the greater ability of the system to co-ordinate the activities undertaken by the various health actors or providers in their respective roles.

SOURCE: 'Coming—the era of telemedicine' by R Allen *IEEE Spectrum* p 33 (December 1976)

1964   MICROELECTRONICS (DIP) (Dual-In-Line Package)   B Rogers (USA)

Bryant (Buck) Rogers fostered the invention of the DIP while at Fairchild Semiconductors in 1964. It originally had 14 leads and looked just as it does today.

SOURCE: 'Solid state devices—packaging and materials' R L Goldberg *Electronic Design* vol 24, p 127 (23 November 1972)

1964   'ETCH-BACK' TECHNIQUE IN PRINTED WIRING   Autonetics (USA)
PLATED THROUGH HOLES

The interconnection of the internal layers of circuitry is made at the area where the drill penetrates through the copper pad exposing a cylinder of copper equivalent to the diameter of the drilled hole times 0.0044 (as times thickness of one ounce of copper). This small area of exposed copper can also be contaminated with epoxy smeared onto it during the drilling operation which can affect the resultant adhesion of the copper to the electroless copper deposit. Therefore, a smoothing process was developed at Autonetics which would expose a greater amount of copper at the interconnection areas to provide a more reliable bond. This, coupled with the fact that the smoothing operation also removes from the copper any smeared epoxy, provides for a more reliable interconnection than the standard T-joint.

SOURCE. 'Electroplating of plated through-hole interconnection circuit board' by L J Quintana *AFS Proc.* p 175 (1964)

1964   WORDPROCESSOR   IBM (USA)

One specialised office application that attracted computers was word processing. IBM, already a dominant manufacturer of electric typewriters, is credited with creating the market in 1964 when it introduced a magnetic-tape typewriter. This unit could store information on magnetic media for later modification and automatic retyping.

SOURCE: *Electronics* p 387 (17 April 1980)

1965   SYNTHESIZER   Moog, Deutsch and Carlos (USA)

Despite a few earlier attempts, the history of sound synthesis (the creation of sounds from electric pulses) did not begin until the early 1950s, with experiments carried out at the University of Bonn, West Germany. The first electronic music studio was set up in 1951 at a West German radio station. Through a complex assemblage of generators and filters, the composers created sounds which they put together manually afterward on magnetic tapes. Because this was a very slow process, the engineer Robert Moog (US) (in collaboration with the composers Herbert A Deutsch and Walter Carlos) had the idea of bringing together all the necessary equipment in one instrument. His research culminated in the Minimoog, which became available in 1965, and that was when the word 'synthesizer' was first used.

SOURCE: *Inventions and Discoveries 1993* edited by Valerie-Anne Giscard d'Estaing and Mark Young (New York: Facts on File) p 135

1965   THE MOUSE   D Englebart (USA)

The mouse is a small device that slides in all directions on a desk and which makes it possible to interact naturally with the computer. Its used was popularised by Apple with the Lisa and the Macintosh models in 1983. However, it was the little-known American inventor Douglas Engelbart who conceived and designed the mouse at the Stanford Research Institute in the mid-1960s. His brilliant idea was to have the computer operator place his or her hand on a small box or mouse. A sphere on the underside of the mouse is used to measure movements which are then transmitted to the computer via a lead—the tail of the mouse. These movements are translated to the cursor on the screen: if the mouse is pushed to the right the cursor goes to the right; if the mouse is pushed away from the user the cursor moves up, and so on. This revolutionary input device, originally found only on Apple computers, was adopted by IBM in 1987.

SOURCE: *The Book of Inventions and Discoveries* Associate Editor Valerie-Anne Giscard d'Estaing (UK: Queen Anne Press, Macdonald & Co.) (1990) p 124

1965   WIEGAND WIRE   J Wiegand (USA)

About 10 years ago, John Wiegand discovered that by properly work hardening a magnetic wire, it is possible, along the exterior 'shell' of the wire, to produce a coercive force significantly greater than the coercive force in the wire's core. By virtue of this magnetic differential, and depending on certain external conditions, the direction of magnetisation in the core of the wire can be the same or opposite to that in the shell. And switching from one state to the other is easily and repeatedly induced at well-defined magnetic-field levels.

Short lengths of wire exhibiting the Wiegand effect can serve as the heart of magnetic pulse generators that have distinct advantages over similar devices, including non-contact operation and a facility for being 'read' by detection devices having virtually no input power. Other important advantages are that pulse signals are not rate sensitive, meaning the amplitude of the pulse signal remains the same regardless of speed of operation; they offer any combination of pulse-generation direction and polarity, that is, unidirectional or bidirectional, unipolar or bipolar. Thus any combination of direction and polarity are available for pulse generation. And such devices are capable of withstanding severe environments, including temperatures from $-95°F$ to $+300°F$. Over the years, Wiegand has developed material composition and work hardening procedures to a point where brief pulses (10^{-4} duration) at levels of 2 milliwatts can be produced. With properly-designed detectors, peak voltages of 500 millivolts in the 50-ohm load have been observed.

SOURCE: 'Wiegand Wire: new material for magnetic-based devices' by P E Wizen *Electronics* p 100 (10 July 1975)

SEE ALSO: 'Wiegand effect pushing its way into new products' *Electronics* p 39 (14 April 1977)

1965   SMOOTH-SURFACED WIRE DRAWING   K M Olsen, R F Jack and E O
                                                        Fuchs (USA)

A technique for producing wire with a very smooth surface by drawing it through dies submerged in an ultrasonically agitated Squid has been devised at Bell Laboratories. The agitated liquid continuously cleans the wire and dies so that the drawn wire is relatively free of embedded particles and surface scratches.

Reduction of surface imperfections in wire improves its properties in some instances. For example, a smooth finish is desirable in those types of magnetic memories that store information on a thin film of metal plated onto a wire. The wire finish should be as smooth as possible so that the film can be deposited evenly.

In this technique, the ultrasonic energy forms extremely minute vapor cavities in the liquid wherever it contacts a solid surface. The expansion and collapse of these cavities—known as cavitation—'scrubs' the wire clean of foreign particles before it enters the dies to be reduced. The ultrasonic agitation keeps the particles suspended in the liquid and prevents them from collecting in the entry area of the dies; thus they do not score the wire as it it drawn through the dies.

SOURCE: 'Very smooth-surfaced wire produced by new drawing technique' *Bell Laboratories Record* p 390 (October 1965)

Copyright 1965, Bell Telephone Laboratories, Inc. Reprinted by permission of Bell Laboratories Record.

1965   SATELLITE—INTELSAT I   (International)

The first internationally owned satellite, INTELSAT I, was put into operation in 1965. It was placed in a geo-stationary equatorial orbit, that is at an altitude of 22 400 miles, in a longitudinal position 30° West for transatlantic operation. It had a mass of 39 kg (85 lb) primary power 45 W from solar cells and was capable of relaying 120 voice circuits or one television channel. The INTELSAT I system was to some extent experimental for two main reason. Firstly, it was to ascertain whether reliable communication could be maintained in spite of the high path loss of 200 dB; however, the Earth stations employed the now well-known parabolic reflector type aerials diameter 85–100 ft—with cryogenic-cooled low-noise amplifiers and in this respect it was a great success. Secondly, it was to determine whether the transmission delay, Earth–satellite–Earth, of 250 ms was operationally acceptable. The decision was taken to continue with satellites in the geostationary orbit and this is used for all internationally owned satellites today.

SOURCE: 'Fixed communications' by A S Pudner *Radio and Electronic Engineer* vol 45, No 10, p 547 (October 1975)

1965   SELF-SCANNED INTEGRATED PHOTODIODE   G P Weckler (USA)
         ARRAYS

The possibility of forming image sensors from arrays of silicon photo diodes on a single silicon chip has been recognised for many years, probably since the inception of integrated circuit technology some fifteen years ago. It was quickly apparent that the array size was limited, not by the number of diodes that could be included on the silicon, but by the number of output leads necessary to form connections to these diodes. To circumvent this problem, it was necessary to scan the diodes, that is, to multiplex them in to a single output lead by means of switching circuitry on the same integrated chip. A second problem was that of detecting the minute photocurrents produced by the necessarily very small diodes. The technique of charge integration had been used in the Vidicon for some years, and it was pointed out by G P Weckler in 1965 that this technique could be used with photodiode arrays, the switching being achieved by m.o.s. transistors. The one step necessary to complete the picture was now to include a shift register on an integrated circuit with the diodes and m.o.s.t's to perform the serial multiplexing function. The first fully self-scanned arrays using this technique were announced in 1967.

SOURCE: 'Applications of self-scanned integrated photodiode arrays' by P W Fry *The Radio and Electronic Engineer* vol 46, No 4, pp 151–60 (April 1976)

SEE ALSO: 'Operation of p-n junction photodetectors in a photon flux integrating mode' by G P Weckler *IEEE J Solid State Circuits* SC-2, No 3, pp 65–73 (September 1967)

'Development and potential of optoelectronic techniques' by P J W Noble *Component Technology 2* No 8, pp 23–8 (December 1967)

1965   PROTON-1 Satellite   (USSR)

Launched 16 July 1965. Investigation of solar cosmic rays. Investigation of the energy spectrum and chemical composition of particles of primary cosmic rays in the eneray range up to 10^{14} eV. Investigation of nuclear interaction of ultra-high energy cosmic rays up to 10^{12} eV. Determination of the absolute intensity and energy spectrum of electrons of galactic origin. Determination of the intensity and energy spectrum of gamma rays of the galaxy with energies over 50 million eV. Solar cells, batteries. Decayed on 11 October 1965.

SOURCE: Table of Artificial Satellites Launched Between 1957 and 1976 (Geneva: International Telecommunication Union) (1977)

1965   **ELECTRONIC TYPEWRITER**   **IBM (USA)**

IBM launched the 72BM in 1965. The 72BM was the first typewriter with a memory, which was stored on magnetic tape. In 1972 Rank Xerox developed the first electronic typewriter with a live memory. This machine also featured the first daisy wheel, which was invented by Dr Andrew Gabor (US). In 1978 the Italian company Olivetti and the Japanese company Casio marketed the first electronic typewriters with rapid-access memories. They featured 'type wheels' rather than balls.

SOURCE: *Inventions and Discoveries 1993* edited by Valerie-Anne Giscard d'Estaing and Mark Young (New York: Facts on File) p 219

1965   **VIRTUAL REALITY**   **Military (USA)**

It is possible today to be 'absorbed' by a computer and move around in an imaginary and synthetic universe. This concept is called virtual reality. It is best known through its applications in the video game industry; however, its origins are in military applications. The concept is simple enough. On the one hand, a computer generates synthetic images, and on the other, the user controls these images through the intermediary of receptors placed in a glove (Dataglove), and visualises the result with stereoscopic glasses that have listening devices (Eyephone). Each movement of the fingers or the head is transmitted to the computer, which consequently interprets and modifies thc surroundings. It is thus possible to touch or displace objects or to change the field of vision.

SOURCE: *Inventions and Discoveries 1993* edited by Valerie-Anne Giscard d'Estaing and Mark Young (New York: Facts on File) p 219

1966   **OPTCAL FIBRE COMMUNICATIONS**   **K C Kao and G A Hockham (USA)**

Another method of providing guidance, and of cunningly circumventing the problem of light travelling in straight lines, is to use a fibre consisting of a glass core having a high refractive index surrounded by a cladding of lower index. As early as August 1964, in an address to the British Association for the Advancement of Science, the author speculated on the use of light and glass fibres in the telephone network, instead of electric currents and wires, but developments did not start in earnest until publication of the classic article of Kao and Hockham of STL in 1966. At the time the problem seemed a formidable one; the attenuation of existing fibres was about 1000 dB/km, the band-width was expected to be low and fibre bundles were fragile. Since then enormous strides have been made resulting in fibre attenuations of 2 dB/km produced as a matter of routine, bandwidths of 1 GHz in a 1 km length of fibre having a diameter of 100 μm, and fibres coated with nylon which are too strong to be broken by hand. Such fibres are flexible and capable of being incorporated into simple but effective forms of cable. The bandwidth of a single fibre is much greater, and the attenuation lower, than existing copper coaxial cables and the diameter is considerably smaller. Thus the capacity of the present telephone network could be very greatly increased, with little additional installation expense, by the gradual introduction of optical fibre cables.

SOURCE: 'Lasers and optical electronics' by W A Gambling *The Radio and Electronics Engineers* vol 45, No 10, p 541 (October 1975)

SEE ALSO: 'Dielectric-fibre surface waveguides for optical frequencies' by K C Kao and G A Hockham *Proc. Instn. Elect. Engrs.* vol 113, pp 1151–8

1966   **BIOSATELLITE-1 Satellite**   **(USA)**

Launched 14 December 1966. To determine the effects of the space environment on various life processes and study the effect of weightlessness on the life processes of certain organisms and the effects of radiation on organisms in weightlessness. Due to failure of the retro-rockets to fire, it was not possible to recover the capsule. Batteries. Decayed on 15 February 1967.

SOURCE: Table of Artificial Satellites Launched Between 1957 and 1976 (Geneva: International Telecommunication Union) (1977)

1966   **ATS-1 Satellite**   **(USA)**

Launched 7 December 1966. Experiments to advance the fields of spacecraft communications (aircraft and ground), meterology (photos, transmission of weather facsimile) and control technology. Number of scientific experiments to measure the orbital environment of the satellite. 22 000 solar cells (185 W) batteries.

SOURCE: Table of Artificial Satellites Launched Between 1957 and 1976 (Geneva: International Telecommunication Union) (1977)

1966   **LUNAR ORBITER-1**   **(USA)**

Launched 10 August 1966. Flying photographic laboratory. Obtained high resolution photographs of various types of surface on the moon to assess their suitability as landing sites for APOLLO and SURVEYOR spacecraft; monitored the meteroids and radiation intensity in the vicinity of the moon; provided precise trajectory information for use in improving the definition of the moon's gravitational field. 10 856 solar cells, battery.

SOURCE: Table of Artificial Satellites Launched Between 1957 and 1976 (Geneva: International Telecommunication Union) (1977)

1966   **NITRIDE-OVER-OXIDE SEMICONDUCTORS**   **F H Horn (USA)**

After more than three years of legal proceedings between GE and International Business Machines. Corporation, the US Patent and Trademark Office has upheld GE's claim to priority of the invention, thus reaffirming GE's right to the patent, number 3597 667.

In one form of the GE inventions a thin film of silicon nitride is placed between the gate and silicon dioxide in metaloxide-semiconductor field-effect transistors (MOSFETs). This structure virtually eliminates the contamination by alkali ions that previously caused widespread failure of the tiny devices.

In another application of the invention, these nitride-overoxide layers are used in standard bipolar transistors as a surface and junction-sealing passivation layers Both the manufacturing yield and reliability of modern integrated circuits and semiconductor devices are 'substantially enhanced' by the GE invention.

The original GE patent application was filed on 1 March 1966, following a discussion of semiconductor instability problems between GE scientists Dr Dale M Brown and Dr Horn, during which Dr Horn suggested silicon nitride over silicon dioxide as a passivation technique for overcoming these problems. The idea was tried and successfully demonstrated shortly thereafter.

SOURCE: GE Public Information Release (GE Research & Development Center, Schenectady, NY 12301) p 2 (11 October 1976)

1966   **TIROS-1 Satellite**   **(USA)**

Launched 3 February 1966. Meteorogical satellite Environmental Survey Satellite. Part of the TIROS Operational System (TOS), Advanced vidicon camera system (AVCS). Switched off on 8 May 1967, 9100 solar cells, batteries.

SOURCE: Table of Artificial Satellites Launched Between 1957 and 1976 (Geneva: International Telecommunication Union) (1977)

1966   **FLIP-FLOP BONDING TECHNIQUE (FLIP-CHIPS)**   **M Wiessenstern and G A S Wingrove (USA)**

The flip-flop bonding structure and method was invented and subsequently patented in 1906 by Wiessenstern and Wingrove. Since that time nearly every semiconductor manufacturer has experimented with various forms of flip-flop bonding for the purpose of assembling integrated circuits, and possibly some discrete components, into larger subsystems. To this day, no successful method of flip-chip bonding has become generally utilized on the open market.

SOURCE: 'A multichip package utilizing In-Cu flip-chip bonding' by A P Youmans, R E Rose and W F Greenman. *Proc. IEEE* vol 57, No 9, p 1599 (September 1969)

SEE ALSO: 'Semiconductor device assembly with true metallurgical bonds' by M Weissenstern and G A S Wingrove *US Patent* No 3256465, 14 June 1966

'Joining semiconductor devices with ductile pads' by L F Miller *3rd Annual Hybrid Microelectronics Symposium* (29 October 1968)

1966   **SURVEYOR-1 Satellite**   **(USA)**

Launched 30 May 1966. Soft landed on the moon on 2 June 1966. Transmitted 11 150 photos up to 13 July 1966. 3960 solar cells (77 W) batteries.

SOURCE: Table of Artificial Satellites Launched Between 1957 and 1976 (Geneva: International Telecommunication Union) (1977)

1967   **LASER TRIMMNG OF THICK FILM RESISTORS**   **(USA)**

The trimming of electronic components started in 1967 when the first experiments were conducted on the trimming of thick film resistors with CO_2 lasers. Two years later 'Q-switched' laser systems were installed in the General Motors Delco Electronics plant in Indiana for the manufacture of thick film voltage regulators for automobiles.

SOURCE: 'Bright future for laser trimming' by W B Cozzens *Electronic Engineering* p 58 (February 1976)

1967   **SOYUZ-1 Satellite**   **(USSR)**

Launched 23 April 1967. Manned spacecraft. Re-entered 24 April 1967 after 17 orbits. Failed to land. Pilot: V Komarov, killed. Solar cells.

SOURCE: Table of Artificial Satellites Launched Between 1957 and 1976 (Geneva: International Telecommunication Union) (1977)

1967   **AUDIO NOISE REDUCTION SYSTEM**   **R M Dolby (USA)**

Utilizing the masking effect, together with signal compression and expansion, the Dolby Laboratories A301 achieves noise reduction (a) by boosting low-level signal components during recording whenever possible (compression), followed by complementary attenuation during playback (expansion), and (b) by the masking effect whenever the signal level is already so high that compression and expansion are not possible.

Since masking is less effective with noise frequencies somewhat removed from the signal frequency, it is necessary to deal with the various portions of the spectrum independently. The noise reduction system then yields a lower—and apparently constant—noise level, the classical hush-hush or swish of normal compression and expansion being absent.

The A301 system splits the audio spectrum into four bands and compresses and expands each of these in an essentially independent manner. Separate bands are provided for the hum and rumble frequency range (80 Hz, lowpass), for the mid-audio range (80 Hz–3 kHz, band-pass), for medium high frequencies (3 kHz, high-pass), and for high frequencies (9 kHz, high-pass). A high-level signal in one band hence cannot prevent noise reduction in another band in which the signal level may be low.

From another point of view, the system effectively produces a recording equalisation characteristic which continuously conforms itself to the incoming signal in such a way as to improve the signal to noise ratio during playback.

SOURCE: Dolby Laboratories Technical Report A301.

SEE ALSO: 'An audio noise reduction system' by R M Dolby *J. Audio Eng. Soc.* vol 15, p 383 (1967)

'Audio noise reduction: some practical aspects' by Ray M Dolby *Audio* magazine (June & July 1968)

'The Dolby noise-reduction system—its impact on recording' by John Eargle *Electronics World* (May 1969)

Dolby noise reducer part 1 (An introduction to the Dolby noise reduction system) by Geoffrey Shorter *Wireless World* pp 200–5 (May 1970)

1967   ION BEAM COATING   K L Chopra and M R Randlett
                                                                (USA)

The history of Ion Beam Coating (IBC) covers a period of more than ten years beginning with metallic coatings reported by Chopra and Randlett. Carbon deposition with 'diamond like' properties was initially reported by Aisenberg and Chabot in 1970 and 1971. Rapid expansion of the field resulted when Spencer Schmidt *et al* showed that essentially any solid material can be deposited when a target is bombarded with an energetic ion.

Although ion beam milling has been accepted for several years as the most desirable technique when compared with chemical or plasma etching for the fabrication of high resolution micron or submicron circuitry in both research and production installations, ion beam deposition is just now being accepted by research laboratories as the single economical process which affords real flexibility in thin film fabrication. High resolution microfabrication process with electron beam or x-ray lithography require improved deposition technology. The answer may be ion beam coatings both for research and production applications.

SOURCE: 'Ion beam coating: A new deposition method' by George R Thompson, Jr *Solid State Technology* p 73 (December 1978)

SEE ALSO: 'Duoplasmatron ion beam source for vacuum sputtering of thin films' by K L Chopra and M R Randlett *Rev. Sci. Instrum.* vol 38, No 8, p 1147 (1967)

S Aisenberg and R Chabot *J. Vac. Sci. Technol.* vol 6, p 112 (1970)

S Aisenberg and R Chabot *J. Appl. Phys.* vol 42, p 2953 (1971)

E G Spencer and P H Schmidt *J. Vac. Sci. Tech.* vol 8, p 368 (1971)

'Deposition and evaluation of thin films by DC ion beam sputtering' P H Schmidt, R N Castellano and E G Spencer *Solid State Technol.* vol 15, No 7, p 39 (1972)

1967   'ROTATOR' CIRCUIT NETWORK   L O Chua (USA)

This paper presents a new linear, reciprocal, active two-port network element called a rotator, of which there are three types: an R-rotator, an L-rotator, and a C-rotator. They have the unique property that whenever a nonlinear resistor, inductor, or capacitor is connected to one port of an R-, L-, or C-rotator, respectively, the resulting two-terminal network behaves as a new resistor, inductor or capacitor whose characteristic curve is that of the original resistor, inductor or capacitor rotated by a prescribed angle about the origin.

The rotator is realizable by either a π-network or a T-network of linear resistances, inductances or capacitances. It can also be realised by a balanced lattice network of linear elements. Operational laboratory models are reported, and experimental data agree remarkably well with theoretical predictions.

The sensitivity, power rating, and stability performances of rotators are considered in detail in this paper and practical stability criteria are given. They are shown to be indispensable building blocks for realising multi-valued elements and some potential applications are described.

SOURCE: 'The rotator—a new 'Network Component'' by L O Chua *Proc. IEEE* vol 55, No 9, p 1566 (September 1967)

1967 **TRAPATT DIODE (Trapped Plasma Avalanche Transit Time Diode)**   **H J Prager, K K N Chang and S Weisbrod (USA)**

The trapatt mode was discovered in 1967 by Prager, Chang and Weisbrod. It has permitted the realisation of high-efficiency solid-state microwave oscillators and amplifiers. Dc–rf conversion efficiencies as high as 60% are obtained at frequencies of 1–2 GHz, and several authors have reported efficiencies as high as 35% at X-band frequencies.

Trapatt action occurs when a rapidly increasing reverse-bias voltage of magnitude greater than the breakdown voltage is applied across the depleted diode. An avalanche zone sweeps rapidly from the junction, through the depletion layer to the substrate, leaving in its wake a dense plasma of holes and electrons and collapsing the electric field. The diode drops into a low voltage high-current state, and the carriers are said to be trapped, as their drift velocities fall well below their saturated values. As the carriers drift slowly out of the active region under low-field conditions, the electric field within the diode recovers. When the electric field has fully recovered and the current returns to essentially zero, the cycle is repeated. The current and voltage waveforms produced are favourable to the production of high efficiencies. The frequency of operation is much lower than in the impatt mode, since the carriers spend a long part of the cycle with drift velocities well below the saturated drift velocity.

SOURCE: 'Design and performance of trapatt devices, oscillators and amplifiers' by C H Oxley, A M Howard and J J Purcell *IEE Solid-State and Electron Devices* vol 1, No 1, p 24 (September 1976)

SEE ALSO: 'High power, high efficiency silicon avalanche diodes at ultra frequencies' by H J Prager, K K N Chang and S Weisbrod *Proc. IEEE* vol 55, p 586 (1967)

1968 **AMORPHOUS SEMICONDUCTOR SWITCHES**   **S R Ovshinsky (USA)**

Switching phenomenon have been noted for a decade but Ovshinsky first attracted international attention by producing a discrete component commercially in 1968, figure 11.27(a) shows an early switch made by his company in which a thin film of undisclosed glassy amorphous material is sandwiched between massive carbon electrodes in a thermister type package. Another possible discrete geometry used in earlier studies is the cross-over sandwich using thin film electrodes (figure 11.27(b). These devices and their more modern counterparts are so-called threshold-switches, and their circuit configuration and resulting $I - V$ characteristics are shown in figure 11.27(c). These devices have the remarkable property of having an initial OFF resistance, corresponding to region (a), of order tens of megohms, which drops to a much lower ON resistance, region (c) of around 100 Ω when the voltage across it exceeds a certain threshold value V_{th} typically around 10 V. The switching, which occurs along a load line corresponding to the load resistor, region (b), happens in a time typically of order nanoseconds, but there is a delay time which can be as long as 10 μs, between the application of a switching voltage and the onset of switching. If the current in the On-state is allowed to fall below a minimum holding value, I_h, the device reverts to its high resistance state, along region (d). In this sense, the device behaves as a monostable element. Notice that the switch is insensitive to the polarity of the supply; the $I - V$ curve for negative voltages is a mirror image of that for positive values.

SOURCE: 'Amorphous semiconductor devices and components' by J Allison and M J Thompson *The Radio and Electronic Engineer* vol 46, No 1, p 12 (January 1976)

SEE ALSO: 'Reversible electrical switching phenomena in disordered solids' by S R Ovshinsky *Phys. Rev. Lett.* vol 21, p 1450 (1968)

1968 **IRIS (ESRO-1) Satellite**   **ESRO (Europe)**

Launched 17 May 1968. First European Space Research Organisation satellite. Measurements of the solar and cosmic radiations. 3456 solar cells, batteries. Decayed on 8 May 1971.

SOURCE: Table of Artificial Satellites Launched Between 1957 and 1976 (Geneva: International Telecommunication Union) (1977)

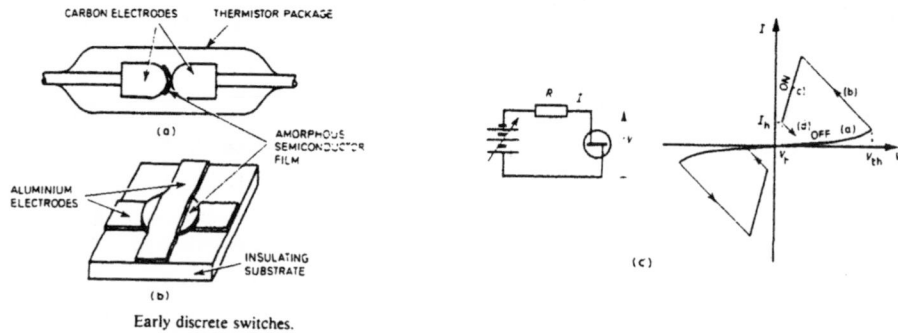

Early discrete switches.

Figure 11.27. Amorphous silicon switch.

1968   **C-MOS (Complementary Metal-Oxide-Semiconductor)**   **Westinghouse, GT & E Labs., RCA**
       INTEGRATED CIRCUIT   **Sylvania (USA)**

At the October 1968 International Electron Devices Meeting in Washington, the Westinghouse Molecular Electronics division announced two experimental MOS field effect transistor schemes that may show the way to more successful solid state memory systems. One, the MONTOS (metal-oxide-nitride-oxide-semiconductor) element for read-only memory applications, is a non-volatile sandwich of silicon nitride and silicon dioxide. Turned on by a negative voltage, MONOS could lead to circuits having all their active, passive and storage elements on the same chip, according to Hung C Lin, manager of advanced techniques development.

The other development, a complementary MOS-bipolar structure, gets around the fact that MOSFETs are usually limited to driving only low-capacitance loads by placing vertical and lateral non bipolar transistors on the same chip to drive higher loads. The resulting devices are 10 times faster than most MOSFETs Lin says, and are made with standard MOS manufactur ing techniques, Potentially, he adds, the structures could be used not only as large random-access memories but as logic and shift-register elements. And the use of bipolar transistors at the output end could reduce the interface problems between MOS devices and others such as transistor-transistor logic or diode-transistor logic.

The performance of complementary metal oxide semiconductor logic is so much better than p-channel logic that there must be a good reason why its not used more often. There is. It has been very difficult to fabricate n and p channels in the same substrate.

Paul Richman, who with Walter Zloczower described the complementary MOS circuit at the International Electron Devices Meeting in 1968, says that their difficulties with the extremely resistive substrate have been economic rather than technological. The material is easy enough to make, but the only source is a chemical firm in Germany.

Richman feels that the GT & E approach is the first simplified method of making complementary MOS ICs. The RCA and Westinghouse fabrication techniques have disadvantages, he says.

RCA's method of forming conventional n and p channels in the same substrate requires extremely careful control of the diffusion process and results in a rather high threshold voltage.

Westinghouse uses an elaborate procedure of etching pits, filling them with epitaxial p-type material, then etching back to form the p channels. This process involves critical mechanical operations and results in relatively slow circuits.

Neither GT & E nor Sylvania has immediate plans for marketing complementary MOS IC's. But Richman hopes that the new fabrication method will open up the memory applications for which complementary MOS integrated circuits are so well suited.

SOURCE: 'Electronics review—integrated electronics' *Electronics* p 49 (28 October 1968)

1968   **THE 'TRINITRON' Colour Cathode Ray Tube**   **Sony (Japan)**

A new colour CRT employing a single lens in-line gun was developed by Sony Corporation. The single large diameter lens minimizes electron beam aberration, resulting in a high quality image. Electrostatic plates attached to the top of the gun effectively converge the side beams at the phosphor screen. The unique arrangement of the electron optics of the gun permits modulating the electron scanning velocity, giving rise to further improvement of the picture image. A new colour selection mechanism called the Aperture-Grill, which has a great number of slits instead of the holes or slots found in a conventional shadow mask, is incorporated in the CRT. Since the Aperture-Grill has a greater beam transparency than a shadow mask, the CRT yields a brighter picture. The vertically continuous phosphor stripes produces a high resolution image limited only by the electron beam diameter. Additional advantages are that the Aperture Grill is less sensitive to terrestrial magnetism and that it is free from Moiré patterns. The cylindrical face plate whose vertical curvature is almost infinite, reduces the ambient light problem. Sony Corporation has manufactured more than 20 million of this unique high-quality CRT, ranging from 5V" to 30V", the world's largest, with a variety of deflection angles.

SOURCE: 'The 'TRINITRON'—a new colour tube' by S Yoshida, A Ohkoshi & S Miyaoka. *IEEE Trans.* vol BTR-14, p 19–27 (July 1968)

SEE ALSO: 'The 'TRINITRON'—a new colour tube' by S Yoshida, A Ohkoshi & S Miyaoka *Electronics & Radio Technician* vol 3, No 4 (December 1969)

'A wide-deflection angle (114°) TRINITRON colour picture tube' by S Yoshida, A Ohkoshi & S Miyaoka. *IEEE Chicago Spring Conference on BTR* (12 June 1973)

'25V inch 114 degree TRINITRON colour picture tube and associated new developments' by S Yoshida, A Ohkoshi & S Miyaoka *IEEE Chicago Spring Conference on BTR* (10 June 1974)

1968   **'BARITT' DIODE**   **G T Wright (UK)**

In 1968 Wright described a new negative resistance microwave device based on the principle of barrier controlled injection and transit time delay the BARITT diode. His simple analysis suggested that the device should operate at moderate power and low noise level. In the same year, independently, Ruegg presented a paper on the simplified large-signal theory of a similar punch-through structure giving considerably optimistic prospects—an estimated efficiency of the order of 20% and power output of 10–100 W at 10 GHz. These theoretical works were confirmed experimentally in 1970 when Sultan and Wright achieved negative resistance in npn silicon structures, and subsequently oscillations in pnp structures and in 1971 when Coleman and Sze reported oscillations in metal-semiconductor-metal structures. Several experimental papers have since been presented, comparing the properties of different BARITT diode structures and pointing out the reliable and low-noise operation of the device at moderate power levels.

SOURCE: 'Large-signal analysis of the silicon pnp-BARRIT diode' by M Karasek *Solid State Electronics* vol 19, p 625 (1976)

SEE ALSO: G T Wright *Elect. Lett.* vol 4, 543 (1968)

H W Ruegg. *IEEE Trans.* ED-15, 577 (1968)

1968   **INTEGRATED CIRCUIT ALUMINIUM**   **R Noyce (USA)**
         METALLISATION

An insight into Noyce's style of technical leadership is provided by Gordon Moore, a chemical physicist who was one of the eight founders of Fairchild Semiconductors, and who quit in 1968 to join Noyce in starting Intel, where he is now president and chief executive officer. 'Bob was certainly the idea man in the group. I can think of two things that at the time impressed me even more than what he did for the integrated circuit. One was the use of aluminium for transistor contacts. I remember struggling with all kinds of complex alloys to find one metal that would make contact with both the emitter and the base. One day Bob said, 'Why don't you try aluminium'.

'So I tried aluminium and it worked beautifully. That really got the double-diffused transistor out of the laboratory as a practical device. Then there was the use of nickel to fabricate junctions with good electrical contacts. Bob suggested it one day and it worked. These were both cases where he proposed something I really thought wouldn't work and which then got us past significant barriers.'

SOURCE: 'The genesis of the integrated circuit' by M F Wolff *IEEE Spectrum* p 49 (August 1976)

1968   'MUTATOR' CIRCUIT NETWORK   L O Chua (USA)

The basic problem of synthesising a nonlinear resistor, inductor, or capacitor with a prescribed $i - v$, $\phi - i$, or $q - v$ curve is solved by introducing three new linear two-port network elements, namely the mutator, the reflector, and the scalor. The mutator has the property that a nonlinear resistor is transformed into a nonlinear inductor, or a nonlinear capacitor, upon connecting this resistor across port two of an appropriate mutator. The reflector has the property that a given $i - v$, $\phi - i$, or $q - v$ curve can be reflected about an arbitrary straight line through the origin. The scalor is characterised by the property that any $i - v$, $\phi - i$, or $q - v$ curve can be compressed or expanded along a horizontal direction, or along a vertical direction. Using these new elements as building blocks, it is shown that any prescribed single-valued (which need not be monotonic) $i - v$, $\phi - i$, or $q - v$ curve can be synthesized. Active circuit realisations for each of these new elements are given. Laboratory models of mutators, reflectors, and scalors have been built using discrete components. Oscilloscope tracings of typical mutated, reflected and scaled, $i - v$, $\phi - i$, and $q - v$ curves are given. The experimental results are in good agreement with theory at relatively low operating frequencies. The practical problems that remain to be solved are the stability and frequency limitation of the present circuits.

SOURCE: 'Synthesis of new nonlinear network elements' by L O Chua *Proc. IEEE* vol 36, No 8, p 1325 (August 1968)

SEE ALSO: 'Additional types of mutators and active RC synthesis using mutators' by T Murata *Int. J. Electronics* vol 42, No 1, p 33 (1977)

1968   HIGH DEFINITION TELEVISION   Nippon Broadcasting Corporation (Japan)

In 1968 the NHK (Nippon Broadcasting Corporation), that is to say Japanese television, began research into high definition television and in 1974 were joined by Sony in this work. Sony engineers worked with those of the NHK in the development of a new system called Hi-Vision which would give television a quality of image comparable to that at the cinema images of 1125 scanning lines on 60 hertz. However this would need a change in production cameras and the whole stock of current television sets. As far as Europeans were concerned this expense was excessive. They have launched a counter-attack under the Eureka programme a high definition system with images of 1250 lines that is compatible with existing networks (Pal Secam NTSC) and in particular with the standard D2-Mac Pacs (1984). So the 600 million or so television sets in service in the world won't have to be replaced. This system has been available since 1988.

Doubtless it will take another ten years before high definition television systems, be they Japanese Hi-Vision or the European and American (ACTIV Advanced Compatible TV), become fully operational. The first live high definition re-transmission organised by the national Japanese channel NHK took place at the time of the Opening ceremony of the Seoul Olympics on 17 September 1988.

SOURCE: *The Book of Inventions and Discoveries* Associate Editor Valerie-Anne Giscard d'Estaing (UK: Queen Anne Press, Macdonald & Co.) p 238 (1990)

1969   INTERNET   Arpanet (USA)

Towards the end of September 1969, four US academic institutions—the University of Utah, the Stanford Research Institute, the University of California at Santa Barbara and at Los Angeles—were linked by an experimental computer network, ARPAnet, funded by the Advanced Research Projects Agency (ARPA) of the US Department of Defense (DoD). With the benefit of a quarter of a century of hindsight, this military-inspired development can now be seen as one of the seminal events in the history of communications. Out of the ARPAnet has emerged the extraordinary phenomenon of the Internet—a computer network that already links tens of millions of users worldwide, while growing at a rate that, unchecked, would cover the entire population of the earth by 2001.

The first distinguishing feature of the Internet is that it is based on packet switching. Traffic carried across the network is broken up into conveniently sized chunks of data, which are then augmented by addressing and a variety of quality-control information before being committed to the network for transmission.

Packet switching has two distinct advantages over the circuit-switching regime used in conventional telephony. First, it is highly efficient in the use of transmission bandwidth. Bandwidth is only allocated to a connection if it is generating packets, otherwise the bandwidth is available to other connections.

Secondly, a packet switched network is tolerant of failures or breaks in the network. If one transmission link fails, then packets can simply be rerouted along some alternative path through the network. The perceived robustness of packet switching was an important consideration behind DoD funding for ARPAnet.

Aside from being a packet-switched network, the Internet is distinguished by being organised as a multiplicity of tens of thousands of interconnected subnetworks. This explains why the Internet has been able to grow so fast. It has not been built from scratch; rather, it has been largely bolted together from existing subnetworks.

The individual subnetworks cover a variety of networking technologies, such as Ethernet local-area networks and X.25 wide-area networks. These technically disparate subnetworks are bound together by the intellectual glue of the IP (Internet Protocol) and the TCP (Transmission Control Protocol), generally known by their joint label TCP/IP.

An Internet transmission path comprises three basic elements: the host computers at either end of the path, the subnetworks, and the gateways that link the subnetworks (figure 11.28). The IP resides at level 3, the network layer, of the OSI seven-layer model so that IP functions have to be provided within the terminal machines and within every gateway.

Given the shortness of its history, the progress of the Internet has been remarkable. However, its most radical achievement may yet lie ahead—the undermining of the dominant role of the telecoms operators. Conventional wisdom holds that the multimedia future will be created around telecoms networks and the standards of the Broadband ISDN. The message of the Mbone is that such cosy assumptions 'ain't necessarily so'.

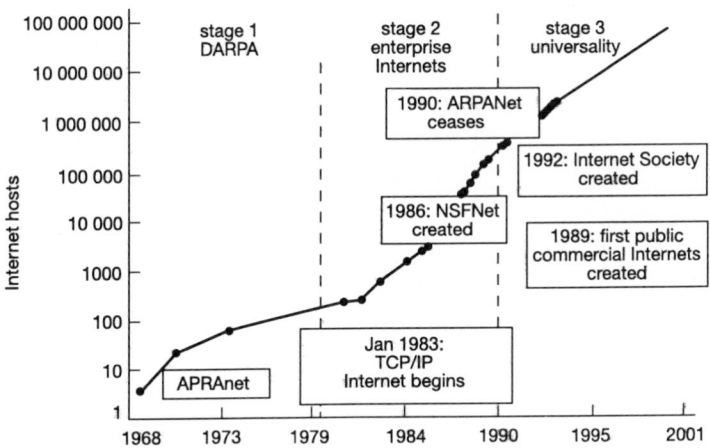

Figure 11.28. The three ages of the Internet.

SOURCE: 'The net effect' by Roger Dettmer *IEE Review* p 67 (March 1995)

1969   **AERIAL MATCHING UNIT**   **Eur Ing A M Gordon (UK)**

In order to match a random-length aerial to a transmitter it is necessary to transform the aerial impedance to the 50 or 75 ohms which will allow efficient power transfer. The ideal way to match impedances is with a transformer of the correct ratio.

In such a transformer the bottom end of each winding may be connected to earth. It is a small step to realise that a tapped coil with one end connected to the transmitter, the other to the aerial and the appropriate tap earthed will make a variable ratio transformer.

In the implementation, there were 3 turns (on a 2-in dia former) to the first tap, 12 taps 5 turns apart and 2 turns to the end of the coil.

This was the basic matching unit but a variable capacitor was added, together with a 5 way, 2-gang switch so that as well as the variable-ratio transformer with the capacitor across the primary one could have a parallel network, an L-network, a series capacitor or straight-through connection.

SOURCE: Technical Topics column, Radio Communications (Journal of the RSGB) (May 1969)

SEE ALSO: *Amateur Radio Techniques* by Pat Hawker, G3VA, Edition 3 (and subsequent editions) (RSGB) pp 241, 242

1969   **MAGNETIC BUBBLES**   **A H Bobeck, R F Fischer, A J Perneski, J P Remeika and L G Van Uitert (USA)**

A magnetic material usually consists of arrays of discrete localised volumes of material, defined as domains, Each domain separated from its neighbours by domain walls has a preferential orientation of the magnetisation vectors of all of the atomic magnetic moments within its volume, Domains may have different orientations with respect to each other depending on their net energy content and the force vectors acting upon them. In some cases, domains can be produced and moved about in a thin plate or layer in a reproducible way. This is the case for a number of ferromagnetic and ferrimagnetic materials. Figure 11.29 shows a typical domain structure in a magnetic garnet at zero magnetic field with the random, worm-like domain patterns maintained, in the steady state, by the inherent uniaxial magnetic anisotrophy of this material, If an increasing magnetic field is applied perpendicular to the plane of the plate in figure 11.29 then the unfavourably oriented domains (with respect to the applied field) may be made to shrink and then finally to collapse into cylindrical domains which look like 'magnetic bubbles' if they are observed in polarized light under a microscope; they can be displaced in the direction of an applied magnetic gradient and their presence or absense at a certain position of the plate constitutes the binary-coded information stored in the memory.

SOURCE: 'Magnetic domain bubble memories' by J L Tomlinson and H H Weider *The Radio & Electronic Engineer* vol 45, No 12 p 727 (December 1975)

SEE ALSO: 'Application of orthoferrites to domain-wall devices' by A H Bobeck, R F Fischer, A J Perneski, J P Remeika and L G Van Uitert *IEEE Trans. on Magnetics* MAG-5, pp 544–54 (1969)

1969   **SKYNET-A Satellite**   **(UK)**

Launched 22 November, 1969. Government communication satellite to be placed in synchronous orbit at 45°E longitude over Indian Ocean. Spin stabilized 7236 solar cells, batteries, Launched by NASA.

SOURCE: Table of Artificial Satellites Launched Between 1957 and 1976 (Geneva: International Telecommunication Union) (1977)

1969   **PARCOR**   **NTT(Japan)**

A new method of speech analysis and synthesis, in which the speech spectra are expressed with the use of the partial autocorrelation (PARCOR) coefficients was proposed and developed by Nippon Telegraph and Telephone Public Corporation in 1969.

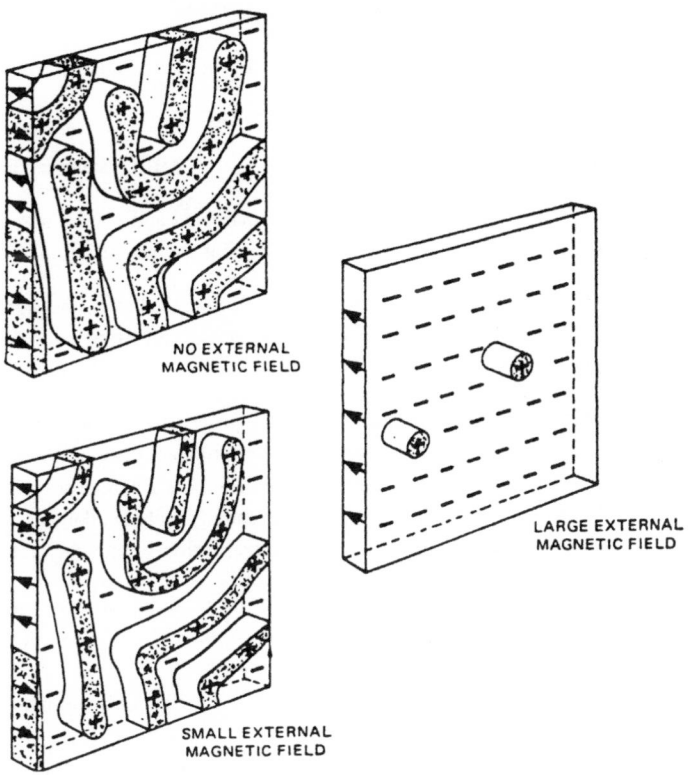

Figure 11.29. Oppositely oriented magnetic domains shown with and without an applied external magnetic field having indicated polarity and direction. (After Bobeck and Scovil.)

In order to obtain a speech of high quality, it is necessary to extract the spectral envelope parameters and the driving source parameters and to reproduce these features of the original speech as accurately as possible. In extracting these parameters efficiently, PARCAR coefficients were introduced.

Speech signals are sampled at every 125 μs through a PARCAR analyzer. The signal amplitude just before and just after the sampling time is predicted with a set of n samples by the least square method. Then the deviation between the predicted value and the real value is measured. The extracted eight informations obtained during a certain time (one frame 15 ms) are the PARCOR coefficients. Speech informations are compressed to 57 bits which consist of the PARCOR coefficients of $8 \times 5 = 40$ bits (5 bits to a PARCOR coefficient) and driving source parameters of 17 bits. This corresponds to 3800 bits/s. The speech compression by the PARCOR method is about one fifteenth of the 56 000 bits/s of the PCM method and is very efficient. Speech synthesis is just an inverse process of the speech analysis mentioned above. The PARCOR type speech analysis and synthesis system is superior to the conventional parameter editing and synthesis method. When the PARCOR synthesizer is composed of the recent high-speed logic elements (LSI IC), it is possible to respond simultaneously to many telephone circuits.

This speech analysis and synthesis system based on PARCOR coefficients has many response words compared with that of the conventional recording and editing system and various flexible speech response services will be possible. Moreover, this analysis method is expected to produce new services such as the automatic speech recognition and the perception of speaking voices, etc.

SOURCE: 'Speech analysis and synthesis system based on partial auto correlation coefficients' by F Itakura and S Saito *Meeting Record of the Acoustical Society of Japan* 2-2-6 (October 1969) (In Japanese)

SEE ALSO: 'Digital filtering techniques for speech analysis and synthesis' by F Itakura and S Saito

Conference Record 7 Int. Cong. Acoust. 25C1, Budapest (1971)

'New speech analysis and synthesis system PARCOR' by Fumitada Itakura *Nikkei Electronics* vol 2, No 12, pp 58–78 (1973) (In Japanese)

1969   'BUCKET-BRIGADE' DELAY CIRCUIT   F L G Sangster and K Teer (USA)

The general principle is that the signal to be delayed is sampled and stored in a cascade of capacitors interconnected by switches operated at the same frequency as the signal sampler.

As a new signal sample can evidently not be stored in a capacitor before the signal sample present is completely removed, only half the number of capacitors actually do store information at any moment, the others being empty.

In the past only rather complicated circuitry has been proposed for this function, so that even in integrated form there was no chance for an inexpensive compact design. A much simpler solution presents itself when signal sample transfer is not established by a charge transfer in the direction of signal travel but in the opposite direction, by what is essentially a charge deficit transfer. This principle leads to a much simpler resistorless circuit suitable for realization in integrated-circuit form.

SOURCE: 'Bucket-Brigade electronics—possibilities for delay, time-axis conversion and scanning' by F L J Sangster and K Teer. *IEEE Journal of Solid State Circuits* vol SC-4 No 3, p 131 (June 1969)

1969   MICROELECTRONICS (Bipolar)   Bell Laboratories (USA)

CDI (Collector Diffusion Isolation)   Ferranti (UK)

In 1970 manufacturers began to investigate bipolar processes which seemed to offer prospect of being competitive with m.o.s. For example there was the c.d.i. process (collector diffusion isolation) developed first at Bell Labs and then by Ferranti, the Isoplanar process of Fairchild, the Process IV which was suggested at Plessey's research centre at Caswell, and the Dutch Locos process developed by Philips. All of these were compatible with circuits which could operate in excess of 1.5GHz and all of then had the advantage of using less surface area than earlier processes. The c.d.i. system for example, started with a slice of 10 to 20Ω cm p-type silicon into which n-layers were diffused. These were later to be the collectors of transistors formed in a Lyle cm p-type epitaxial layer put down on top of them. The n^+ diffusions were made through the epitaxial layer to make contact with the now buried n^+ layers laid down at first. These not only acted as collector contacts but isolated the area within. In this base area the n^+ emitter diffusion is made, as well as any second emitter for a Shottky diode. After the oxide has been deposited and holes cut in it to gain access to the electrodes, silicon is grown in the holes to the s same level as the oxide, thus giving a flat surface.

SOURCE: 'The semiconductor story' by K J Dean *Wireless World* p 170 (April 1973)

SEE ALSO: 'Collector k-diffusion isolated integrated circuits' by B T Murphy V J Glinski, P A Gary and R A Pedersen *Proc. IEEE* vol 57, No 9, p 1523 (September 1969)

1969   'MAGISTOR' MAGNETIC SENSOR   E C Hudson IBM (USA)

The Magistor, invented by E C Hudson, Jr, is a dual-collector planar transistor operating below avalanche breakdown with beta values in the range of 30 to 100. In effect, this beta appears to amplify a typical Hall voltage. The sensitive axes are orthogonal to the substrate surface.

SOURCE: 'A magnetic sensor utilising an avalanching semiconductor device' by A W Vinal *IBM J. Res. Dev.* vol 25, No 3, p 196 (May 1981)

SEE ALSO: 'Transistor responds to magnetic fields' ed Robert H Cushman *Electronic Design News* pp 73–8 (15 February 1969)

1969   **SEMICONDUCTOR MEMORY SYSTEM**   **B Agusta, R D Moore and G K Tu**
                                                            (USA)

Since the first disclosure, by Agusta and Ayling *et al*, of the actual application of semiconductor memory in a computer system, the rapid development of silicon technology has led to the gradual (and probably eventually complete) replacement of magnetics by semiconductors in the memory field due to the speed, density, cost and power advantages of the latter.

SOURCE: 'Nonvolatile semiconductor memory devices' by J J Chang *Proc. IEEE* vol 64 No 7, p 1039 (July 1976)

SEE ALSO: 'A 64 bit planar double-diffused monolithic memory chip' by B Agusta ISCC *Digest of Tech. Papers* p 38 (February 1969)

'A high-performance monolithic store' by J K Ayling, R D Moore and G K Tu *ISCC Digest of Tech. Papers* p 36 (February 1969)

1970   **TUNG-FANG-HUNG (CHINA 1) Satellite**   **(People's Republic of China)**

Launched 24 April 1970. First satellite of the People's Republic of China.

SOURCE: Table of Artificial Satellites Launched Between 1957 and 1976 (Geneva: International Telecommunication Union) (1977)

1970   **VIDEO CASSETTE RECORDERS**   **Various (Japan, Europe)**

In addition to professional video recorders, manufacturers also designed models intended for the mass market. It was for this purpose that video cassette recorders were developed (as opposed to tapes).

Towards the end of the 1960s Matsushita, JVC and Sony together developed the standard U-Matic. The first models were launched on the market in 1970. Subsequently the standard U-Matic gained in prestige to the point where, today, it is considered to be the standard professional recorder.

In October 1970 Philips launched its VCR an apparatus aimed at the mass market.

SOURCE: *The Book of Inventions and Discoveries* Associate Editor Valerie-Anne Giscard d'Estaing (UK: Queen Anne Press, Macdonald & Co.) p 117 (1990)

NOTE: Philips (Holland) were the first to produce a machine with its own tuner and timer (i.e. suitable for home viewers) in 1973.

SOURCE: Private communication from E Davies, London

1970   **UNIX**   **Bell Labs, University of California**
                                                            (USA)

The Unix operating system grew out of research done at Bell Laboratories and at the University of California at Berkeley. It was orginally designed for minicomputers by Ken Thompson and Dennis Ritchie. Today there are versions of it for almost every sort of machine, from portables to supercomputers. The name Unix itself dates from 1970, and the first version was marketed in 1975. The main advantage of this rather complex operating system on a microcomputer is that it is multi-task and multi-user, which explains its growing share of the market. The two main versions for micros are called Unix System V and Xenix. AT&T (to which Bell Laboratories belong) have invested considerable sums in the development of the Unix operating system. At the time when its Unix System V version 3 seemed set to become the world standard, seven major builders representing 40 per cent of world computer science announced that they were setting up a foundation which would develop its own Unix standard. The Open Software Foundation was born in May 1988. Formed of IBM, HP, DEC, Bull, Siemens, Nixdorf and Apollo, it seems all set to fight the monopoly that AT&T was building up.

SOURCE: *The Book of Inventions and Discoveries* Associate Editor Valerie-Anne Giscard d'Estaing (UK: Queen Anne Press, Macdonald & Co.) p 117 (1990)

1970   **CHARGE COUPLED DEVICES**   **W S Boyle & G E Smith (USA)**

Storing charge in potential wells created at the surface of a semiconductor and moving the charge (representing information) over the surface by moving the potential minima.

Principle of operation: creation of potential wells.

Consider the application of an increasingly positive voltage to the gate of the m.o.s. structure shown in figure 11.30(a). As is to be expected, the influence of the gate on the underlying semiconductor closely resembles that of the gate of an m.o.s. transistor, i.e. a small positive gate bias causes the repulsion of majority carriers (i.e. holes) from the semiconductor immediately beneath the gate (figure 11.30(b)), whilst a gate bias in excess of the threshold voltage, V_{th}, makes it possible for an inversion layer to form at the oxide–semiconductor interface (figure 11.30(c)). Unlike the situation in an m.o.s. transistor, however, the existence of a gate voltage in excess of V_{th} does not necessarily mean that an inversion layer will form immediately in the structure of figure 11.30(a). This is because, whereas in an m.o.s.t. there is a source diffusion capable of supplying, almost instantaneously, a large number of minority carriers to form the channel, no such source exists adjacent to each c.c.d. electrode.

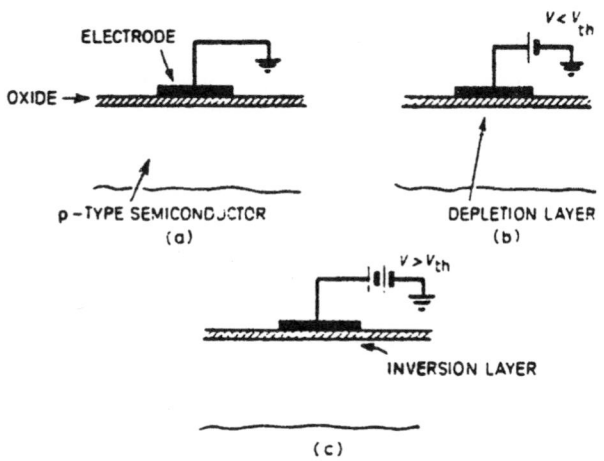

Figure 11.30. Single c.c.d. electrode showing the creation of depletion and inversion layers under the influence of an increasingly positive electrode voltage.

SOURCE: 'Charge coupled devices—concepts, technologies and applications' by J D A Benyon *The Radio & Electronic Engineer* vol 45, No 11, p 647 (November 1975)

SEE ALSO: 'Charge coupled semiconductor devices' by W S Boyle and G E Smith *Bell System Tech. J.* vol 49, p 583 (1970)

1970   **X-RAY LITHOGRAPHY FOR BUBBLE DEVICES**   **E Spiller, E Castellani, R Feder, L Romankiw, J Topalian and M Heritage (USA)**

X-rays have been used for several decades to obtain images of objects, First proposals to use x-rays for the fabrication of microelectronic devices were made in 1970. X-ray lithographic systems are very simple, they have a high throughput because many wafers can be exposed simultaneously and they have a resolution which is at least as good as that of electron beam systems. In 1972 Spears and Smith produced the first devices using proximity printing with x-rays and demonstrated the high resolution capability of x-ray lithography. In particular, x-ray lithography lends itself to the fabrication of high resolution devices requiring no alignment capability, such as magnetic bubble devices.

SOURCE: 'X-ray lithography for bubble devices' by E Spiller, R Fedor. J Topalian, E Castellani, L Romankiw and M Heritage *Solid State Technology* p 62 (April 1976)

SEE ALSO: 'X-ray projection printing of electrical circuit patterns' by R Feder *IBM Report* TR22, 1065 (August 1970)

'High resolution pattern replication using soft x-rays' by D L Spears and H I Smith *Electron. Lett.* vol 8, p 102 (1972)

'Evolution of bubble circuits processed by a single mask level' by A H Bobeck, I Danylchuck, F C Rossol and W Strauss *IEEE Trans. on Magnetics* vol MAG-9, p 474 (1973)

1970   'FLOPPY DISC' RECORDER   IBM (USA)

The concept of floppy discs is rapidly becoming accepted in many areas of data processing. Indeed, it has become a very much in-device when talking about data entry and data communications systems. And yet the very concept of floppy discs appears to have started almost by accident.

The first, designed and developed by IBM, was simply a component of the IBM 370 when introduced in 1970—and a pretty obscure one at that. In its original form it was part of the diagnostic system for the 370/155 and 165 and formed part of the controller for the IBM 3300 disk drive. At that stage it was a read-only device and enabled diagnostic programs to be introduced quickly, to identify faults and help to reduce maintenance time.

Indeed, it was not until 1973 that it can be properly said to have formed an integral part of an IBM system. That was the 3740 data entry system ... which promptly established an industry standard and, because it provided three times the storage and much faster access, rendered what competition there was virtually obsolete overnight.

SOURCE: "Accident' became the floppy discs' by W Boffin, *Electronics Weekly* p 11 (12 May 1976)

1970   NATO-1 Satellite   NATO (International)

Launched 20 March 1970. Telecommunication satellite. Geostationary space craft stationed over the equator at approximately 18°W longitude. Hundreds of communications of various types (voice, wideband data, telegraph and facsimile data). More than 7000 solar cells, batteries.

SOURCE: Table of Artificial Satellites Launched Between 1957 and 1976 (Geneva: International Telecommunication Union) (1977)

1971   BUMPED TAPE AUTOMATIC BONDING   T S Liu and H S Fraenkel (USA)

The concept began with Tape Automated Bonding which automated the bonding and assembly of packaged IC devices, initially using patterned etched copper foil lead frames and IC chips with raised contact pads or bumps. BTAB modified this approach by providing the bumps at the end of each lead— making it now possible to gang bond to conventional IC chips. These bumps provided the requisite bonding material and the physical standoff to prevent lead/chip shorting. The potential for volume production and cost savings motivated the development of essential equipment, materials, processes and tooling. However, because of competitive and proprietary aspects, much development was carried out independently and thus often duplicated.

SOURCE: 'BTAB's future—an optimistic prognosis' *Solid Stae Technology* p 77 (March 1980)

SEE ALSO: 'Bumped tape automated bonding (BTAB) applications' by R F Unger, C Burns and J Kanz *Proc. International Microelectronic Conference* Anaheim, CA 27, pp 71–7 (February–March 1979)

'Application of tape automated bonding technology for hybrids' by R G Oswald and W R Rodrigues de Miranda *Solid State Technology* pp 33–8 (March 1977)

1971   HOLOGRAM MATRIX RADAR   K Iizuka V K Nguyen and H Ogura
                                                 (Canada)

The concept of hologram matrix is proposed. This concept was incorporated into the design of a novel radar which, unlike conventional radars, determines the distance by the spatial distribution of the scattered wave rather than by the lapse of times. The radar based upon this principle was developed and built for the purpose of mapping ice thickness in the range of 0.5 ~ 5 m but it has potential applications in other fields.

Such a radar has real-time processing capability resulting from an amalgamation of the antenna and computer subsystems. The programability of the radiation pattern by software of the processing simplifies the construction of the radar. Capability of dual focussing of the transmitter and receiver eliminates the necessity of either pulsing, or frequency modulation of the transmitting signal. Superior performance in the short range, with high resolution, is particularly advantageous for measuring lossy ice.

These features were substantiated by experimental results obtained from the field operation of the system.

SOURCE: 'A hologram matrix radars' by K Iizuka, H Ogura, J L Yen, Van-Khai Nguyen and John R Weedmark *Proc. IEEE* vol 64, No 10, p 1495 (October 1976)

SEE ALSO: 'Review of the electrical properties of ice and HISS down-looking radar for measuring ice thickness' by K Iizuka, V K Nguyen and H Ogura. Presented at the Aerospace Electronic Symposium in Toronto, Canada on March 16 1971. Text is published in *Can. Aeronaut. Space J.* vol 17, No 10, pp 429–30 and pp M28–M33 (December 1971)

'Hologram matrix and its application to a novel radar' by H Ogura and K Iizuka. *Proc. IEEE (Lett.)* vol 61, pp 1040–1 (July 1973)

1971 CARRIER-DOMAIN MAGNETOMETER   B Gilbert (USA)

A carrier-domain device, as conceived by Gilbert, consists of an elongated bipolar transistor, within which emitter-current flow is restricted to a small region known as a domain. The domain can be moved within the device, subject to an external signal. Using this concept novel devices, whose functions arise directly from their geometry, can be designed. One form of carrier-domain device proposed by Gilbert is a magnetic-field sensor in which two domains are caused to rotate together around a circular device by the application of a magnetic field normal to the silicon surface. Output current pulses are produced at a rate proportional to the magnetic-flux density. The first successful carrier-domain magnetometer (c.d.m.) based on this design has been fabricated and operated by the authors, and brief details of its operation have been published.

SOURCE: 'The carrier-domain magnetometer: a novel silicon magnetic field sensor' by M N Manley and G G Bloodworth *Solid-State and Electron Devices* vol 2, No 6, p 176 (November 1978)

SEE ALSO: 'New planar distributed devices based on a domain principle' by Gilbert B *IEEE ISSCC Technical Digest* p 166 (1971) (Ref. 1: p 183)

'Novel magnetic field sensor using carrier domain rotation: proposed device designs' by Gilbert B *Electron. Lett.* vol 12, pp 608–10 (1976) (Ref. 2: p 183)

'Novel magnetic field sensor using carrier domain rotation: operation and practical performance' by M H Manley, G G Bloodworth and Y Z Bahnas *Electron. Lett.* vol 12, pp 610–11 (1976) (Ref. 3: p 183)

1971 ELECTRONIC DIGITAL WATCH   Time Computer Corporation (USA)

The first electronic digital watch was introduced in the fall of 1971: the Pulsar, retailing for $2000 with an 18-carat gold case bracelet. Touch a button and the light-emitting diodes showed the time. (It took about a year for Pulsar to add the day and date.)

SOURCE: *Electronics* p 401 (17 April 1980).

1971 FAMOS (Floating-Gate Avalanche-Injection   D Frohman-Bentchkowsky (USA)
Metal-Oxide-Semiconductor) INTEGRATED
CIRCUIT

Famos describes the floating-gate avalanche-injection metal-oxide semiconductor transistor that Dov Frohman-Bentchkowsky developed at Intel Corp. in 1971.

The Famos device is essentially a silicon-gate MOS field-effect transistor in which no connection is made to the floating silicon gate. Instead, charge is injected into the gate by avalanches of high-energy electrons from either the source or the drain. A voltage of −28 volts applied to the pn junction releases the electrons.

Data is stored in a Famos memory by charging the floating-gate insulator above the channel region. The threshold voltage then changes, and the presence or absence of conduction is the basis for readout.

The Famos cell has generally been considered more reliable than nitride storage mechanism used in reprogrammable metal-nitride-oxide-semiconductor memories. In MNOS memories carriers tunnel through a thin oxide layer into traps at the oxide-nitride interface. But a partial loss of stored charge during readout limits the number of readout cycles to approximately 10".

In Famos memories, on the other hand, there is no loss of charge due to reading. Moreover, over time, the loss of stored electrons is negligible, less than one per cell per year, and information retention is excellent.

SOURCE: 'The Famos principle' *Electronics* p 109 (3 March 1977)

1971 **DSCS-1 Satellite** (USA)

Launched 3 November 1971. Defense Satellite Communication System. Synchronous satellite carrying multichannel communications payload. Four antennae, two for wide earth coverage and two with narrow beams for ground controlled direction beaming for high-volume communications. Capacity: 1300 circuits.

SOURCE: Table of Artificial Satellites Launched Between 1957 and 1976 (Geneva: International Telecommunication Union) (1977)

1971 **CERAMIC CHIP CARRIER** 3M Co. (USA)

A popular IC package was the ceramic chip-carrier. About one third the size of a comparable DIP, it originated in 1971 at the 3M Co., in St Paul, Minn. It was a square, multilayered ceramic package whose bottom periphery contained a pattern of gold bumps on 40- or 50-mil centres. The chip was bonded to a gold base pad inside a cavity within the ceramic. The small hermetically sealed package could be easily attached or removed from pc boards and hybrids.

SOURCE: *Electronics* p 389 (17 April 1980)

1971 **SALYUT-1 Satellite** (USSR)

Launched 19 April 1971. Objectives: scientific research and testing of on-board systems and units. Control by remote command or by crew. Visited by crews of Soyuz-10 and Soyuz-11. The latter spent 23 days in Salyut. Decayed on 11 October, 1971.

SOURCE: Table of Artificial Satellites Launched Between 1957 and 1976 (Geneva: International Telecommunication Union) (1977)

1971 **LIQUID CRYSTAL STUDY OF OXIDE DEFECTS** J M Keen (UK)

One of the most convenient techniques was first reported by Keen. It consists of introducing, between the oxidised silicon and a tin oxide coated glass slide, a thin film of negative nematic liquid crystal. On applying a voltage across this 'capacitor' structure, defects can be seen as highly turbulent regions of liquid crystal. For plane electrodes without an oxide layer the same turbulence is present everywhere. In the case of oxides containing defects, however, the turbulence is particularly violent making location of a defect easy.

SOURCE: 'Polarity dependent oxide defects located using liquid crystals' by A K Zakzouk, W Eccleston and R A Stuart *Solid State Electronics* vol 19, p 133 (1976)

SEE ALSO: J M Keen *Electron Lett.* vol 7, No 15, p 432 (1971)

1971   **MICROPROCESSOR**   **M E Hoff (USA)**

In 1971 Marcian E Hoff, then working for Intel, developed the first microprocessor which he baptised the 4004. Hoff brought together the elementary functions of a computer on a single electronic component (an integrated circuit). It contained the equivalent of 230 transistors and was a four-bit processor.

SOURCE: *The Book of Inventions and Discoveries* Associate Editor Valerie-Anne Giscard d'Estaing (UK: Queen Anne Press, Macdonald & Co.) p 119 (1990)

1972   **MICROCOMPUTER**   **Intel (USA)**

For three years a great revolution has been taking place in digital electronics. Since 1972, when the first microcomputer was introduced by Intel, these devices have been very successfully used in a wide range of applications, including process control, data communications, instrumentation and commercial systems. The key to this success is due to the price/performance ratio enhancement that occurs when microcomputers are used in a system when compared with more traditional approaches.

SOURCE: 'The microcomputer comes of age' by H Kornstein *Microelectronics* vol 8, No 1, p 17 (1976)

1972   **VIDEO GAMES**   **Magnavox (USA)**

First to market a video game consumers could buy and take home was Magnavox, in 1972. However, their original Odyssey game was not an immediate sensation, perhaps because it had no automatic score-keeping feature, lacked sounds, and required a plastic overlay on the TV screen to simulate the net, goals and boundaries of a playing field (static electricity held the overlay in place). Since then, Odyssey has evolved through four model changes and is now offered with automatic serve, digital scoring and sound for $89.95. Also, all stationary playing-court features are electronically generated. Magnavox's game circuits are produced by Texas Instruments and the General Instrument Corp.

The second major milestone in the evolution of home video games was established just over a year ago when Atari and Sears teamed up to produce and market Hockey Pong for the 1975 Christmas season. The product had many of the important features associated with Atari's successful line of coin games and sold briskly. By working closely with one of its integrated circuit suppliers (American Microsystems, Inc.) Atari was assured a supply of dedicated, proprietary game chips. These chips greatly reduced the parts count, and costs, associated with Atari's original coin-game products that were constructed with hundreds of standard logic circuits. The coin games normally sell for $1000–$3000 each—but. consumer acceptance of the add-on TV version was expected to peak somewhere under $100 retail.

Meanwhile, with Odyssey more streamlined and the Atari/Sears venture a proven success, the lucrative aspects of consumer TV games became obvious to other entrepreneurs. For many, the first opportunity to enter the market came in March 1976 when the General Instrument Corp. (GI) Hicksville, NY, announced production of its AY-3-8500 TV-game integrated circuits. Product acceptance was swift and favourable. New companies anxious for an early shot at the home video game sweepstakes quickly snapped up GI's projected 1976 production capacity of game chips.

SOURCE: 'Electronic Gamesmanship' by D Mennie *IEEE Spectrum* p 27 (December 1976)

SEE ALSO: 'Video games: perishable or durable' *J. Electronics Industry. Japan* vol 23, No 10, p 38 (October 1976)

1972   **LANDSAT-1 (ERTS-1) Satellite**   **(USA)**

Launched 23 July 1972. Earth Resources Technology Satellite. Objectives: to obtain coverage of the United States and other major land masses with multispectral, high spatial resolution (60 m) images of solar radiation reflected from the earth's surface. These images will be used in agricultural, geological, geographical, hydrological and oceanographical research.

SOURCE: Table of Artificial Satellites Launched Between 1957 and 1976 (Geneva: International Telecommunication Union) (1977)

1972   **MICROELECTRONICS (V-MOS technique)**   **T J Rodgers (USA)**

Because of the work of a 27-year-old research engineer, American Microsystems Inc., in Santa Clara, California, is on the verge of committing itself in a big way to V-MOS—an n-channel metal-oxide semiconductor technology that will compete with the new, faster and denser bipolar static designs and processes.

The engineer is T J Rodgers, who, as a doctoral candidate in electrical engineering at Stanford University in nearby Palo Alto, invented the V-groove MOS process (*Electronics* 18 September p 65). His goal was to push MOS technology to its limits so it would achieve bipolar speeds as well as high speed power products and high packing densities in read-only and static random-access memories, random logic and microprocessor designs.

SOURCE: 'Young EE's ideas to alter AMI's direction' *Electronics* p 14 (22 January 1976)

1972   **NITROGEN-FIRED COPPER WIRING**   **J D Grier (USA)**

In late 1971, market researchers at Owens-Illinois Inc., Toledo, Ohio, decided that the rising cost of precious metals made the time right for research on non-noble conductors for thick-film microelectronics. A young chemist, John D Grier, who had been with the firm for five years, was assigned as program manager.

Grier decided to concentrate on creating a workable nitrogen-fired copper paste. There had been earlier research on copper pastes, but these compositions used 100-micrometre copper particles to produce conductors with poor peel strength and low conductivity.

He went to 3 to 5 μm copper particles for the functional phase of the ink and found both a glass binder and vehicle that could survive firing at about 800°C. The late 1972 result was a patented, practical, screenable copper paste that had good peel strength and conductivity. At that point Grier correctly predicted that the new copper paste would be suitable for microstrip and thick-film hybrid applications.

SOURCE: *Electronics* p 121 (28 October 1976)

1972   **X-RAY SCANNER**   **EMI (UK)**

The skull surrounds the brain and provides a very good protection for this most delicate and vital organ; it also heavily attenuates diagnostic X-rays. The brain is a relatively homogeneous organ, when imaged by X-rays, which does not have much contrast to show up its structure. These two problems make imaging of the brain by conventional X-radiography of very limited diagnostic value. Contrast techniques can be used to improve the imaging but they do involve some risk to the patients and the need for hospitalisation. They are expensive.

In 1972, EMI Limited introduced computerised axial tomography to overcome these limitations. This radical new technique was developed at the Central Research Laboratories of EMI. Clinical trials rapidly showed that this was a major advance in diagnostic imaging.

In computerized axial tomography the patient is scanned by a tightly collimated narrow beam of X-rays. The transmitted beam is detected and converted to an electric signal after passing through the patient. Another detector is used in the reference mode to measure the primary X-ray beam.

The frame, carrying the X-ray source and detectors traverses linearly across the patient, a large number of readings of X–ray intensity are taken and stored as it traverses, the gantry is then indexed round by a small angle and the process is repeated. This series of verses and angular movements is repeated until a large matrix of data has been acquired.

The computer then uses this data to calculate the X-ray absorption coefficient map of this cross section of the anatomy. This can then be displayed as a brightness modulation map on a cathode ray tube or printed as a map of X-ray absorption numbers by a line printer. The computation cancels out the effects of absorption in other parts of the anatomy so that the problem of shadowing by the skull or bone structure is overcome.

SOURCE: 'Section by section' by Shelley Stuart *Electronics Weekly* p 16 (7 April 1976)

1972   AUTOMATIC CONTROL OF CRYSTAL GROWTH   W Bardsley, G W Green, C H Holliday and D T J Hurle (UK)

In this note, we describe a novel, alternative method of automatic diameter control (or, more strictly, control of cross-sectional area, since the crystals may be of non-circular section) for which certain advantages can be claimed. Put simply, the method comprises 'weighing' the growing crystal by means of an industrial weighing cell from which the pull rod is hung. The method requires that there are no constraints to the vertical motion of the pull rod, and this is achieved by a gas bearing where the rod enters the growth chamber. Normally, some of the ambient gas escapes through the gas bearing, but for the initial evacuation and flushing before growth, the rod is sealed by a constrictable rubber sleeve. The pull rod has a self-aligning bearing at its upper end to provide a connection to the weighing cell, and is rotated by a low friction pin and fork arrangement.

The electrical signal from the weighing cell is compared with a signal from a rectilinear potentiometer driven from the leadscrew nut and any difference is amplified and used to adjust the crucible heating power in that direction which minimises the difference signal. The desired diameter is predetermined by setting electrically the magnitude of the potentiometer output voltage per unit distance of pull rod travel. The initial growth out from the diameter of the seed crystal to the final diameter has also been automatically controlled by introducing a non-linear element in series with the potentiometer output circuit.

SOURCE: 'Automatic control of Czochralski crystal growth' by W Bardsley, G W Green, C H Holliday and D T J Hurle *J. Crystal Growth* vol 16, p 277 (1972)

SEE ALSO: 'Developments in the weighing method of automatic crystal pulling' by W Bardsley, B Cockayne, G W Green, D T J Hurle, G C Joyce, J M Roslington, P J Tufton and H C Webber *J. Crystal Growth* vol 24/25, p 369 (1974)

1972   1024 BIT RANDOM ACCESS MEMORY   Intel (USA)

The RAM-father of them all, the 1103 from Intel started the stampede to semiconductor memories. It was the first time that more than 1000 bits of read/write memory could be supplied on a single semiconductor chip in a low-cost MOS configuration.

SOURCE: 'Special report—semiconductor RAM's land computer mainframe jobs' by L Altman *Electronics* p 64 (28 August 1972)

1972   INTEGRATED INJECTION LOGIC   K Hart and A Slob (Holland)

Logic gates suitable for large-scale integration (LSI) should satisfy three important requirements. Processing has to be simple and under good control to obtain an acceptable yield of reliable IC's containing about 1000 gates. The basic gate must be as simple and compact as possible to avoid extreme chip dimensions. Finally, the power-delay time product must be so high that operation at a reasonable speed does not cause excessive chip dissipation.

Multicollect or transistors fed by carrier injection proved to be a novel and attractive solution. A simplified (five masks) standard bipolar process is used resulting in a packing density of 400 gates/mm^2 with interconnection widths and spacings of 5 μm. The power-delay time product is 0.4 pJ per gate. An additional advantage is a very low supply voltage (less than 1 V). This, combined with the possibility of choosing the current level within several decades enables use in very low-power applications. With a normal seven-mask technology, analog circuitry has been combined with integrated injection logic (I^2L).

SOURCE: 'Integrated injection logic—a new approach to LSI' by Kees Hart and Arie Slob *IEEE Journal of Solid State Circuits* vol SC-7 No 5, p 346 (October 1972)

SEE ALSO: 'Super integrated bipolar memory devices' by S K Wiedman and H H Berger. Presented at the *IEEE Int. Electron Devices Conference* (11–13 October 1971)

1972 **DEEP PROTON-ISOLATED LASER**   **J P Dyment, L A D'Asaro, J C North, B I Miller and J E Ripper (USA)**

Proton bombardment as a means of isolation is now widely used for a number of semiconductor devices. It was first demonstrated for (GaAl)As/GaAs heterostructure lasers in 1972.

High peak-power lasers with junctions at a depth of 40 μm from the surface are required for applications where fibreoptics, with 25 μm square fibres, are used to couple the output of several lasers to form a high brightness source.

SOURCE. 'Deep proton-isolated lasers and proton range data for InP and GaSb *Solid-State and Electron Devices* vol 3, No 1, p 1 (January 1979)

SEE ALSO: 'Proton-bombardment formation of stripe-geometry heterostructure lasers for 300 K c.w. operation' J C Dyment, L A D'Asaro, J C North, B I Miller and J E Ripper *Proc. IEEE* vol 60, pp 726–8 (1972)

1972 **VIDEO DISCS**   **Philips (Holland)**

Philips Gloeilampenfabrieken demonstrated a long-playing video disk in September 1972. It was a dramatic improvement over an AEG-Telefunken/Decca black-and-white video disk that had been demonstrated in 1970. Because the Teldec disks had grooves and mechanical tracking, they suffered from short playing time (an 8-inch disk played for only 5 minutes) and high record wear. The Philips disks held 30 to 45 minutes of colour material and instead of grooves had submicrometer pits molded into a spiral track; a laser-generated light spot read the patterns.

SOURCE: *Electronics* p 409 (17 April 17 1980)

1973 **DRY ETCHING**   **Mitsubishi Electric Co (Japan)**

The application of dry etching techniques using plasma chemistry to semiconductor processing was introduced by Mitsubishi Electric Co., Japan. It shows that gas plasma containing fluorine species are able to etch silicon and its compounds (SiO_2, Si_3N_4). In the experiments, the gas used was CF_4. A plasma was produced by rf (13.56 MHz) discharge and a barrel type plasma reactor was used. The etching mechanism was principally considered to be a chemical reaction between silicon and the fluorine radicals in plasma. However, the details of the etching characteristics and the etching mechanisms were not known in those days.

This technique promised a number of advantages over wet etching methods in terms of improved precise pattern control, problems of etchant preparation and disposal, and cost. It was expected to play an important role in the fabrication of Si integrated circuits (SLI, VLSI).

SOURCE: 'Etching characteristics of silicon, and its compounds by gas plasma' by H Abe, Y Sonobe and T Enomoto *Jap. J Appl. Phys.* vol 12, No 1 (1973)

1973 **SCANNING ACOUSTIC MICROSCOPE**   **C F Quate (USA)**

In 1973 Professor Quate conceived an approach of elegant simplicity to produce a microscope that would use sound, rather than light, in order to form images. This achievement which had been the aim of applied scientists for more than fifty years, led to the extremely rapid development of a microscope which already exceeds the resolution of optical microscopes.

The key idea, which was the recognition of the fact that velocities of acoustic waves in some solids can be as much as seven times greater than the velocity in water, resulted in the production of a lens which could focus a beam of sounds on its axis, without significant aberrations. Whilst such a lens cannot image a complete field, Professor Quate recognised that the axial focus was enough for the realisation of a mechanically-scanned microscope in which the image was reproduced point by point. The scanning acoustic microscope has opened up a completely new field of microscopy which permits the direct imaging of biological specimens and the examination of silicon integrated circuits and other solid objects.

SOURCE: 'Major prizes for opto-electronics inventions' *The Radio and Electronic Engineer* vol 52, No 3, p 107 (March 1982)

SEE ALSO: 'Seeing acoustically' by R K Mueller and R L Rylander *IEEE Spectrum* p 28 (February 1982)

'Thermal imaging via cooled detectors' by D B Webb *The Radio and Electronic Engineer* vol 52, No 1, p 17 (January 1982)

'Recent developments in scanning acoustic microscopy' by D A Sinclair, I R Smith and H K Wickramasinghe *The Radio and Electronic Engineer* vol 52, No 10, p 479 (October 1982)

1973   **LOGIC-STATE ANALYSER (Displaying Binary**   **C H House (USA)**
       Notation in 1s and 0s)

       LOGIC-TIMING ANALYSER (for Recording,   **B J Moore (USA)**
       Displaying and Analysing Complex Timing
       Relationships)

One problem, two men, two solutions: yet both designers were right; both of their designs were needed. So Charles H House of Hewlett-Packard Co's Colorado Springs (Colo.) division, and B J Moore of Biomation Corp. in Cupertino, California, developed two markedly different diagnostic instruments that were the first such electronic tools for studying, designing and troubleshooting complex digital logic circuits and systems.

The 160 1L is a plug-in unit for HP's 180 series oscilloscopes, giving a 12-channel, 16-word-memory logic-state analyser for 10-megahertz operation. The Biomation 10-MHz 810-D digital logic recorder stores 256 logic states on each of eight channels, displaying waveform-like timing diagrams on an oscilloscope.

SOURCE: 'Logic-analyser originators cited for testing innovation 1977 award for achievement' *Electronics* p 83 (27 October 1977)

1973   **SKYLAB-1 Satellite**   **(USA)**

Launched 14 May 1973. Manned orbital research laboratory. Objectives: to determine man's ability to live and work in space for extended periods: to extend the science of solar astronomy beyond the limits of earth-based observation; to develop improved techniques for surveying earth resources, to make various investigations requiring a constant zero gravity environment.

SOURCE: Table of Artificial Satellites Launched Between 1957 and 1976 (International Telecommunication Union, Geneva) (1977)

1974   **IT (Information Technology)**   **Various (Worldwide)**

Information and communications technologies are changing the way work, study, do research, and educate our children and ourselves. They are influencing the way we do our banking, pay our bills, entertain ourselves and do business. New options are being provided for us in the field of health care, education, environmental protection, culture, and business. A more direct and open rapport between private individuals and public administrations is becoming increasingly possible.

The impact of this information revolution on our society cannot yet be fully measured or predicted at this time. The combination of new and rapidly developing interactive multimedia computers and applications with electronic networks will require a restructuring of our traditional approach to strategic planning and organisational structure. It will also mean a considerable change in the way we interact with each other, with business and with government.

Moreover, it has the potential to overcome the marginalising effects of distance and geography. It could enable regional economies to be revitalised, and consumers and businesses in rural and remote areas to be re-integrated into mainstream economic and cultural activity.

For each individual citizen, the information society also means greater choice and new opportunities, sharing of cultural knowledge and experiences and the creation of new markets and employment opportunities.

SOURCE: *I & T Magazine* No 17, p 1 (July 1995)

1974 ELECTRON BEAM LITHOGRAPHY   Bell Laboratories (USA)

Electron-beam lithography was the key to making the masks for the optical lithography units. Without its ability to make masks with micrometre-wide lines, no LSI lithography technique based on the use of either masks or reticles would have been possible.

One of the first electron-beam systems came from Bell Laboratories in 1974. Called the Electron-Beam Exposure System, it made masks by using a raster-scanned beam aimed at a continuously moving table. Wafer alignment with the beam was controlled by a laser interferometer.

SOURCE *Electronics* p 388 (17 April 1980)

1974 CATT (CONTROLLED AVALANCHE   S P Yu, W R Cady and W
TRANSIT-TIME TRIODES)   Tantraporn (USA)

The use of avalanche and transit-time effects in microwave transistor-like structures for increased gain and higher-frequency operation has been proposed by us and others in recent publications. We have previously described the basic principle and large-signal theory of the controlled-avalanche transit-time triode (CATT) and have also reported some initial experimental results in the 1–3 GHz region. The purpose of the present paper is to discuss more fully certain aspects of CATT design and operation resulting from the avalanche-multiplication process and those whose importance has become clearer through our further investigations.

SOURCE: 'Avalanche multiplication in CATTs' by J R Eshbach, S P Yu and W R Cady *IEE Solid State and Electron Devices* vol 1, No 1, p 9 (September 1976)

SEE ALSO: 'A new three-terminal microwave power oscillator' by S P Yu, W R Cady and W Tantraporn *IEEE Trans.* vol ED-21, p 736 (1974)

'The third terminal in microwave devices' by J E Carroll *Proc. European Solid-State Device Research Conf.* Nottingham (1974)

'Transistor improvements using an impatt collector' by A M Winstanley and J E Carroll *Electron. Lett.* vol 10, p 516 (1974)

1974 PRESTEL System   S Fedida (UK)

Mr Fedida invented the concept of viewdata whilst working at the Post Office Research Centre in the early 1970s. It combines a modified television set, a telephone line and a computer: a push button control panel calls up a 'page' of the information required by a subscriber on to a television screen using a telephone line link routed into a computer data bank. The simplicity of operating the system provides the potential for the mass marketing of information on a wide range of general and technical subjects.

SOURCE: '1979 MacRobert award for software system inventor' S Fedida *The Radio and Electronic Engineer* vol 50, No 1/2, p 10

1974 WESTAR-1 Satellite   (USA)

Launched 13 April 1974. First United States domestic communication satellite placed in synchronous orbit over equator at 99°W. Can transmit 12 colour television channels or up to 14 400 one-way telephone circuits through five earth stations located close to New York, Atlanta, Chicago, Dallas and Los Angeles.

SOURCE: Table of Artificial Satellites Launched Between 1957 and 1976 (Geneva: International Telecommunication Union) (1977)

1974   **16-BIT SINGLE CHIP MICROPROCESSOR**   **National (USA)**

The semiconductor industry's first 16-bit, single-chip microprocessor is soon to be introduced by National Semiconductor Corp. Called PACE (for processing and control element), the device will handle 16-bit instructions and addresses, and either 16-bit or 8-bit data. It is being built with p-channel silicon-gate MOS technology because, the company says, p-MOS is a more predictable and better established technology than n-MOS and meets both of PACE's main requirements: 10-microsecond execution time for instructions, and enough density to fit the entire circuit on a single chip.

PACE requires only two power supplies, $+5$ V and -12 V, instead of the three required with n-channel fabrication.

SOURCE: 'National to show 16-bit processor on single chip' *Electronics* p 35 (28 November 1974)

1974   **BAR CODES (The Uniform Product Code)**   **Ad Hoc Committee of the Grocery Industry (USA)**

A historic moment came at 8.01 am on 26 June 1974 when Clyde Dawson, director of research and development for Marsh Supermarkets, bought a 10-pack of Wrigley's Chewing Gum in his company's store in Troy, Ohio—the first purchase made in the first store to be fully equipped with scanners which could read the new Uniform Product Codes (UPC). The UPCs, or barcodes, the patterns of black and white lines printed on the packaging of groceries and other merchandise, signalled a new stage in the development of the grocery industry in the United States, a transformation brought about by the use of the UPC.

What was novel about the process of innovation leading to be barcode was the way in which the people involved in this development were organized. A committee, the Ad Hoc Committee of the Grocery Industry, was formed to decide whether a code was needed and once they had established the need, they went on to develop specifications for both a code and the equipment to read it. From these specifications various manufacturers then designed the actual equipment.

The barcode system is an early, perhaps the first, example of a new way of handling technological change on a large scale that it affected very rapidly, not just one firm or sector but the whole of a major international industry. Essentially, the barcode was an innovation produced to order—a 'free enterprise' version of central planning—and the organization that carried out this project, the Ad Hoc Committee, came out of the particular structure and conditions found in the US food industry in the 1960s. The great achievement of the committee was not just to devise the actual code system but to facilitate the future development of the whole of the food industry.

SOURCE: 'Packaging history: The emergence of the uniform product code (UPC) in the United States, 1970–75' by A Morton *History and Technology* vol 11, p 101 (Amsterdam: Harwood Academic Publishers) (1994)

1975   **THE GYROTRON**   **A G Gapanov *et al* (USSR)**

Although the foundations for high-power-gyrotron development were laid sometime previously, the first reported real breakthrough describing a working device was in 1975. In this device, the electrons are made to rotate at a cyclotron frequency (which is also near the operating frequency or a sub-harmonic of it) of the static magnetic field. Hence it is sometimes called a 'cyclotron resonance maser'. All devices mentioned here are based on developments arising from the concept of the cyclotron resonance maser.

The impetus for the development of gyrotrons came from Russian workers in the nuclear-energy field, who hoped to be able to heat dense plasma, confined by a powerful magnetic field, by adsorbing microwave energy at the cyclotron resonance frequency. If sufficient power could be absorbed in this way, then temperatures approaching those required for fusion could be reached. The power level required is in the region of 10 to 20 MW for several seconds. The frequency required for typical magnetic fields used in plasma containment machines (such as Tokamaks) is 50 to 100 GHz for magnetic fields of 2.0 to 4.0 T.

SOURCE: 'The gyrotron' by M J Smith *Electronics and Power* p 389 (May 1981)

SEE ALSO: Gapanov A V *et al Radiofizika* 18, p 280 (1975)

Gapanov A V *et al* 'A device for cm–mm and sub-mm wave generation' Copyright No 223931 with priority of 24 March 1967 (Official bulletin KD10 of SM USSR (11) p 200, 1976)

1975 LOCMOS (Locally Oxidised Complementary Metal-Oxide-Semiconductor) INTEGRATED CIRCUIT — Philips (Netherlands)

LOCMOS is an acronym for locally-oxidized CMOS, a process invented by Philips Research Laboratories which produces a high performance, high density CMOS that costs no more than standard CMOS. The LOCMOS 4000 range which is pin-for-pin compatible with other popular 4000 ranges, needs less chip area per function and thus enables full buffered circuitry to be built into every device.

Features of LOCMOS 4000 include high noise immunity, standardised outputs and increased system speeds. The increased voltage gain, due to buffering, gives almost ideal transfer characteristics, and every device will give a guaranteed output of 400 μA from a 5V power supply. Output impedance and propagation delay are independent of input pattern and reduced sidewall capacitance results in higher speed.

SOURCE: 'Mullard announce LOCMOS 4000' Mullard Press Information Sheet, p 1 (September 1975)

1975 VIKING-1 Satellite — (USA)

Launched 20 August 1975. Objectives: to explore the surface and atmosphere of the planet Mars. Includes an orbiter and a lander separating on approach to Mars.

Orbiter: spacecraft arrived at Mars in June 1976.

Lander: landed on Mars on 20 July 1976.

SOURCE: Table of Artificial Satellites Launched Between 1957 and 1976 (Geneva: International Telecommunication Union) (1977)

1975 MICROELECTRONICS (Integrated Optical Circuits) — F K Reinhart and R A Logan (USA)

For the first time, scientists have combined a laser with components such as modulators, filters, and lightguides in a single crystal microcircuit, just as multiple components are fabricated in an integrated electronic circuit.

The devices, integrated optical circuits measuring usually about 6 by 15 mils, operate within the structure of a semiconductor injection laser.

This type of circuit represents an alternative to hybrid integrated optics where components—often fabricated from different material systems are interconnected on a base. By contrast, the new monolithic optical circuit 'contains' many of the required components within the same single crystal.

Franz K Reinhart and Ralph A Logan of Bell Laboratories, Murray Hill, New Jersey, developed the new circuit.

SOURCE: 'Integrated optical circuits: another step forward' *Bell Labs. Record* p 349 (September 1975)

1975 SILICON ANODISATION — R Cook (ITT) (USA)

The discovery that silicon itself can be anodised opens an unexpected path to cheaper, denser, faster integrated circuits. The low-temperature process produces in one step the dielectric needed to isolate the active elements on a chip, thus adding the advantages of dielectric isolation to any semiconductor technology, whether bipolar or metal-oxide-semiconductore

Direct silicon anodisation was discovered quite by chance. An anodizing voltage was accidentally increased beyond the point required to anodise aluminium. The aluminium was destroyed, but the silicon substrate beneath the aluminium, when examined under a microscope, was seen to have been transformed into a porous dielectric layers. Further experiment revealed that the dielectric on the silicon surface could be tailored to almost any desired thickness simply by adjusting the anodising process.

SOURCE: 'Anodizing silicon is economical way to isolate IC elements' by R Cook *Electronics* (13 November 1975) p 109

1975   4096-BIT RANDOM ACCESS MEMORY   Fairchild (USA)

In a significant development, Fairchild Semiconductor has applied its oxide-isolated Isoplanar technology to an injection-logic configuration. The result: the industry's first 4096-bit I^2L random-access memory. The part has a nominal access time of 100 nanoseconds, making it more than twice as fast as today's n-MOS 4-kilobit dynamic RAMs. The device will be ready for selective prototyping late this summer.

SOURCE: 'Fairchild develops first 4K RAM to use I^2L' *Electronics* p 25 (26 June 1975)

1975   THIN FILMS—DIRECT BONDED COPPER   J F Burgess, C A Neugebauer, G
PROCESS   Flanagan and R E Moore (USA)

In the direct copper to ceramic bonding process, bonding is accomplished by heating Al_2O_3 or BeO substrates in contact with the Cu foil. Foil thicknesses from 250 to 1 mil can be used. The gas atmosphere consists principally of inert gas such as argon or nitrogen with a small addition of oxygen, typically of the order of a few hundredths of a percent. The length of time required for bonding is typically a few minutes. The temperature for bonding is critically important. Bonding does not take place unless the temperature exceeds 1065°C, but it must be below 1083°C, which is the melting point of copper.

SOURCE: 'Hybrid packages by the direct bonded copper process' by J F Burgess, N A Neugabauer, G Flanagan and R E Moore *Solid State Technology* p 42 (May 1975)

1975   VHS RECORDER   JVC (Japan)

The VHS format (Video Home System) was launched by JVC in October 1975 and marketed as from 1976. The VHS now holds a dominant position in the world market, with more than 80 per cent of sales.

A variation of the VHS model, the VHS-C was launched by JVC in 1982. It was then intended for the portable video and uses reduced size video cassettes which can be re-read on a traditional VHS recorder thanks to an adaptor.

In July 1985 JVC launched the VHS HQ (High Quality). In March 1987 JVC brought out a Super VHS in Japan with an increase of horizontal lines on the image from 240 to 400.

In 1988 a technological agreement was made between ten Japanese companies (such as Sony and Matushita) for the purposes of developing a Super-8mm capable of competing with JVC's Super VHS.

SOURCE: *The Book of Inventions and Discoveries* Associate Editor Valerie-Anne Giscard d'Estaing (UK: Queen Anne Press, Macdonald & Co.) p 239 (1990)

1975   LASER PRINTER   IBM (USA)

The first laser pinter as introduced by IBM in 1975. It was an extremely expensive and bulky machine, designed for high-speed printing. In 1978 the IBM 3800 was followd by the ND2 from Siemens and the 9700 from Xerox, but it was not until 1984—with Hewlett-Packard's Laserjet—that the laser printer began to expand into the world of microcomputers It works on a principle similar to that of offset printing: a laser beam 'paints' the letters onto a roller, and the sheets of paper are printed by rotation. In 1988 colour laser printers came on to the market.

SOURCE: *Inventions and Discoveries 1993* edited by Valerie-Anne Giscard d'Estaing and Mark Young (New York: Facts on File) p 218

1975 **STATIC INDUCTION THYRISTOR** **J-I Nishizawa (Japan)**

The Static Induction Thyristor (SIThy), I proposed the three types of fundamental structures in 1975, has been developed to the various high power devices with the high efficiency. Figure 11.31 shows the top surfaces of cathode electrode of the two SIThys. These are able to turn-off at 4000V within one or two micro-seconds. The continuous current of the small SIThy (34 mmϕ is 200 A, the other (62 mmϕ) is 350 A. The large one is the reverse conducting SIThy that is composed with the reverse direction diode.

Figure 11.31. Static induction thyristor.

These SIThys have been developed as the power switching devices of the motor inverter of the electric railway vehicle by Toyo Electric Mfg. Co., Ltd. Conventionally, one of the latest motor cars has been adopted GTO (Gale Turn-off Thyristor) which has been switching in 400 Hz. The high switching frequency of the SIThy, in this case 2000 Hz, brings more smoothness and better controllability to the train.

SOURCE: Personal communication from Em. Professor Dr Jun-ichi Nishizawa, President of Tohoku University.

1975 **BETAMAX VIDEO RECORDER** **Sony (Japan)**

Invented by the Japanese company Sony, the Betamax was launched in 1975. Today Betamax has 10 per cent of the world market. In January 1985 Sony launched a new version of the Betamax format, called High Band, with a better quality image.

SOURCE: *The Book of Inventions and Discoveries* Associate Editor Valerie-Anne Giscard d'Estaing (UK: Queen Anne Press, Macdonald & Co.) p 238 (1990)

1976 **SPREAD-SPECTRUM COMMUNICATION** **R C Dixon *et al* (USA)**
TECHNIQUES

Spread-spectrum communication techniques are increasingly being used, particularly for satellite communications. A spread-spectrum system is one in which the transmitted signal is spread over a wide frequency band which is many times greater than the bandwidth of the original modulation. Various code mixing techniques are used, e.g. direct sequences (DSK), 'frequency hopping' (which can also be 'time-hopping' or 'time-frequency hopping') and chirp, or frequency sweeping. At the receiver, the modulation (and its bandwidth) is recovered by reversing the transmitting code process. The deleterious effects of interfering can be significantly reduced. Because the system is based on coding, privacy is obtained by the choice of suitable spreading codes.

SOURCE: Private communication from J R Guest, Malvern Wells, UK

SEE ALSO: *Spread Spectrum Systems* by R C Dixon (New York: John Wiley) (1976)

'Spread spectrum com. system uses modified PPM' *Electronic Design* (21 June 1961)

'Poisson, Shannon & the radio amateur' by J P Costas *Proc. IRE* (December 1959)

1976   **MICROELECTRONICS (16 384 bit Random Access   Intel (USA)
        Memory)**

A triumph of semiconductor device technology, the 16 384-bit random access memory has arrived. Its bit density is unprecedented and springs from an enhanced n-channel silicon-gate technique, in which a double level of poly silicon conductors shrinks the memory cell to 400 micrometres square. That is less than half the cell size in the densest 4096-bit RAM.

SOURCE: 'Enter the 16 384 bit RAM' by J B Coe and W G Oldham *Electronics* p 114 (19 February 1976)

1976   **AMORPHOUS SILICON SOLAR CELL   RCA (USA)**

A new type of solar cell has been developed at RCA Laboratories using amorphous silicon (a-Si) deposited from a glow discharge in silane (SiH_4). These solar cells utilize ~ 1 μm of a-Si and have been fabricated in heterojunction, p-i-n, and Schottky-barrier structures on low-cost substrates such as glass and steel.

Discharge-produced a-Si has optical and electronic properties that are ideally suited for a solar cell material. The optical absorption coefficient is significantly larger than that of crystalline Si over the visible light range and therefore most of the solar radiation with $\lambda < 0.7$ μm is absorbed in a film ~ 1 μm thick.

SOURCE: 'Properties of amorphous silicon and a-Si solar cells' by D E Carlson, C R Wronski, J I Pankove, P J Zanzucchi and D L Staebler *RCA Review* vol 38, p 211 (June 1977)

SEE ALSO: 'Amorphous silicon solar cell' by D E Carlson and C R Wronski *App. Phys. Lett.* vol 28, p 671 (1976)

1976   **POLYSILICON RESISTOR LOADED RAMs   Mostek (USA)**

In 1976, Mostek introduced its Poly R process with the MK4104, a 4-K-by-1-bit static RAM. The part diverged from the usual static RAM designs in that it replaced the depletion-mode MOS transistor loads in its cell with ionimplanted polysilicon resistors. The design not only saved chip area but also greatly lowered power dissipation. Since the polysilicon resistors are actually laid over the four transistors, the cell of the 4104 shrank to 2.75 mil² —roughly half the size of conventional cells.

The power is reduced because the high resistivity of the polysilicon loads—typically, 5000 megohms, accurately controlled by ion implantation squeezes the current flow down to less than 1 nanoampere per bit. Another feature of the Poly R loads is their negative temperature coefficient which automatically compensates for increased leakages that normally occur at elevated temperatures. Moreover, the polysilicon loads allow data retention in the cells even at greatly reduced supply voltages.

SOURCE: 'Concepts for a dense new RAM' *Electronics* p 1320 (27 September 1979)

1976   **COMPUTER (One Board with Programmable I/O)   Intel Corp. (USA)**

A complete general-purpose computer subsystem that fits on a single printed-circuit board has been a major goal all through the steady evolution of LSI technology. Such a computer, consisting of a central-processing unit, read/write and read-only memories, and parallel and serial input/output interface components, could satisfy most processing and control applications needed by original-equipment manufacturers. A single board computer could greatly extend the range of computer applications by providing a single solution to three problems that have often precluded the use of conventional computers.

The primary reason for use of a single assembly of LSI devices rather than a multiboard subsystem is economic. Extra board assemblies are costly in themselves and need related equipment, such as backplanes and housing, that also adds to cost.

Compactness and low power consumption are often prerequisites for products. Using LSI for all key computer functions reduces power consumption and provides a higher functional density than conventional subsystem designs. This new class of LSI devices—programmable input/output interface chips enables an 8-bit computer to be built as a subsystem on one printed-circuit board.

SOURCE: 'The 'super component': the one-board computer with programmable I/O' by R Garrow, J Johnson and M Maerz *Electronics* p 77 (5 February 1976)

1976   MARISAT-1 Satellite   (USA)

Launched 19 February 1976. Maritime communications satellite positioned at 15°W over the Atlantic Ocean.

SOURCE: Table of Artificial Satellites Launched Between 1957 and 1976 (Geneva: International Telecommunication Union) (1977)

1976   MICROELECTRONICS (Versatile Arrays)   Philips (Holland)

A simple variation of standard silicon-gate technology has produced extremely versatile arrays that make novel analog-to-digital converters, analog type displays and light-pattern scanners. The arrays consist of devices similar to standard metal-oxide-semiconductor elements, except that a resistive electrode structure replaces the normal metal insulated gate. This structure permits a voltage gradient to be set up across the ends of the gate and then manipulated to control the transistors either singly or in groups.

SOURCE: 'Resistive insulated gates produce novel a-d converters, light scanners' by M V Whelan, L A Daverveld and J G deGroot *Electronics* p 111 (18 March 1976)

1977   MRI: MAGNETIC RESONANCE IMAGING   G Houndsfield (UK)

A major electronic technique is non-invasive diagnosis by computerised scanning, using nuclear, ultrasonic, fluoroscopic and x-ray equipment. MRI (Magnetic Resonance Imaging) is a medical imaging technique that relies on the response of hydrogen atoms to a magnetic field to distinguish between various types of soft tissue. Computerised axial tomography can provide 'slices' of patients' anatomy and yield valuable diagnosis, enabling physicians to visualise anatomic structures of live patients. This was the first equipment to provide detailed pictures of the body's soft tissues.

SOURCE: *Electronic Inventions and Discoveries* 4th edn, chapter 9 (Bristol: Institute of Physics Publishing) (1997)

1977   CCD ANALOG-TO-DIGIAL COVERTER   G E Corp. (USA)

For the industrial marketplace, the Research and Development Center of General Electric Corp, in Schenectady, NY, has fabricated the first CCD analog-to-digital converter as a p-channel MOS chip providing a resolution equivalent to 10 to 12 binary bits.

The GE converter chip is big, measuring 240 by 180 mils, but it contains a comparator, voltage reference, clock, counter and all necessary control logic—in fact, it even has decoder/drivers for gas-discharge displays. Operating speed, though, is slow: about 20 milliseconds total for a 10-bit conversion. The device, which runs at a clock frequency of up to 500 kilohertz, resolves analog inputs to within better than 1 millivolt, digitising them to an accuracy of ±0.5 least significant bit.

Chip operation relies on the transfer of fixed-size charge packets from one site to another, with conventional digital circuitry controlling the conversion process.

SOURCE: 'CCD's edge towards high volume use' *Electronics* p 74 (17 March 1977)

1977   ANISOTROPIC PERMANENT MAGNET   Matsushita Electric (Japan)

Matsushita Electric has developed what they claim to be the world's first anisotropic permanent magnet for practical use. The new magnet is made of manganese, aluminium and carbon materials which are available in abundance instead of cobalt and nickel, which are scarce and expensive. It has higher magnetic energy than the ordinary 'alnico' type magnet which contains cobalt and nickel and has good mechanical strength and machinability, allowing it to be shaped and drilled. Samples should be available from June 1977.

The new magnet consists of manganese, aluminium and carbon. The basic composition of the magnet was developed by Matsushita in 1967; however, the magnet produced was of 'isotropic' type, in which the direction of magnetisation was distributed at random. With the anisotropic version, the company has succeeded in aligning the direction in which it can be magnetised, so increasing the magnetic energy to more than five times that of the isotropic magnet, giving a maximum energy product of 7 MG Oe.

SOURCE: 'First anisotropic permanent magnet' *Electronic Equipment News* p 10 (June 1977)

1977 POCKET TV RECEIVER Sinclair Radionics (UK)

What is claimed to be the world's first pocket television set was launched in London early this year by Sinclair Radionics, the British company which pioneered the revolution in miniature electronic calculators four years ago.

The result of a 500 000 12-year research and development programme, the Microvision, which has a 2-inch screen, is now in production at Sinclair's new assembly plant in St Ives, Huntingdon.

The set is 4 inches wide, 6 inches from front to back, just 12 inches deep and weighs $26\frac{1}{2}$ oz. Yet, operated by internal rechargable batteries or direct from the mains, it produces a sharp black and white picture which, when viewed at a distance of one foot, is of equivalent size and brilliance to that of normal domestic compact portables at 6 feet and 24 inch models at 12 feet.

SOURCE: 'Television with two inch screen' *National Electronics Review* vol 13, No 4, p 74 (July/August 1977)

1977 TRIMOS (Triac + MOS) Stanford University (USA)

Another innovation in power at ISSCC comes from California's Stanford University, which has developed a way to put signal and power devices on one and the same piece of silicon.

Called Trimos, the new technology permits integrating an insulated-gate triac with metal-oxide-semiconductor components, inviting a host of new applications in crosspoint switching, output stages, and power control.

Trimos is actually a merged device based on double-diffused MOS technology—two high-voltage D-MOS transistors are merged around a common drain. Contact is made to the source and diffused channel of each D-MOS device, forming symmetrical anode and cathode contacts. The shared gate metal forms the unit's control electrode.

In its on state, the Trimos device exhibits a dynamic resistance of less than 10 ohms and can pass currents on the order of amperes. A simple shunt switch, in the form of a conventional MOS transistor, can be fabricated adjacent to the Trimos unit for switching it out of its on state or inhibiting it from triggering. Without such a bypass structure, the Trimos device typically has turn-on and turn-off times on the order of 200 nanoseconds, and its single pulse dv/dt capability exceeds 1000 volts per microsecond.

SOURCE: 'Trimos combines triac, MOS devices' *Electronics* p 42 (2 March 1978)

1977 FLAD (Fluorescence-Activated Display) Institute for Applied Solid State Physics, Freiburg (Germany)

Display-system designers will soon have a new device to work with the fluorescence-activated display (FLAD). Invented at the Institute for Applied Solid State Physics in Freiburg, West Germany, the device uses a layer of plastic material appropriately doped with fluorescent organic molecules. In this layer, ambient light is collected, guided, and then emitted at the segments of the display's digits.

The FLAD dissipates the same power as a liquid-crystal display, but its light intensity is said to be much higher. Moreover, the light can be any color in the spectrum between green and red, the inventors say.

West Germany's Siemens AG will produce the FLAD display first, initially for use in battery-operated digital tabletop and alarm clocks, Later, FLADs will be used in pocket calculators, portable instruments, and scales that indicate price and weight.

SOURCE: *Electronics* p 55 (17 March 1977)

1977 MICROELECTRONICS H-MOS   Intel (USA)

To achieve their new high-performance process called H-MOS, Intel has chosen the direct device-scaling method for two reasons. First, it evolves directly out of standard silicon-gate processing and so requires neither new device structures not complex circuit schemes (either requirement would make yields and fabricating costs too unpredictable to guarantee their usefulness over a wide range of semiconductor products). Second, it fits in with the trend to smaller and smaller circuit patterns, as photolithographic methods grow more refined and electron-beam wafer-fabrication techniques stand ready to take over.

SOURCE: 'H-MOS scales traditional devices to higher performance level' by R Pashrey, K Kokonnen, E Boleky, R Jecmen, S Liu and W Owen *Electronics* p 94 (18 August 1977)

1978 LASER-ANNEALED POLYSILICON   Texas Instruments (USA)

Researchers at Texas Instruments Inc.'s Central Research Laboratories have succeeded in fabricating MOS devices in laser-annealed polysilicon on silicon dioxide. Not only will the devices have the speed and density of those fabricated on sapphire, but the all-silicon construction could lead to true three dimensional circuitry.

Unannealed polysilicon is made up of randomly oriented crystal grains on the order of 500 Å across. Carrier flow is impeded at each grain-to-grain boundary and the resulting low mobility would yield a poor device at best.

Armed with a pulsed frequency-doubled neodymium-yttrium-aluminum garnet laser, TI scans the polysilicon surface to induce localized melting. The grains recrystallise with much larger dimensions so the number of interfaces is reduced and mobility is enhanced. With this set up, researchers have observed grains as large as 10 micrometres across.

To build what it calls silicon-on-insulator MOS FETS, TI begins with a single-crystal p-type silicon substrate. On this, it grows a 1 μm-thick oxide layer and then deposits a 0.5 μm film of undoped polysilicon film. The samples then go onto an $X - Y$ translation stage, which is heated to 350°C while a stepping motor moves it in synchronisation with the pulsed laser.

Then the polysilicon is selectively etched down to the oxide level to isolate islands for each transistor. Boron ions are implanted to form the channel, which is covered with a thermally grown 500 Å gate oxide. A polysilicon gate is deposited on the gate oxide and implanted with phosphorus. It also is used for a self-aligned arsenic ion implantation of the source and drain.

SOURCE: 'All-silicon devices will match SOS in performance' by John G Rosa *Electronics* p 39 (22 November 1979)

1978 ISL (INTEGRATED SCHOTTKY LOGIC)   J Lohstroh *et al* (Philips) (Holland)

Consider a pair of bipolar technologies: low-power Schottky transistor–transistor logic aiming at high speed for medium-scale parts like the 7400 family, and integrated injection logic, which merges transistors specifically for the high packing density needed in large-scale integration. What if the attributes of both could be found in a high-speed, low-power logic suitable for LSI. Apparently they can—in ISL, a newly developed technology that stands for integrated Schottky logic (see figure 11.32).

Developed by Jan Johstroh and colleagues at the Digital Circuitry and Memory Group of Philips Gloeilampenfabrieken in Eindhoven, the Netherlands, ISL has already performed admirably in 'kit' parts—flip-flops, oscillators, and the like. Such devices have exhibited gate propagation delays of about 3.5 nanoseconds (half that of low-power Schottky and a quarter that of I²L), with each gate drawing only about 400 microamperes. An ISL D-type flip-flop toggles comfortably at 60 megahertz, as compared with a limit of about 33 MHz for a similar low-power Schottky device.

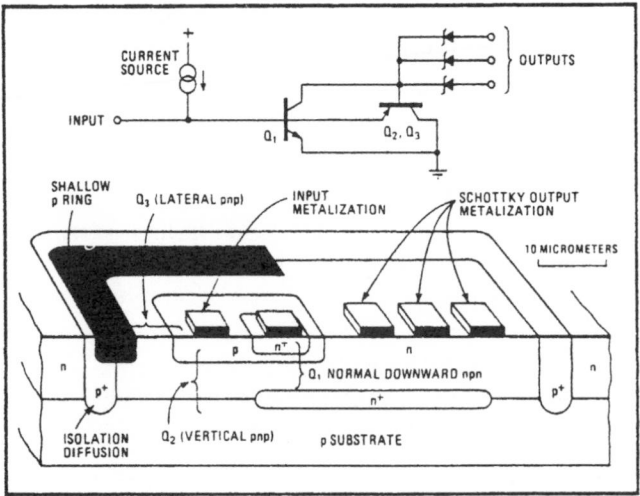

New logic. In ISL, the normal downward npn transistor inherently adds a vertical pnp device. A p ring parallels the vertical transistor with a lateral pnp one.

Figure 11.32. Integrated Schottky logic.

SOURCE: : 'Two popular bipolar technologies combine in Philips' device' *Electronics* p 41 (8 June 1978)

1978   **LIGHT BUBBLES**   **W Ruehle, V Marello and A Onton**
                                                          IBM (USA)

Mobile 'light bubbles', which appear to be electrical analogs to magnetic bubbles, have been generated in magnesium-doped zinc-sulfide thin films, IBM scientists reported at the Electronic Materials Conference in Santa Barbara, California. The 1 μm bubbles appear when 10 kHz, 190 V ac current is applied to the film via sets of parallel 1-mm-wide metallic lines orthogonally placed on each side of the film. When voltage is applied to a pair of intersected electrodes, *the intersected area will emit light bubbles that appear to move in discrete steps*—and move faster when the frequency rises to 50 kHz. The alternating-current thin-film electroluminescence, or actel, lasts as long as an hour and can be used to form images on the film by stimulating different areas with light or an electron beam or by applying voltages across the film. However, before the technique becomes practical, much work needs to be done to direct the bubbles.

SOURCE: IBM scientists report observation of 'light bubbles' *Electronics*, p 33 (6 July 1978)

SEE ALSO: 'Electrical analogy to magnetic bubbles' *Electronics Weekly* p 7 (12 July 1978)

1978   **LIGHTWAVE POWERED TELEPHONE**   **Bell Laboratories (USA)**

Bell Laboratories has developed a telephone powered by light carried to it by a fibre-optic cable. Such a development brings the totally fibre-optic phone system closer to reality, but it is more than a major technological advance.

The major problem was making the phone run off the power available from the fibre cable. The biggest headache here, according to De Loach, was the conventional telephone ringing mechanism.

So a new ringer was designed that worked perfectly well at a couple of volts. It uses an electroacoustic tone generator with a thin piezoelectric active element. 'This device has some exceptional specs,' says De Loach. 'For example, its overall efficiency from the optical input to the acoustic output is more than 33%'.

With the ringer problem solved, the power requirements of the rest of the phone were readily satisfied. This power, as well as the ringer drive, comes from a Bell-developed photodetector, which converts light to electrical pulses. The same device can act as a photodiode, too.

The photodetector's narrow-bandwidth conversion efficiency at 0.81 micrometre is 56%—the highest ever reported. It is a double-heterostructure device with layers of gallium aluminum arsenide sandwiching one of gallium arsenide and grown on a single crystal substrate. A hole etched in the substrate exposes some of either GaAlAs layer, and the glass fibre butted at that point couples light to and from the photodetector.

But the phone handset must send as well as receive, so Bell has designed the photodetector to generate light as well, at a wavelength different from the incoming light. With time-sharing and automatic switching, the phone sends signals back to the central office at a 0.9 μm wavelength. It operates in a duplex mode, taking further advantage of fibre's bandwidth capability.

The laser light coming from the central office is on for about 95% of the time. It is pulse-width-modulated by the voice or data. Since the modulation bandwidth is small compared with the carrier wavelength, the change in pulse width is easily controllable.

SOURCE: 'Bell Labs develops telephone powered by lightwaves alone' *Electronics* p 39 (23 November 1978)

SEE ALSO: 'De Loach built the fibre-optic phone' by H J Hindin *Electronics* p 231 (25 October 1979)

1978	**OMIST (Optical Metal Insulator Silicon Thyristor)**	**A G Nassibian, R B Calligaro (Australia) and J G Simmons (Canada)**

Recently a novel metal-tunnel oxide n/p$^+$ silicon device with I/V characteristics similar to those of a silicon control rectifier (metal-insulator-silicon-thyristor, the MIST), but with added advantage of being of a much simpler structure which is also compatible with LSI techniques, has been described. The device has been shown to have wide ranging digital and analogue circuit applications, including oscillators RAMS and ROMS. Furthermore, the device has been shown to be light sensitive.

This device has an advantage over the conventional analogue light-sensitive devices in that it is a true digital-optical switch. When the incident light is above a certain threshold intensity the device switches between two well-defined states. In optical systems, such a device can perform both optical transduction and thresholding, thereby greatly simplifying the receiver system.

SOURCE: 'Digital optical metal insulator silicon thyristor (o.m.i.s.t.)' by A G Nassibian, R B Calligaro and J G Simmons *Solid-State and Electron Devices* vol 2, No 5, p 149 (September 1978)

SEE ALSO: 'Bistable impedance states in m.i.s. structures through controlled inversion' by H Kroger and H A R Wegener *Appl. Phys. Lett.* vol 23, pp 397–9 (1973)

1978	**ANALOGUE ALL-ELECTRONIC CLOCK FACE**	**Hosiden Electronics and NEC (Japan)**

Electronic parts manufacturer Hosiden Electronics and Nippon Electric Company (NEC) have developed a 60-pole analogue fluorescent displace tube for use in clocks and the drive circuitry to go with it. The new inventions now make it possible to introduce electronics on a full scale to analogue clocks (which use hands).

Since the clock face has been turned into a fluorescent display tube, it is now possible to achieve a range of colours from yellow to blue using filters. Ordinary fluorescent tubes are direct-head 3-electrode vacuum tubes with anodes coated with phosphor, and they use the electro-luminescence principle of light emission. This kind of tube was first developed some ten years ago in Japan since which time it has undergone improvements in performance. It radiates a green light which is easily visible. It has a low voltage low power consumption and high response speed making it suitable for use with MOS LSI chips. Other features include its dependability, long life and mass production capability.

Analog systems have been around for a long time in timepieces and meters, and they have become an irreplaceable part of every-day life. The all-electronic analog clock has, therefore, a bright future now that this new fluorescent display tube and accompanying drive circuitry have been developed.

SOURCE: 'Analogue clocks can now go electronic' *Journal of Electronics Industry* (Japan) p 42 (April 1979)

1978   **LASER OPTICAL RECORDING SYSTEM**   **Philips (Netherlands)**
(COMPACT DISC)

An ultra-compact diode laser optical recording system, the world's first, has recently been introduced by Philips. It allows high-density recording and retrieval of up to 10^{10} bits of data, equivalent to about 500 000 typewritten pages, on a pregrooved 30 cm disk. This capacity represents an improvement of ten times compared with the most advanced magnetic disk pack systems currently available. The system offers direct read-after-write with random access; any address can be reached in a mean time of 250 ms, providing virtually instant access to 5×10^9 bits (the capacity of one side of the disk).

The system uses similar techniques to those developed for VLP (Video Long Play). The real breakthrough, however, has come with the development of a suitable miniature diode laser and matching recording material. The laser used is of the AlGaAs DH type and employs a 0.1 mm square semiconductor chip housed in a transistor-sized encapsulation. Despite its small size, the device develops a pulsed light output power equivalent to that of a large gas laser and its associated modulator.

SOURCE: 'World's first diode laser optical recording system' *Electronic Components and Applications* vol 1, No 2, p 128 (February 1979)

SEE ALSO: 'An optical disk replaces 25 mag tapes' by G C Kenney, D Y K Lou, R McFarlane, A Y Chan, J S Nadan, T R Kohler, J G Wagner and F Zernike *IEEE Spectrum* p 33 (February 1979)

'Consumer electronics: personal and plentiful' by D Mennie *IEEE Spectrum* p 62 (January 1979)

1978   **TAMED FREQUENCY MODULATION**   **N Wiedenhof and J M Waalwijk**
(Philips) (Holland)

Philips Research Laboratories in Eindhoven, have designed a different system of frequency modulation for transmitting digital information. Using this method a very narrow spectrum is obtained while the quality of detection is almost equal to the maximum that can be obtained with digital transmission.

The new method has been given the name 'tamed frequency modulation' (TFM). Because of its properties tamed FM is eminently suitable for digital radio communication.

SOURCE: 'Tamed FM for efficient digital transmission via radio' Philips Research, Philips, Eindhoven (7810/0920/186E PR+PREL)

1978   **LCP (LASER COLD PROCESSINGS OF**   **Quantronix Corporation (USA)**
SEMICONDUCTORS

A new and exciting technology has appeared in the last two years that could be of major importance in producing tomorrow's very large-scale integrated circuits as well as in raising the yields of today's LSI devices. It is the use of a laser as a heat source for some of the many high-temperature process steps in semiconductor manufacture—for instance, annealing a wafer to eliminate crystal damage due to ion implantation, diffusing a wafer with dopants, and growing crystalline material from amorphous or polycrystalline material.

In this new technique, an intense laser beam heats a semiconductor surface to a temperature at which some desirable physical or chemical change takes place in the material. The main practical advantage of this process is that the laser spot limits irradiation to specific areas and the short pulse limits heating to a small depth while the rest of the material stays near ambient temperatures. Hence the name of the technique—laser cold processing, or LCP.

SOURCE: 'Laser cold processing takes the heat off semiconductors' by R A Kaplan, M G Cohen, and K C Kiu *Electronics* p 137 (28 February 1980)

1978   **ONE MEGABIT BUBBLE-MEMORY**   **Intel, Texas Instruments (USA)**

There can no longer be any doubt that bubble-memory systems are a reality—1978 saw the introduction of not one but a pair of million-bit bubble memory chips. Intel Magnetics Inc., Santa Clara, Calif., was first with its 7110 chip, a 4-square-centimetre device organized into 256 4096 bit loops, which when operated at a 100-kilohertz field frequency provides an average access of about 20 ms. The organization of Texas Instruments' megabit chip is 512 loops of 2048 bits each; with shorter loops, it has an access of about 10 ns.

SOURCE: *Electronics* p 133 (25 October 1979)

SEE ALSO: 'Megabit bubble-memory chip gets support from LSI family' by D Bryson, D Clover and D Lee *Electronics* p 1059 (26 April 1979)

1978   **INTEGRATED OPTOELECTRONICS**   **A Yariv *et al* (USA)**

Today a single optical fibre can transmit billions of bits of information per second over many kilometers. In fact, fibres can handle more bits of information per second than conventional optical sources—lasers or light emitting diodes—can transmit and detectors can receive through them. Combining lasers, detectors, and active electronic devices for modulating the light on a single-crystal chip of GaAs—a means for reducing an electronic bottleneck—was first suggested 10 years ago by Amnon Yariv of the California Institute of Technology in Pasadena. Research efforts since then indicate that the time is ripe to put the idea into practice

SOURCE: 'Integrated optoelectronics' by N Bar-Chaim, I Ury and A Yariv *IEEE Spectrum* p 38 (May 1982)

SEE ALSO: 'Integration of an injection laser with a Gunn oscillator on a semi-insulating GaAs substrate' by C P Lee, S Margalit, I Ury and A Yariv *Appl. Phys. Lett.* vol 32, p 806 (1978)

1979   **TWO-LAYER RESIST TECHNIQUE FOR**   **Bell Laboratories/MIT (USA)**
        SUBMICROMETRE LINES

In the two-level resist method, a very thin, 1500-to-2000-angstrom amorphous upper layer of selenium and germanium is sputtered or evaporated onto a polymer layer. The latter, about 2 μm thick, may be made of any of the standard resist polymers without the silver that renders them photosensitive.

This beginning layer is thick enough to compensate for silicon's microscopic roughness, which often makes fine geometries impossible because of depth-of-field or optical-interference effects, Thus the S Ge layer is an almost perfectly flat optical surface which also cuts the reflections and refractions that make for uncontrolled line widths.

To make the thin upper layer photosensitive, the team soaks the wafer in a room-temperature potassium silver selenide solution for 30 seconds. Then it exposes patterns with ultraviolet light, usually at a 4300 Å wavelength, but sometimes as short as 3250 Å.

Kai notes that the upper layer has more than twice the contrast of conventional resists and that the amorphous material is finely grained, thus reducing the optical-dispersion effects that can impair resolution. In fact, ultimate resolution is finer than the minimum spot or line sizes possible with available UV sources.

SOURCE: 'Two-layer resist technique produces submicrometer lines with standard optics' by James B Brinton *Electronics* p 47 (14 February 1980)

1979   **CCD COLOUR TV CAMERA**   **Sony (Japan)**

For improved sensitivity and resolution, the new camera uses two imager chips, one to generate the green signal and the other to generate the red and blue signals. Sensitivity is high because an entire chip is used for green—a stripe filter is not required—and green light generally contains most of the energy in images.

The two chips are offset horizontally by one half the horizontal pixel pitch, and sophisticated signal-processing techniques are used to increase resolution to a value almost as high as what could be obtained with a single chip having twice as many pixels along each horizontal line. Thus a measured optical resolution of 280 test-pattern lines per picture height is obtained even though there are only 245 pixels across the width of each sensor. The measured vertical resolution is 350 lines from 492 pixels.

SOURCE: 'Color TV camera using CCD imager chips gets first sale' *Electronics* p 79 (14 February 1980)

SEE ALSO: *Electronics* p 67 (5 July 1979)

Electronics p 33 (9 January 1978)

Electronics p 63 (20 July 1978)

Electronics p 68 (28 September 1978)

1979   **AMORPHOUS SILICON LIQUID-CRYSTAL**   **RSRE and Dundee University (UK)**
     DISPLAY

A small experimental liquid-crystal display that is addressed by a matrix of amorphous silicon thin-film transistors has been developed by researchers at Dundee University, Scotland, with funds from Britain's Royal Signals and Radar Establishment. The display panel is far from complex—it measures 1.6 by 2.2 cm (0.6 by 0.9 in) and consists of a five-by-seven array of display elements each 2 mm square. But its development, which began four years ago, points to the potential of amorphous silicon thin-film transistors as a means of overcoming the addressing limitations of LCDs at low cost. The electrical performance of individual devices looks acceptable, with an on current of 5 μA, a 10^5 on-to-off current ratio, and a response time for each liquid-crystal element of less than 100 μs. The work is to be described in a paper at the Sixth European Solid State Device Research Conference at York, England, 15–18 September 1980, together with one from Plessey's Allen Clark Research Centre on device physics of amorphous silicon transistors—work also funded by the RSRE.

SOURCE: 'British address LCD with amorphous silicon thin-film transistors'. *Electronics* p 67 (28 August 1980)

SEE ALSO: *Electronics* p 69 (21 June 1979)

1979   **LASER-ENHANCED PLATING AND ETCHING**   **IBM (USA)**

Recently, an interesting and novel application of lasers was discovered in which a laser beam impinging on an electrode is used to enhance local electroplating or etching rates by several orders of magnitude. It has also been discovered that with the aid of the laser it is possible to produce very highly localized electroless plating at high deposition rates, to greatly enhance and localise the typical metal-exchange (immersion) plating reactions, to obtain thermo-battery-driven reactions with simple single-element aqueous solutions, and to greatly enhance localized chemical etching. Since laser beams can be readily focused to micron-sized dimensions and scanned over sizeable areas, the enhancement scheme makes it possible to plate and etch arbitrary patterns without the use of masks.

SOURCE: 'Laser-enhanced plating and etching: mechanisms and applications' *IBM J. Res. Dev.* vol 26, No 2, p 136 (March 1982)

SEE ALSO: 'Laser enhanced electroplating and maskless pattern generation' by R J von Gutfield, E E Tynan, R L Melcher and S E Blum *Appl. Phys. Lett.* vol 35, p 651 (1979)

'Laser enhanced electroplating and etching for maskless pattern generation' by R J von Gutfield, E E Tynan and L T Romankiw. Extended Abstract No 472 *Electrochemical Soc.* pp 79-2 (1979) (156th Meeting of the Electrochemical Society, Los Angeles, CA)

1979   **SATELLITE ECHO-CANCELLING CIRCUIT**   **Bell Laboratories (USA)**

By the beginning of 1980, telephone communications by satellite will have undergone a drastic technical change. The voice-garbling echoes that occur along the 45 000-mile-long satellite paths will be just about eliminated by a new integrated circuit that was developed by researchers at Bell Laboratories.

The chip—dubbed an echo canceller—could by itself double the number of satellite circuits used by AT & T in its telephone network. So far, because of echo problems, the company transmits transcontinental telephone calls only one way by satellite.

The echo for one-way transmission is controllable by the present echo suppressors, which open the transmission paths when the echo's amplitude becomes too high. But these devices do not meet high enough two-way transmission standards for AT & T. The new digital device removes an echo signal from a circuit by sampling it electronically as it occurs, making a replica of it, and adding the replica to the original signal so that the two cancel each other. It is ideal for connection to Bell's all-digital telephone network, which will be completed over the next 20 years.

SOURCE: 'No more echoes' *Electronics* p 228 (25 October 1979)

SEE ALSO: 'Bell's echo-killer chip' by Donald L Duttweiler *IEEE Spectrum* p 34 (October 1980)

'Silencing echoes on the telephone networks' by Man Mohan Sondhi and David A Berkley *Proc. IEEE* vol 68, No 8, p 948 (August 1980)

1979 FLOTOX (Floating-Gate Tunnel Oxide) PROCESS   Intel (USA)

Flotox resembles the Famos structure except for the additional tunneloxide region over the drain. With a voltage V_g applied to the top gate and with the drain voltage V_d at 0 V, the floating gate is capacitively coupled to a positive potential. Electrons are attracted through the tunnel oxide to charge the floating gate. On the other hand, applying a positive potential to the drain and grounding the gate reverses the process to discharge the floating gate.

Flotox, then, provides a simple, reproducible means for both programming and erasing a memory cell.

SOURCE: '16-K EE-PROM relies on tunneling for byte-erasable program storage' by W S Johnson, G L Kuhn, A L Renninger and G Perlegos *Electronics* p 113 (28 February 1980)

1979 COMPACT DISC   Philips (Holland)

(See also page 238.)

The invention of the compact disc (CD) was the result of research carried out on the video disc by the Dutch electronics company Philips NV. Under a joint licensing agreement by Philips and the Japanese company Sony, the CD was first developed in 1979. A process of digital recording is used, rather than the analogue recording process used for the microgroove. The signal is coded in binary form, using the series 0 and 1. The conventional groove has therefore disappeared and has been replaced by millions of microcells known as pits: approximately 4 million per second. The sound is reproduced by a laser beam. The compact disc has a diameter of 5 inches and can hold 75 minutes of music or sound on one side.

The CD was first marketed in 1983, and by 1991 had outstripped both traditional forms of recorded music—records and tapes—in terms of unit sales and values. In the space of a few years, the CD has achieved incredible success, and its applications are many and varied. In 1984 Philips and Matsushita brought out the prototypes of decoders that enabled fixed images, which had been stored on CDs alongside an audio signal, to be viewed on television. In 1985 the extensive storage capacity of CDs was applied to computers. CD players now have the capability of running a disc at twice the normal speed, which makes it possible to record an hour-long disc onto a cassette in just 30 minutes.

SOURCE: 'Inventions and Discoveries 1993' edited by Valerie-Anne Giscard d'Estaing and Mark Young *Facts on File* New York p 138

1979 SEVEN-COLOUR INK-JET PRINTER   Siemens (Germany)

Engineers of the Teletype and Data-Transmission group at Siemens AG in Munich have developed a prototype colour ink-jet printer that can produce characters of seven different colours at the rate of 200 per second in both directions across normal paper. The unit consists of the company's older PT80 printer whose single-colour ink-jet mechanism has been replaced by a multinozzle unit for red, green, and yellow ink. Proper mixing of these inks yields the colours blue, magenta, cyan, and black. Colour and mixture data are stored in a floppy-disk memory whose program is derived from the output of a separate colour scanner. One of the biggest problems involved, the Siemens engineers say, was developing nonsmearing inks that would stay liquid in the nozzles yet become dry right after they hit the paper. If customer reaction is favorable, the colour printer will go into production. It will make possible multicoloured graphic representations.

SOURCE: 'Siemens develop seven-colour ink-jet printer' *Electronics* p 71 (13 September 1979)

1980   **OUTDOOR LARGE SCREEN COLOUR DISPLAY**   **Mitsubishi Electric Corporation**
        SYSTEM   **(Japan)**

A large screen colour display system for use in outdoor video system has been developed and introduced by Mitsubishi Electric Corp. The display screen consists of a matrix array of small, high brightness light emitting tubes and can present sharp colour pictures even in full daylight. Many video display systems have been installed for various outdoor video services such as in sports stadiums, racetracks, or as an advertising media. However, these previous systems consist of an array of incandescent lamps, and have many problems especially in displaying colour pictures such as insufficient colour quality, high power consumption and short operating life.

To overcome these problems, Mitsubishi has developed the high brightness light emitting tube (LET). The LET is a small flood beam CRT (28mm in diameter) having a single phosphor of red, green, or blue for each tube, and works for a single picture element in the screen. Brightness of the LET is 8000 bits for a green tube (over 20 times brighter than the usual TV picture tube).

SOURCE: 'AURORA VISION'. A large screen colour display system' by K Kurahashi, K Yagishita, T Tomimatsu and H Kobayashi (In Japanese) *Technical Report of Inst. of TV Eng. (Japan)* IPD49-3, p 319 (1980)

SEE ALSO: 'An outdoor large screen colour display system' by K Kurahashi, K Yagishita, N Fukushima and H Kobayashi *SID Symp. Digest of Technical Papers* vol 12, 13, 1 (p 132) (1981)

1980   **MAT (MAGNETIC AVALANCHE TRANSISTOR)**   **IBM (USA)**

A new semiconductor device for sensing uniaxial magnetic fields has been realized. The device is basically a dual-collector open-base lateral bipolar transistor operating in the avalanche region, and is referred to as a magnetic Avalanche Transistor. It exhibits high magnetic transduction sensitivity compared to traditional Hall-effect and conventional nonlinear magnetoresistive devices. Several hundred experimental devices have been designed, fabricated, and tested over the past two years. Many structural and some process parameters were varied. The magnetic sensitivity of a typical device was found to be proportional to substrate resistivity. A sensitivity of 30 volts per tesla was measured for devices which used 5-ohm-cm p-type substrates. The output signal measured between collectors is differential and responds linearly with field magnitude and polarity. A typical signal-to-noise ratio is 20 000 per tesla. The bandwidth is known to extend well beyond 5 MHz. The sensitive area is calculated to be on the order of 5 amp. This communication describes the basic structure, fabrication, and characteristics for the magnetic avalanche transistor.

SOURCE: 'A magnetic sensor utilizing an avalanching semiconductor device' by A W Vinal *IBM J. Res. Dev.* vol 25, No 3, p 196 (May 1981)

1980   **256K DYNAMIC RAM**   **NEC-Toshiba Musashino Electrical**
                                                          Communication Labs. (Japan)

NEC-Toshiba Information Systems Inc. and NTT-Musashino Electrical Communication Laboratory, both of Tokyo, each present 256-K-by-1-bit dynamic RAMs. The NEC-Toshiba chip has a 160-nanosecond access time and a 350-ns cycle time, while the NTT device, which uses molybdenum to speed signal propagation, accesses in just 100 ns and cycles in double that. NEC-Toshiba's chip consumes 225 milliwatts of active power and 25 mW on standby, while NTT's device uses 230 mW but only 15 mW on standby. Both designs require a 256-cycle/4-millisecond refresh.

NEC-Toshiba uses two levels of polysilicon, 1. 5 μm direct-step-on wafer photolithography, and all-dry processing to build its RAM. The device is already being shown in a 16-pin package, the pinout of which meets the Joint Electron Device Engineering Council's standard with the eighth address line on pin 1. The oblong die, measuring 191 by 338 mils, contains two 128-K arrays. Each array is further split into two 128-by-512-bit sections, separated by 512 sense amplifiers that run the length of the chip. Various techniques including silicone coating, are used to keep the mean time between failure due to alpha radiation below 30 000 device hours.

NTT-Musashino's RAM is built with electron-beam direct writing, dry processing, and three interconnection levels: molybdenum word lines, aluminum bit lines, and polysilicon for storage capacitor electrodes and gates of non-array devices. NTT's die is organized just like NEC-Toshiba's but with one interesting difference: each 128-K array has attached to it a 2-K block of redundant cells and a dummy sense circuit. The extra cells, connected via four pairs of spare bit lines and two spare word lines, are replaced by electrically programming on-chip poly-silicon resistors during wafer probing. The additional circuits take up no more than 10% of NTT's nearly square 230-by-232-mil die.

SOURCE: 'ISSCC: a gallery of gigantic memories' *Electronics* p 138 (14 February 1980)

| 1980 | **FIBRE-OPTIC LASER DRIVEN SUPERHETERODYNE** | **S Saito, Y Yamamoto and T Kimura (Japan)** |

Researchers at Nippon Telegraph & Telephone's Musashino Laboratories have demonstrated for the first time the use of classical superheterodyne detection in an optical-fibre transmission system. The technique represents a significant step in the practical realisation of optical communication systems, and could lead to the more efficient use of the low-loss frequency window recently opened by advances in optical-fibre technology, since finer separation of carrier frequencies will become possible. Improvements will also be made in optical signal reception.

The main signal was supplied by an AlGaAs laser, emitting at 820nm, the drive current being directly frequency modulated by an rf signal. This was matched to a local-oscillator signal from a similar laser by temperature control and direct-current adjustment. No feed-back stabilisation system was used, The mixed signal was aligned on a photodetector and the resulting fm intermediate-frequency signal fed to a frequency discriminator for conversion to an am output. The spurious signal depth was very small and easily filtered out.

Digital (at 100Mbit/s) and analogue (at 300MHz) signals were transmitted satisfactorily by the system, which is described in the 23 October issue of Electronics Letters. The system is claimed to have many advantages over conventional amplitude-modulation systems, such as low power, simplicity and better signal/noise ratio, and is expected to be improved in the future by the use of recently developed frequency-stabilisation feedback systems.

SOURCE: 'Classical steps in optical fibres' *Electronics and Power* p 855 (November/December 1980)

SEE ALSO: 'Fibre optics adopts superheterodyne principles' *Electronics* p 73 (20 November 1980)

| 1980 | **MCZ (Magnetic Field Method of Silicon Crystal Growths)** | **Sony (Japan)** |

Sony has developed a new method for producing very high-quality single crystals of silicon with greater uniformity and lesser defect generation, which greatly reduces the wafer warpage and distortion, through the application of a high magnetic field in the silicon pulling process.

The new silicon crystal growth method, called 'MCZ (magnetic-field CZ) method', has been developed to meet the requirements of the coming age of ultra-high-density semiconductor devices, including CCDs and super-LSIs, which integrate tens of hundreds of thousands of elements into several-millimetre-square chips. Sony's MCZ method is the world's first of its kind, which applies a high magnetic field, instead of zero gravity, to mass produce very high-quality crystals of silicon for industrial applications.

SOURCE: 'Sony develops magnetic-field method for high-quality silicon crystals' *Journal of the Electronics Industry, Japan* p 42 (August 1980)

1980 **FIBRE-OPTIC SUBMARINE CABLE**   **Standard Telephones and Cables (UK)**

"Tomorrow (14 February) weather permitting, Standard Telephones and Cables will start laying what is probably the first purpose-built fibre optic submarine cable in the world.

The British Post Office cableship Monarch will lay the trial system, a five nautical mile loop of armoured cable made by STC, in Loch Fyne at Inveraray, Scotland.

Results of the trial will be monitored by STC which will be looking at 'all possible parameters', and the British Post Office. It is considered vitally important since STC is a major exporter of submarine cable and the Post Office and STC need experience of a realistic sea situation to face competition from foreign companies, especially those in the USA and Japan.

The armoured cable is about two inches in diameter and has four fibres although it has been designed to take up to eight. In about a year's time, two regenerators will be inserted by Cableship Monarch. The housing for these regenerators has already been incorporated and the cable will be lifted and the regenerators placed within the housings."

SOURCE: 'PO first with seabed optic link' by D Clark *Electronics Weekly* p 1 (13 February 1980)

1981 **MS-DOS**   **W Gates (USA)**

In 1981 IBM asked Microsoft (a small company which has since expanded greatly) to provide them with an operating system for their future microcomputer, the PC. Bill Gates, who was Microsoft's young owner at the time, then bought Seattle Computer Products' Tim Patterson's 16-bit operating system SCP-DOS. Having adapted it, he christened it MS-DOS and delivered it to IBM (who call it PC-DOS). MS-DOS has since been considerably improved and is today the most widely used for professional microcomputers, since it is used by all IBM compatibles.

The irony of the story is that Bill Gates first advised IBM to go to his competitor, Digital Research, the producers of CP/M. Digital Research refused to sign the promise of secrecy required by IBM, thus losing the market and, at the same time, taking the first step on their path to decline.

SOURCE: *The Book of Inventions and Discoveries* Associate Editor Valerie-Anne Giscard d'Estaing (UK: Queen Anne Press, Macdonald & Co.) p 117 (1990)

1981 **HYDROPLANE POLISHING OF SEMICONDUCTORS**   **J V Gormley M J Manfra, A R Calawa (Massachusetts Institute of Technology) (USA)**

A new polishing technique for semiconductor materials promises very smooth surfaces free of mechanical defects, faster polishing, and, by implication, improved yield and throughput. Called hydroplane polishing by its developers at the Massachusetts Institute of Technology's Lincoln Laboratory in Lexington, Mass., the system may beat today's mechanical and chemical polishing approaches. *So far used on gallium arsenide and indium phosphide, the technique removes material at up to 30* μm *per minute*, as much as 60 times faster than other methods, and the surfaces produced are flat to within 0.3 μm and free of mechanical damage. In the new process, semiconductor wafers are mechanically suspended about 125 μm above the surface of a smooth, spinning disk coated with continually replenished etchant solution; the wafers thus hydroplane just above the disk's surface. The new method was developed to satisfy the stringent surface-quality requirements of molecular-beam epitaxy.

SOURCE: 'Hydroplaning could yield smoother IC wafers' *Electronics* p 34 (15 December 1981)

SEE ALSO: 'Spinning etchant polishes flat, fast' *Electronics* p 40 (13 January 1982)

1981 **PLANE POLARIZED LIGHT OPTICAL FIBRE**   **T Suganuma (Hitachi) (Japan)**

The big difference in the new fibre is the presence of an 80-by-26 μm elliptical jacket surrounding the cladding and itself surrounded by the support. Boric oxide in the jacket material increases its temperature coefficient of expansion far above that of the support. Thus as the fibre drawn from the four-part preform (core, cladding, jacket, and support) cools down from the 2000°C temperature at which it is fabricated, differences in the thickness of the jacket material exert anisotropic forces on the core along the major and minor axes of the ellipse. The direction with the higher compression, along the short axis of the ellipse, has the higher index of refraction.

Because the index of refraction is highest along the shorter of the ellipse's perpendicular axes and lowest along the longer one, a single-polarized wave of light launched into the cable along either axis will be transmitted unchanged. Even after transmission over 1 km, the polarisation, as measured by the extinction ratio, will be better than 30 decibels—that is, the conversion of energy from one plane into the orthogonal one is less than 0.1%. For a standard fibre under the best of conditions the ratio would probably not exceed 10 dB. And the slightest vibration of a standard fibre, which causes anisotropic mechanical pressure on the core, can reduce the ratio to about 3 dB.

SOURCE: 'Fibre transmits plane-polarized wave of light' *Electronics* p 77 (28 July 1981)

SEE ALSO: 'Elliptically cross-sectioned fibre' *Electronics* p 67 (30 August 1979)

1981   **HYDROGENATED AMORPHOUS SILICON FILMS**   **V Grasso, A M Mezzasalma and F Neri (Italy)**

The preparation of hydrogenated amorphous silicon films was carried out by a new method consisting in mixing evaporated silicon from an electron-beam source with a stream of ionized hydrogen produced by an Ion Tech low energy source. This source is a cold cathode device which operates at lower pressure than conventional cold cathode sources.

SOURCE: 'A new evaporation method for preparing hydrogenated amorphous silicon films' by V Grasso, A M Mezzasalma and F Neri *Solid State Communications* vol 41, No 9, p 675 (1982)

1982   **FISSION TRACK AUTORADIOGRAPHY**   **AERE (UK)**

Since the 'soft error' was diagnosed in 1978, very-large-scale integrated (VLSI) circuit memory manufacturers have sought a method of detecting the minute amount of naturally-occurring radioactive impurities which, if present in VLSI circuit materials, can disrupt circuit performances.

Now, however, Harwell has developed an extremely sensitive technique known as fission track autoradiography (FTA) which can detect the presence of uranium in concentrations as small as 2 parts in 10^9. This provides manufacturers with a quality control enabling them to assess raw material, and components, thereby reducing the risk of component failure.

The 'soft error' effect is produced by alpha-particle emissions from radioactive impurities present in any part of the VLSI circuit assembly. The energy possessed by an alpha-particle can produce an electric charge which may change the content of a single memory location, giving rise to computational errors. Because of this, semiconductor manufacturers are now specifying alpha-particle emission rates of less than 0.001 particles/cm^2/hour for their memory device materials.

It is not possible to detect such emission levels directly. The Harwell FTA technique exploits uranium-235, the fissile isotope present as 0.72% of natural uranium; prepared specimens of semiconductor material are coated with a polyimide film solid state nuclear track detector (s.s.n.t.d.) and irradiated with thermal neutrons in Harwell's Materials Testing Reactor, DIDO. On irradiation, the U-235 undergoes fission and the resulting fission particles are registered as tracks on the s.s.n.t.d. Afterwards the polyimide film is chemically etched to develop the fission tracks which can then be examined by optical microscopy. From the information gained it is possible to determine precisely the amount of uranium present, down to 2 parts in 10^9 (or a surface distribution of 3×10^{-6} μg/cm^2) and thus to calculate alpha-emission rates of as little as 0.0002 particles/cm^2/hour.

SOURCE: 'Fission track radiography for checking V.L.S.I. circuits' *The Radio and Electronic Engineer* vol 52, No 5, p 200 (May 1982)

1982   **RECRYSTALLIZATION SILICON PROCESS**   **Texas Instruments (USA)**

Using a moving graphite heater to create a thin layer of single-crystal silicon atop an oxide insulator, Texas Instruments Inc., is developing what it believes will be a practical alternative to expensive silicon-on-sapphire substrates for high-density, high-speed complementary-MOS integrated circuits (see figure 11.33).

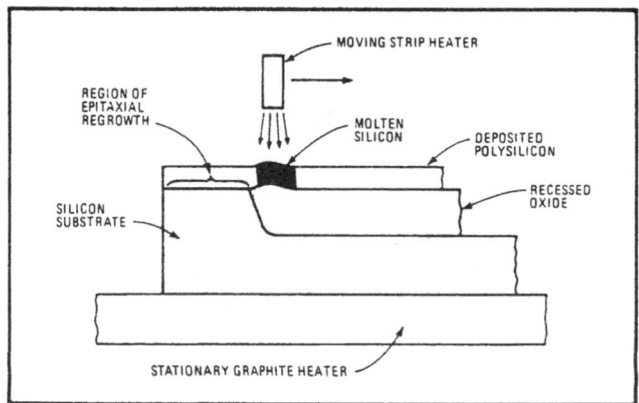

Hot spot. Strip heater that Texas Instruments moves across deposited polysilicon produces a region of epitaxial regrowth above the single-crystal substrate. Recrystallization continues above the insulator region, making a device-quality single-crystal layer.

Figure 11.33. Recrystallisation silicon process.

SOURCE: 'Oxide insulator looks the equal of sapphire for C-MOS ICs' *Electronics* p 45 (2 June 1982)

1982   **AMORPHOUS PHOTOSENSORS**   **Sanyo (Japan)**

Sanyo Electric recently succeeded in developing photosensors made of amorphous semiconductor material for the first time in the world, and announced that they would be released in spring. Based upon the development of the amorphous photosensors, Sanyo also developed one-chip full-colour sensors and intends to expand their application to line sensors. In the case of visible light total spectra photosensors, the new amorphous photosensors cost only half of conventional silicon monocrystal photosensors.

The amorphous sensor developed by Sanyo is made by forming a transparent conductive layer on a glass base, by forming PIN amorphous silicon on it through the continuous separation forming method using silicon carbide in the P layer, separating it into chips after attaching lead wires, and molding it into a resin package after performing face-down-bonding/lead-frame bonding, The mono-colour and full-colour sensors are then made by providing appropriate filters; even in the case of the full-colour sensor, it can be made as a single chip, taking advantage of the large electric resistance between the lead wires.

SOURCE: 'Sanyo develops amorphous photosensors' *JEI, Japan* p 50 (April 1982)

1982   **BUBBLE-JET PRINTING**   **Canon (Japan)**

Canon has made a major breakthrough in ink-jet technology which, it is claimed, is likely to become the printing system of the future. It is a unique 'bubble-jet', which can produce a copy in just six seconds—at least 120 times faster than any other International Standard facsimile ink-jet printers.

The new 'bubble-jet' operates on completely different principles from conventional ink-jet systems. As the name suggests, it ejects ink by means of thermally generated bubbles, rather than piezo-electricity used with other ink-jet printers.

Yet, like many new processes, the 'bubble-jet' system was discovered by accident. In Canon's research laboratories in Japan, a soldering iron was placed, inadvertently, next to a full ink syringe—and a Canon scientist noticed the heat forced out a droplet of ink. Unlike earlier ink-jet systems, which are restricted to using either a single nozzle, or from five to seven multi-nozzles, the new 'bubble jet' ejects ink through nearly 2000 nozzles (covering A4 size) to enable printing to be completed at spectacular speeds.

The 'bubble-jet' nozzles can also be increased to cope with larger copies, and it is expected that even faster speeds will be possible in the future, simply by adding more printer modules.

SOURCE: Canon Press Release, 23 February 1982

1982   CAMCORDER   Sony (Japan)

The Betamovie was the first camcorder: a video camera and recorder combined. Presented at the Japan Electronic Show in Tokyo in October 1982 by Sony, it uses normal Betamax cassettes allowing up to 3 h 35 min recording time.

SOURCE: *The Book of Inventions and Discoveries* Associate Editor Valerie-Anne Giscard d'Estaing (UK: Queen Anne Press, Macdonald & Co.) p 239 (1990)

1984   DIGITAL OPTICAL DISC   ATG (France)

In 1984 the French firm ATG marketed its numerical optical disc system, called Gigadisc: It is designed for large scale storage of all sorts of documents: texts, photos, sound, computer data. The information is read using a laser beam. The Gigadisc can store up to two thousand million bytes, which is the equivalent of 600 000 pages of typed text.

SOURCE: *The Book of Inventions and Discoveries* Associate Editor Valerie-Anne Giscard d'Estaing (UK: Queen Anne Press, Macdonald & Co.) p 125 (1990)

1985   HARD DISC CARD   Plus Development Corp. (USA)

In 1985 the American company Plus Development Corporation revolutionised mass memory technology by launching an extension card designed for IBM PCs equipped with an extra-flat hard disk of 10 MB. Before this, hard discs (large capacity fixed magnetic media) were awkward and fragile. The Hardcard, as it is called, is only 2.5 cm, 0.985 in, thick. Many other producers have followed the lead.

SOURCE: *The Book of Inventions and Discoveries* Associate Editor Valerie-Anne Giscard d'Estaing (UK: Queen Anne Press, Macdonald & Co.) p 124 (1990)

1985   DIGITAL VIDEO RECORDER   Sony (Japan)

In May 1985, at the Symposium of Montreux, Sony introduced the first video recorder with a digital recording facility. This model was solely for professional use. Since then digital video recorders have become commonplace. The effects which can be produced are very varied. For instance, it is possible to superimpose a reduced secondary image on to a corner of the screen. It is also possible to choose the speed of the tape with faster and slower options, etc.

SOURCE: *The Book of Inventions and Discoveries* Associate Editor Valerie-Anne Giscard d'Estaing (UK: Queen Anne Press, Macdonald & Co.) p 238 (1990)

1985   CD-ROM   Philips (Holland)

The CD-ROM, invented by Philips and promoted throughout the World in collaboration with Sony is similar to those used in hi-fi systems, but adapted to computing uses. It has the advantage of containing a thousand times more data than a diskette. Its disadvantage is that the data on it can be read, but new data cannot be written onto it. Developed in 1985, the CD-ROM began to take off in 1988.

SOURCE: 'Inventions and Discoveries 1993' edited by Valerie-Anne Giscard d'Estaing and Mark Young *Facts on File* New York p 218

1985   **WINDOWS**   **Microsoft (USA)**

This was launched for the first time in November 1985 by Microsoft. It required more than 50 man years of work. In April 1987 it was adopted as the standard integrator by IBM. Windows is multi-task (in some conditions) and will to a certain extent accept applications modules not specifically designed for it. The Windows interface may make it possible for IBM compatibles to become more user-friendly like the Macintosh.

SOURCE: *The Book of Inventions and Discoveries* Associate Editor Valerie-Anne Giscard d'Estaing (UK: Queen Anne Press, Macdonald & Co.) p 117 (1990)

1985   **TACTILE SCREEN**   **Zenith (USA)**

In 1985 Zenith (USA) presented the first tactile screen system, based on surface acoustic wave technology, which is even easier to use than the mouse or light pen: all the user has to do to give a command is touch a section of the screen.

SOURCE: *The Book of Inventions and Discoveries* Associate Editor Valerie-Anne Giscard d'Estaing (UK: Queen Anne Press, Macdonald & Co.) p 124 (1990)

1986   **SCANNING TUNNELLING MICROSCOPE**   **IBM (USA)**

The latest in microscopes is so small, you need a microscope to see it. Built on a silicon chip, it is only 200 micrometres across. It was made by Noel MacDonald and his students Yang Xu and Scott Miller at the National Nanofabrication Facility at Cornell University.

The device is basically a scaled-down version of the scanning tunnelling microscope, which earned Heinrich Rohner and Gerd Binnig of IBM's Research Division a Nobel Prize in 1986. STMs rely on running a fine needle, which narrows to a single atom at its point, over the surface of a sample. Electrons 'tunnel' between the needle and the surface to create a minute electric current as the tip moves to and fro across the surface. STMs can sense features as small as a single atom.

A typical modern STM is roughly the size of a thumb, but MacDonald says his version is the size of a human hair. Two tiny electrostatic motors move the silicon tip to and fro, while a third can raise or lower it.

There might not be a lot of use for such a diminutive device had scientists at IBM's Almaden site in California not shown in 1989 that scanning probe tips can pick atoms up, move them and deposit them somewhere else with great precision. To prove this they spelt out their company's logo in 35 atoms of xenon deposited on a silicon surface.

Using little piles of atoms to represent bits of data in a new kind of computer memory was a logical extension of the work. In write mode, an STM could build up the piles or flatten them; in read mode, it would run over the surface to see where the piles are. But the full-size STM was still too big for an electronic engineer's purposes. Because the motors that move the tip around in the miniature version are so tiny, the device experiences hardly any inertia, so it can scan up to 10 000 times faster than a full-scale probe.

'The bigger picture in the next 10 years is putting thousands of them on single chip,' says MacDonald. The little microscopes could then move thousands of atoms around simultaneously. And with all the control circuitry alongside the moving parts on a fully integrated silicon chip, MacDonald believes that the data handling capacity would be enormous. 'You would be able to take something like a hundred or a thousand of your normal computer disks and put them onto a chip the size of your fingernail,' he says.

But MacDonald stresses that their breakthrough is just the first step in a long process.

SOURCE: 'Smallest microscope in the world' by R Pease *New Scientist* p 21 (20 May 1995)

NOTE: Two articles (Van Nostrand Scientific Encyclopaedia, and Engineering Science and Educational Journal) state that the Scanning Tunnelling Microscope was invented by Binning and Bohrer in 1982.

SOURCE: Private communication from E Davies, London.

1987   **DAB (Digital Audio Broadcasting)**   **Eureka 147 (Europe)**

DAB is the result of a standard devised and developed by a group of European broadcasters and consumer electronics industries and their research institutes, including the BBC's Research & Development Department. The European project group is collectively known as Eureka 147—part of a wider European communications and technologies initiative. Since 1987, Eureka 147 has developed DAB to provide a reliable, multi-service digital sound broadcasting system for reception by mobile, portable and fixed receivers, using a simple rod aerial. It is a rugged, and yet a highly spectrum efficient sound and data broadcasting system that can be used in any usual broadcasting band and on terrestrial (land-based transmitters), satellite or cable networks.

SOURCE: BBC Leaflet 1995 (at '100 Years of Radio', IEE Conference, September 1995)

SEE ALSO: 'The COFDM modulation system: the heart of digital audio broadcasting' by P Shelswell *Electronics and Communication Engineering Journal* p 127 (June 1995)

NOTE: BBC Engineering Information state that their broadcasts, commencing in September 1995, were the first.

SOURCE: Private Communication from E Davies, London

1987   **COMPACT DISC VIDEO**   **Philips (Holland)**

In 1987 there was a new development. The CDV, the compact disc video, was brought out by Philips and developed in conjunction with Sony. It enables video pictures to be shown on a television screen while laser quality sound is produced simultaneously on stereo. The new readers can reproduce both sound and picture. They will read standard compact discs, while the gold CDVs of the same format (i.e. 5 inches) play pictures and sound for six minutes as well as 20 minutes of music, the 8-inch CDVs offer 40 minutes of pictures and sound and the 12-inch CDVs last for a maximum of two hours, and give additional backing to films and operas. The CDV is one of the answers to the competition offered by DAT (Digital Audio Tape), an audio-digital cassette.

SOURCE: 'Inventions and Discoveries 1993' edited by Valerie-Anne Giscard d'Estaing and Mark Young *Facts on File* New York p 218

1987   **ADVANCE PROGRAMME CONTROL OF VIDEO**   Matsushita (Japan)
       RECORDERS

This is a Panasonic invention (Japanese Matsushita group). It is a system of monthly advance programming using a light pen and infra-red transmission. The pen is used to scan the bar codes which hold all the information about television programmes and the video recorder is ready to record on the specified date at the exact moment required.

SOURCE: *The Book of Inventions and Discoveries* Associate Editor Valerie-Anne Giscard d'Estaing (UK: Queen Anne Press, Macdonald & Co.) p 238 (1990)

1988   **FIBRE OPTIC TRANSATLANTIC CABLE**   **DGT, BT and AT&T (France, UK,
                                                     USA)**

The first fibre optic transatlantic cable, the TAT-8, has linked the United States with Great Britain and France since 1988. The cable is 6620 km, 4114 miles, long and carries television, telephone and data processing signals. The partners in this venture are DGT (France), American Telegraph and Telephone (AT&T), and British Telecom International. The total cost is estimated to be 220 million.

SOURCE: *The Book of Inventions and Discoveries* Associate Editor Valerie-Anne Giscard d'Estaing (UK: Queen Anne Press, Macdonald & Co.) p 244 (1990)

1988   **VIDEO WALKMAN**   **Sony (Japan)**

Like its audio predecessor, this is also a Sony invention. Marketed in Japan since August 1988, it comprises a case weighing 1.1 kg, 2 lb 7 oz, containing a televusion receiver with liquid crystals and an 8 mm format video recorder.

SOURCE: *The Book of Inventions and Discoveries* Associate Editor Valerie-Anne Giscard d'Estaing (UK: Queen Anne Press, Macdonald & Co.) p 238 (1990)

1991 **VERY HIGH DENSITY DISKETTE** Insite Peripheals (USA)

Developed by the American company Insite Peripherals the Floptical Disk Drive can store 20.8 megabytes on standard 3.5-inch diskettes. Moreover, it is fully compatible with earlier formats: 720 kilobytes and 1.44 megabytes. This achievement is due to the use optical recording techniques. The Floptical has been available in the United States since March 1991.

SOURCE: 'Inventions and Discoveries 1993' edited by Valerie-Anne Giscard d'Estaing and Mark Young *Facts on File* New York p 218

1991 **PLASTIC ELECTRONICS** Philips (Holland)

While the electronics industry strives to make cheaper throw-away silicon microchips, Philips research laboratory in Eindhoven has adopted the opposite approach. It is making disposable materials that behave like a silicon chip. Ultimately this might mean shop assistants can total up your shopping while it sits in the trolley.

De Leeuw has developed plastic coatings that could be incorporated into smart packaging that would, for example, tell a supermarket checkout point what goods were in a passing shopping trolley—or even in a shoplifter's pocket. The plastic paint could also be used to make other disposable devices such as telephone payment cards.

Conventional plastic polymers, for example polyethylene, polypropylene, and polyvinyl chloride, are all insulators. They are built from long chain molecules surrounded by side branches. De Leeuw heated these materials in a furnace to just below their burning point and then added organic solvent to the residue.

This process knocked off the side branches leaving the long central molecules. When these materials are used as a printing ink the molecules line up and the coating behaves like a semiconductor, passing an electric current when a voltage is applied. Last week de Leeuw gave the first demonstration of his plastic electronics, showing that the coating could generate a signal that could be displayed on an LCD screen.

To make smart packaging, a layer of semiconductor ink would be printed on the top surface of ordinary card or plastic. This two-layer structure would behave like a field-effect transistor, where the current flows through the coating when a control or switching voltage is applied.

If the coating is applied in a pattern of connected areas, the plastic acts as a string of FETs—each segment switches on the next FETs in line. Depending on the shape and size of the printed pattern, this domino effect sets up a characteristic oscillating flow of electricity around the FETs. It is this characteristic oscillation that will enable a supermarket checkout to discriminate between the packaging of, say, a carton of milk and a packet of cheese. The oscillations would be triggered by bathing the packaging in an electromagnetic field.

De Leeuw says the plastic coatings could be a much cheaper alternative to silicon for telephone payment cards. He estimates that the minimum cost of implanting a card with a silicon chip is 30 cents, while his technology would cost 'nearly nothing'.

SOURCE: 'Check out the electronic paint' *New Scientist* p 18 (1 July 1995)

1991 **PHOTONIC CRYSTAL** Eli Yablonovitch (USA)

In 1991, Eli Yablonovitch of Bellcore, the telecommunications research company in New Jersey, became the first to mace a photonic crystal. He did it in the simplest way imaginable. Starting with a solid slab of a commercial material called Stycast-12, he used an ordinary workshop drill to bore three sets of long, slanted holes through the top surface of the block, Yablonovitch chose Stycast-12, which is manufactured by the Massachusetts company Emerson and Cumming, because it is transparent to microwaves. The holes he drilled intersected below the surface to produce an intricate, periodic, three-dimensional pattern, which is what forms the photonic crystal. Just drilling the holes turns the material to a perfect mirror to reflect the microwaves.

In the future, photonic crystals could be used to build super-powerful computers based entirely on light. In today's optoelectronic circuits, electronic components still do most of the work. But processing would be much quicker in an optical system. Because optical devices are efficient, they will require less power to operate than their electronic counterparts. And whereas decreasing the size of the circuits in electronic devices tends to increase the resistance, and hence the waste heat they generate, there is no such problem with optical systems. Furthermore, optical circuits can carry many different signals on a single line, so they can process a large amount of information very quickly.

There is as yet no multipurpose optical component like the electronic transistor. But all optical chips, containing devices made from photonic crystals, could eventually provide one.

SOURCE: 'Tricks of the light' *New Scientist* p 26 (25 August 1995)

1995　**LASERCOM—Laser Bridge Across the World**　　**ThermoTrex, Motorola (USA)**

Information 'bridges' could be built in space using infrared lasers, with each beam bouncing a billion bits of information per second between satellites. The fastest radio-based systems on board todays digital TV satellites handle just over 20 million bits per second.

ThermoTrex Corporation of San Diego, California, calls its new system Lasercom. Traditional systems transmit information by bouncing radio waves off satellites and down again to receiving stations. Lasercom will rely on a satellite-to-satellite approach, passing laser signals around the planet between orbiting relay modules known as transceivers. The signals will only come down to Earth at their destination.

The idea of passing information between satellites is not unique. For example, Motorola plans to have its Iridium constellation of satellites—which will carry mobile phone conversations—in place by 2000. But the Iridium satellites will communicate using microwaves, which cannot carry as much data as lasers.

As well as carrying more information than radio or microwaves, infrared lasers should be more secure. The signal travels in a tightly focused beam, rather than scattering like radio signals, so laser transmissions are more difficult for unauthorised receivers to intercept. 'We have improved the engineering on the transceivers to the point where we now feel all the technical hurdles have been passed,' says Scott Bloom, the Lasercom project manager.

In order to pass information reliably between satellites, the on-board transceivers must be able to track each other. Lasercom aims to achieve this with the help of beacons. The transmitting satellite will look for the beacon of the satellite next in line. To cut down interference from ambient light, principally sunlight, ThermoTrex has developed an optical filter that only admits light at the precise wavelength of the beacons. According to Bloom, this will allow the transmitting satellite to track the receiving satellite in broad daylight.

While the lasers should transmit clearly in space, their beams may not be able to penetrate thick cloud cover as the signal passes to and from the Earth at each end of its journey. One option is to rely on radio links between the satellites and the ground. Although they would be slower than laser links, Bloom says that the system can be configured to prevent the radio links from becoming information 'bottlenecks'.

Each satellite will have as many as four laser transceivers, enabling it to communicate with several others simultaneously. If a satellite or a transceiver is disabled, the network could be rapidly reconfigured to keep the information flowing.

Earlier this year, ThermoTrex conducted an Earth-based test of Lasercom when it broadcast several video teleconferences simultaneously by laser between the Naval Research and Development Laboratory on Point Loma in the San Diego Bay and the San Diego Convention Center 10 kilometres away.

ThermoTrex is now negotiating with Teledesic, a company founded by Bill Gates of Microsoft, about launching the transceivers aboard Teledesic's satellites.

The next planned test for Lasercom will be in 1997 when transceivers will be launched aboard US military satellites.

SOURCE: 'Laser bridge across the world' *New Scientist* p 25 (17 June 1995)

1995   INTER-SATELLITE COMMUNICATION   US Air Force (USA)

Two US satellites have passed a message to each other without the aid of a ground station—the first time this has been done. Direct satellite-to-satellite communication could reduce the military's reliance on ground stations, which are vulnerable to jamming during a war, Pentagon officials say. It also heralds the technology that will make global satellite cellphone services possible.

The test message was transmitted from Virginia by the US Air Force and arrived in Hawaii via a 'crosslink' between two Milstar communication satellites. The Milstar satellites were originally designed to relay secret codes for launching nuclear weapons operating after an all-out nuclear war had begun. Since the end of the Cold War, Milstar has come under fire as an unnecessary extravagance. To meet this criticism, the Air Force is promoting Milstar as a way of providing communications in a conventional war.

Other satellites can only talk to each other through ground stations, so any message has to be broadcast from a satellite to a ground station and then back up again to the next satellite in the chain. That makes ground stations a prime target for enemy bombing, says Leonard Kwiatkowski, the general who heads the Milstar programme. Normal transmissions are also vulnerable to eavesdropping. But crosslinked communications several thousand kilometres above the Earth are out of reach to eavesdroppers. 'Someone would have to get right in the pathway of the beam, right between the satellites, to intercept it,' he says.

SOURCE: 'Satellites talk among themselves' *New Scientist* p 7 (6 Jan 1996)

1995   VHS BIT STREAM RECORDER   Hitachi (Japan), Thomson (USA)

In March, Thomson Consumer Electronics Inc. in the US and Hitachi, Ltd successfully demonstrated a prototype of a VHS bit stream recorder that heralds the first major development in VCR technology since RCA, Hitachi and others introduced the analog VHS VCR in 1977.

The first bit stream recording of the DSS® signal was demonstrated using a VHS bit stream recorder prototype and a version of RCA's DSS® receiver.

As co-developers, Thomson and Hitachi are producing the first VHS bit stream recorder that provides the same quality video and audio as is currently being transmitted by DIRECTV® and USSB℠ to an RCA DSS® receiver. The new product is expected to be in US stores by mid-1996.

The standard VHS cassette used by most consumers, as well as the countless software titles on VHS format, are fully compatible with the new VHS bit stream recorder. Owners of digital broadcast system receivers with digital interface will be able to time-shift record superior quality digital picture and sound in addition to regular analog broadcasts.

Thomson and Hitachi are designing the new recorder for possible future applications with other digitally transmitted signals besides the DSS® signal such as the Digital Video Broadcast (DVB) systems in Europe, and digital cable and telco delivery systems.

Digital broadcast recording will spawn a new generation of consumer products that contain a digital interface especially for recording digital video, audio and data.

SOURCE: 'Age of tomorrow 136' Hitachi Ltd, Tokyo, p 19

1995   **BIOLOGICAL MEMORY CHIP**   **Mitsubishi Electric and Suntory**
                                                       (Japan)

One day, 'biological' computer chips may be able to store and process ten thousand times as much information as today's silicon chips, which are approaching the limits of miniaturisation. This dream has moved a step closer to reality with the creation by a Japanese research team of a protein molecule that behaves like a diode.

Diodes are useful in electronics because they only conduct current in one direction: a process called rectification. Two conventional diodes sandwiched back-to-back make a transistor, the basic processing and storage unit in microchips.

The biological diode was built by scientists at Mitsubishi Electric and Suntory using funding from the Japanese Ministry of International Trade and Industry, which is eager to put Japan at the forefront of biocomputing.

A team led by senior researcher Satoshi Ueyama at Mitsubishi's Advanced Technology R&D Centre in Hyogo synthesised the protein from cytochrome c552, a natural protein, and flavin, a vitamin, and called it flavocytochrome. Although the new device is synthetic, both its components occur in nature, so Mitsubishi says it can still be regarded as biological.

To test whether flavocytochrome would behave like a diode, Ueyama's team deposited several of the protein molecules on a thin gold film, which had been grown on a mica substrate. They then probed individual molecules with a scanning tunnelling electron microscope, applying various voltages between the microscope's probe tip and the gold electrode. When a voltage was applied to make the microscope tip positively charged compared to the gold, no current flowed through the molecule. But when the voltage was reversed, and when it reached 900 millivolts, a current of 70 picoamperes flowed through the molecule. This behaviour closely mimics that of an electronic diode.

Ueyama says electrons were found to be flowing from the flavin side of the molecule to the iron-rich haem group in the cytochrome on the other side. 'The electrons only transfer from a high energy level in the flavin to a lower energy level in the haem—just like water flowing from a higher place to a lower place,' he says.

A flavocytochrome molecule is 2.5 nanometres long, and as a memory chip component could be used to create devices that are around 10 000 times the density of today's semiconductor RAM chips, storing several terabits of data. The biggest microchips currently under development at Mitsubishi hold only 256 megabits of data using 0.25-micrometre components.

But like other bioelectronics researchers, Mitsubishi is still a long way from being able to use a biological diode in a circuit. Connecting the molecule to another component, for example, presents huge challenges.

Creating a useful circuit is also a major goal at Israel's Weizmann Institute of Science in Rehovot, where researchers led by Abraham Shanzer are looking at different ways of switching molecules to give 'on' and 'off' settings. Jacqueline Libman, senior staff scientist in Shanzer's team, described Mitsubishi's development as 'a move in the right direction'.

SOURCE: 'Bio-diode could lead to superchip' by P Marks *New Scientist* (27 January 1996)

1995   **GLASS LASER**   **Song-Tiong Ho *et al* (USA)**

The world's most efficient laser is also one of the smallest. It is made of a glass fibre less than half a micrometre across—500 times thinner than a human hair. Its output is measured in microwatts, but it is tens of thousands times more efficient than normal lasers.

All laser light is generated by a cascade of photons passing in phase in a single direction. As the beam travels through the medium, more photons join the cascade and the beam becomes more intense. But in most lasers, for every photon that becomes part of the laser beam, tens of thousands radiate uselessly in other directions.

The trick has been to build on the new art of 'photonics', and use photonic wire, where radiation is suppressed in all directions but one. The photonic wire, developed by Seng-Tiong Ho and his colleagues at Northwestern University, Illinois, is far narrower than the wavelength of the light it carries. Inside, there is only one direction for photons of light to travel: down the axis.

In its present form, the laser is a loop with a diameter of 4 micrometres. The photonic wire is made of silica with a high refractive index, standing proud of a silicon wafer. Embedded in the wire are three 'quantum wells' 0.01 micrometres deep made of indium gallium arsenide. These are the source of the photons. The laser beam would remain trapped inside the tiny ring laser were it not for a second photonic wire, wrapped in a U-shape around the edge of the laser ring. A quantum effect known as tunnelling allows a small fraction of the photons to bleed out of the source laser into the second wire.

This way, 70 per cent of available photons are forced into the laser beam. The high efficiency means that less energy gets wasted as heat. As microelectronic manufacturers try to get more components onto their chips, and look to optical as well as electronic processing of data, they need to minimise heat generation.

SOURCE: 'Tiny glass laser sheds more light than heat' by R Pease *New Scientist* p 23 (23 October 1995)

1995 DVD—DIGITAL VERSATILE DISC   (International)

The DVD format is the result of an unprecedented agreement reached in late 1995 among rival groups of international companies. The competing groups combined the best features of their individual approaches, which had been developed separately. The new breed of optical-disk reader prescribed by the agreement will play both existing CDs and DVDs—discs that, thanks to a variety of design innovations, can store about 14 times more information than current CDs can. In addition, the rate at which the first-generation DVD player plays back data—11 million bits per second—matches that of a fast 9X CD-ROM player, setting a new benchmark for performance.

There are two essential physical differences between CD and DVD discs. First, the smallest DVD pits are only 0.4 micron in diameter; the equivalent CD pits are nearly twice as large, or 0.83 micron wide. And DVD data tracks are only 0.74 micron apart, whereas 1.6 microns separate CD data tracks. So although a DVD is the same size as a CD, its data spiral is upward of 11 kilometres long—more than twice the length of a CD's data spiral. To read the smaller pits, a DVD player's readout beam must achieve a finer focus than a CD player's does. In order to do this, it uses a red semiconductor laser that has a wavelength of 635 to 650 nanometres. In contrast, CD players use infrared lasers with a longer wavelength of 780 nanometres. Also, DVD players employ a more powerful focusing lens—one having a higher numerical aperture than the lens in a CD player. These differences, together with the additional efficiencies of the DVD format described below, account for the huge 4.7-gigabyte capacity of each DVD information layer.

SOURCE: 'Next-generation compact discs' by A E Bell *Scientific American* p 28 (July 1996)

1995 ATOMIC BEAM LITHOGRAPHY   Harvard University, US National
                                                    Bureau of Standards (USA)

Computer chips far more intricate and powerful than those available today may be on the cards with the development of a new technique for 'printing' integrated circuits.

Integrated circuits are usually manufactured by photolithography, in which the circuit pattern is printed onto the chip in a process similar to creating stencil designs. The 'stencil' is a polymer resist that protects certain regions of the underlying layers from etching, coating or implanting operations. Resists are made by cutting away areas of the protective layer using ultraviolet light, which limits the smallest size of any feature to about 100 nanometres, or half the wavelength of the light.

To print smaller features, manufacturers could turn to X-rays, which have a shorter wavelength. Unfortunately, compact, high-energy X-ray sources are not yet available and polymer resists for use with X-rays are still being developed. So until now, the only alternative has been electron-beam lithography, in which a tightly focused electron beam 'writes' the pattern onto the surface like a pen. But this approach is expensive and time-consuming and is not suitable for high-volume manufacturing.

Now researchers at Harvard University and the US National Institute of Standards and Technology have shown that a beam of atoms can be controlled by lasers to create tiny patterns on a resist. They have cut features as small as 40 nanometres and hope to reach the 10 nanometre mark.

Atoms are normally stable when their electrons are in the lowest energy state. In atoms known as neutral metastable atoms, however, electrons can get trapped in higher energy states, from which they cannot decay. When such an atom lands on a surface it deposits its spare energy. The researchers have developed a resist made of a material that is damaged by the energy from neutral metastable argon atoms. When the damaged segments are etched away, a pattern remains.

The resist is a layer of alkanethiolate only one molecule thick applied over a thin gold film. The argon atoms are energised into a metastable state by an electrical discharge and directed by an interference pattern produced by laser beams.

Beams of light interfere with each other in the same way as water waves, creating peaks and troughs of energy. When the light meets the argon atoms, it quenches their internal energy in the brightest regions. The light's magnetic field also exerts a force on the atoms, pushing some into the minimum energy regions. The atoms then create a corresponding pattern in the resist.

SOURCE: 'Atomic beams etch the finest chips' by K Lewotski *New Scientist* p 21 (25 November 1995)

1996   **ATOMIC HOLOGRAPHY**   NEC Tsukuba (Japan)

Three tiny letters, just half a millimetre high, spell out a breakthrough in the art of printing. Scientists at NEC's Fundamental Research Laboratories in Tsukuba, Japan, have written their company's name using a beam of neon atoms. The resulting logo is the atomic equivalent of a hologram, and the technique that created it could one day be used to improve the resolution of microelectronic circuits.

According to quantum mechanics, all particles, including atoms, have wavelike properties. This idea is already exploited in electron microscopes, which use beams of electrons to produce images with a much higher resolution than is possible with light microscopes. More recently, physicists have turned their attention to atoms, bending and diffracting beams of atoms just like beams of light ('Atoms through the looking glass', *New Scientist*, 20 April, p 30).

In conventional holograms, such as the security image on credit cards, light waves pass through the transparent areas between patterns of dark lines. This arrangement acts as a diffraction grating, causing the waves to interfere with one another so that some cancel each other out and others add together to create more intense light. The result is a recognisable image.

To create a diffraction grating for beams of atoms, Junichi Fujita and Shinji Matsui of NEC, working with researchers at the University of Tokyo, punched a computer-generated pattern of holes in a thin sheet of silicon nitride. The pattern was calculated to diffract the neon atoms so that they would write the letters 'NEC' when they hit a surface. In principle, the holes could be punched to create any pattern.

In all, the printed letters were made of just 52 000 atoms, accumulated over two hours, the researchers report in the 29 July issue of *Physical Review Letters* (vol 77, p 802). The edges of the letters were slightly blurred, spread over about 65 micrometres. But with a finer pattern of holes cut into the silicon nitride mask, it should be possible to produce atomic holograms with a resolution of just 10 nanometres—six thousand times sharper.

David Pritchard of the Massachusetts Institute of Technology, a leading expert in atom optics, says that atomic holography could one day be used to create microelectronic circuits. The optical lithography techniques now used to etch circuits are limited by the wavelength of light to a resolution of around 150 nanometres.

SOURCE: 'Quantum hologram says it with atoms' *New Scientist* p 19 (17 August 1996)

1996   ELECTRON-BEAM PROJECTION SYSTEM   Bell Labs (USA)

Thousands more components could be packed onto chips much sooner than chipmakers expected, thanks to an electron beam projection system from Bell Labs in New Jersey. Its developers reckon the system will be working within two years.

Lloyd Harriott, head of the research team, says that SCALPEL, as they have named the technique, can fabricate components just 0.08 micrometres wide—roughly 250 silicon atoms across. 'Features 0.35 micrometres wide are the norm in today's semiconductor plants,' says Harriott. 'Everyone had expected the next steps to be 0.18 micrometres, then 0.13 micrometres, so this is four generations ahead.

The SCALPEL system works in a similar way to the light-based lithography now in use: an electron beam shines high-energy (100 kiloelectronvolts) electrons onto a mask which contains the pattern that must be created on the chip. But instead of blocking the beam with opaque materials in some places to create the pattern, some areas of the SCALPEL mask scatter the electrons while others let them through unscathed. 'We use a high atomic number material that scatters the beam, while in the 'open' areas of the mask the e-beam simply passes through unaltered,' he says.

The e-beam then passes through a magnetic lens, which focuses it on an aperture just 160 micrometres in diameter. The unscattered electrons pass through the aperture, but the vast majority of those that were scattered miss the hole.

A second lens then focuses the beam onto the chip substrate below. Just as in conventional lithography, the energetic electrons bombard the substrate and remove materials to create patterns. The second lens can focus the electrons into an image one quarter the size of the original mask. This means that the mask can be relatively large, which makes it easier to manufacture.

The high energy of the electrons was one stumbling block in previous attempts to make chips with e-beams, says Harriott. 'You can make a mask from material that will stop electrons by absorbing them, but that material quickly heats up as a result, and then becomes distorted,' he says.

An alternative approach is to use a narrow electron beam like a pencil, and write patterns onto the chip. 'This works, but it is like handwriting versus the printing-press speeds of lithography,' says Harriott. 'It's too slow to be practical.'

The laboratory produced its first 0.08 micrometre features this summer, and expects to be etching full circuits next year. 'Commercial use of the machines could come within two years,' says Harriott.

Funding for the project was provided by Bell's parent company Lucent Technologies, a sister company of AT&T, which also provided funding, and by the government's Advanced Research Projects Agency.

SOURCE: 'Chip pioneers tame power of E-beams' by J Beard *New Scientist* p 18 (3 August 1996)

1996   SUPERFAST SWITCH   Argonne Nat. Lab. (USA)

Computing power could be set for another upward surge, by storing data in single molecules that can transmit signals in femtoseconds (10^{-15} seconds). The fastest of today's silicon-based computers rely on switches a million times slower, registering information in nanoseconds.

Some experimental systems have switches that operate in trillionths (10^{-12}) of a second, but the system built by Michael Wasielewski and his colleagues at the Argonne National Laboratory in Illinois is a thousand times as fast. It operates by pumping an electron through a single molecule with two pulses of laser light. 'To the best of our knowledge, it's the first time anyone has developed femtosecond control of a charge shift device,' says Wasielewski.

Each bit of information could potentially be stored in the molecule by a single electron. This is moved from zone to zone with separate laser pulses. The first laser pulse, which has a wavelength of 416 nanometres, pushes the electron from the storage zone to the central, priming zone. The second pulse, with a wavelength of 480 nanometres, shunts the electron within just 400 femtoseconds into the final zone of the molecule, where it could be read.

For this to be possible, the electron would have to be siphoned off into an electrical circuit. Such an action is impossible with the existing molecule, because within 600 femtoseconds, the electron flips back to the original storage zone, which returns the molecule to its maximum stability.

Wasielewski likens the laser pulses to 'pushing an electron up a cliff' in two separate stages. To make reading easier, he hopes to add a further zone that captures the electron in an energy 'well' at the top of the cliff. This would allow more time for the electron's position to be detected.

In practical terms, Wasielewski is attempting to make arrays of the molecules that could serve as miniature electronic devices. One possibility is to anchor the molecules to metal or glass surfaces, which would feed electrons into electrical circuits.

David Vass, leader of the applied optics group at the University of Edinburgh, describes the Argonne experiment as 'extremely elegant'. But he does not expect practical applications to be possible for several years. 'It requires parallel developments in laser technology', he says. The light sources must be fast-acting enough to keep up with the speed of the electrons. 'You are left with the problem of handling the positions of microscopic, femtosecond-long spots of laser light,' he says.

The research is reported in the *Journal of the American Chemical Society* (vol 118, p 8174)

SOURCE: 'Hopping electron promises superfast switch' by A Coghian *New Scientist* p 24 (28 September 1996)

1996 **SURFACE FLAT CHIPS**   **IBM (USA)**

Silicon wafers may look smooth to the naked eye, but under the microscope, they reveal terraces like ancient rice fields. Researchers at IBM and at Cornell University in Ithaca, New York, have devised a way to burn off the irregularities, creating an ultraflat surface suitable for manufacturing the next generation of semiconductors.

The surface of a silicon wafer consists of a flat base of crystalline silicon, from which rise irregular terraces of atoms about 1.5 nanometres high. Semiconductors are manufactured by depositing layers of various materials on top of this silicon base.

For the current generation of semiconductors, the substrate does not have to be absolutely flat because the layers are thick compared to the imperfections and so are not seriously affected by them. But future generations of devices will rely on thinner layers of material, so levelling the imperfections will be crucial. 'Imagine that you're trying to put a carpet on a floor that has ripples a foot high,' says Jack Blakely, professor of materials science and engineering at Cornell. 'You would want to get those ripples out before you put down the carpet. That's what we're doing here.'

Blakely's team and researchers from IBM's T J Watson Research Center in Yorktown Heights, New York, describe their levelling technique in *Applied Physics Letters* (vol 69, p 1235). First they use electron beam lithography to create a criss-cross pattern of ridges on a silicon wafer, dividing it into small squares with sides about 10 micrometres long. The squares are needed to make the smoothing process practical. Creating a level surface within each square is much faster than trying to create a single level across the entire wafer. Next, the wafer is heated rapidly to 1250 °C to drive off oxide contamination from its surface.

The final, key stage involves keeping the wafer at between 1020°C and 1150°C for about 30 minutes. At that temperature, silicon atoms gradually evaporate from the edges of the terraces, where they are not as strongly held by the crystal lattice as atoms that are surrounded on all sides. As the atoms evaporate one by one, the terraces gradually retreat towards the sides of each square, leaving a totally smooth surface behind. Blakely says the temperature is not high enough to drive off atoms from this surface, and evaporation stops at the ridge created by the lithography.

SOURCE: 'Flat is beautiful for future chips' by Vincent Kiernan *New Scientist* p 24 (21 September 1996)

1996   **DIRECT LASER WRITING**   **Mikroelektronik Centre, Lyngby**
                                                              (Denmark)

A Danish group has etched structures in polysilicon and amorphous silicon deposited on silicon oxide by direct laser writing. The patterns can be written with high resolution and transferred to the underlying material by reactive ion etching (RIE). Three-dimensional structures can be obtained by multiple exposure of the silicon mask. The fast turnaround times of the direct writing process enable the technique to be used for the rapid prototyping of large scale structures.

For deep structures and via holes, H Millenborn *et al* of the Mikroelektronik Centre, Danmarks Tekniske Universitet, Lyngby, Denmkar, deposited a 2 μm thick polysilicon layer by low-pressure CVD onto a 10 μm thick oxide deposited by plasma-enhanced CVD on a silicon substrate. Alternatively, a 2 μm thick boron-doped amorphous silicon layer was deposited by dc magnetron sputtering instead of polysilicon. For structures with a subitem resolution, a thin ($\sim$100 nm) phosphorous-doped low-pressure CVD polysilicon layer was deposited on 20 nm of thermal oxide.

Laser direct writing (i.e. 1.5 W of 488 nm light focused to an 8 μm spot) was used to pattern the polysilicon or amorphous silicon layer. Direct etching was used, based on local melting of silicon in chlorine. Solid silicon, oxides and nitrides are not etched, even at high temperatures. The sample was translated at speeds of up to 100 mm/s during etching to produce continuous trenches that can be interrupted by on-the-fly beam switching. There was no time for reflow or diffusion of materials during the fast etching scan. The silicon layer could be locally removed in a single scan to expose the underlying oxide that acts as an etch stop.

The silicon pattern was then transferred to the oxide layer by RIE using a CF_4/CHF_3 plasma that provides much higher etch rates for oxides than for silicon. Stepped oxide structures are produced by iterative patterning of the polysiclion or amorphous silicon mask and transferred by RIE (figure 11.34). Alternatively a 3-D structure can be etched directly into the silicon layer and then transferred to the oxide in a single step. The resulting oxide structure has been applied as a 3-D eroding etch mask for the RIE of bulk silicon using a SF_6/O_2 plasma.

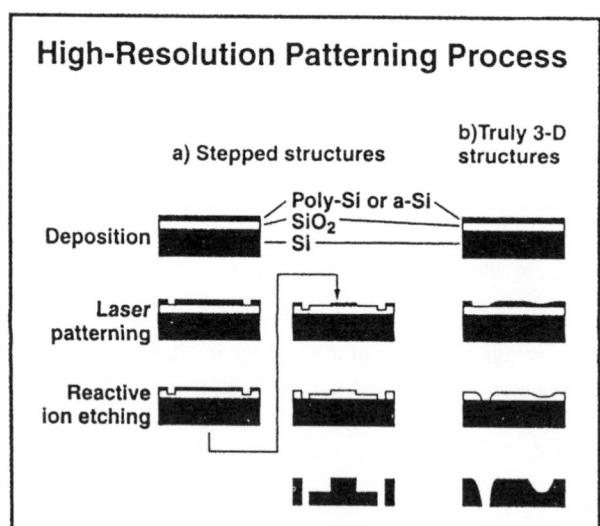

1. The process sequence for prototyping of large high-aspect-ratio 3-D structures in bulk silicon. a)Iterative laser and reactive ion etching, b) single-step pattern generation and transfer.

Figure 11.34. High resolution patterning process.

SOURCE:'Etch techniques achieves rapid prototyping' by B Dance *Semiconductor International* vol 19, No 10, p 60 (September 1996)

Chapter 12

Electronics Acronyms and Abbreviations

Because of very lengthy descriptive technical words necessary in modern electronics, e.g. TRAPPAT diode (Trapped plasma avalanche transit time diode), acronyms and abbreviations have become part of today's electronics language. Arising from the preparation of this book and from a perusal of current technical literature, the following list covering a wide field has been prepared in the hope that it will be useful for reference.

N.B. In cases where the same acronym is used for different interpretations, both are included.

a–d	Analogue to Digital
a-Si	Amorphous Silicon
ACIA	Asynchronous Communication Interface Adapter
ADA	Computer Language (named after Ada Augusta)
ADC	Analogue-to-Digital Converter
ADS	Address Data Strobe
ADP	Automatic Data Processing
AGC	Automatic Gain Control
AIM	Avalanche-Induced Migration
ALGOL	ALG-Orithmic, Computer Language
ALU	Arithmetic/Logic Unit
AM	Amplitude Modulation
ANSI	American National Standards Institute
AOI	AND/OR Invert
AOI	Automated Optical Inspection
APL	A Programming Language developed by Iverson
AQL	Acceptance Quality Level
ASCH	American Standard Code for Information Exchange
ASIC	Application Specific Integrated Circuits
ASPR	Automatic Satellite Position Reporting System
ASTM	American Society for Testing and Materials
ASQC	American Society of Quality Control
ATAB	Area-Array Tape Automated Bonding
ATC	Address Translation Cache
ATE	Automatic Test Equipment
ATM	Asymchronous Transfer Mode
ATS	Automatic Test System
AVC	Automatic Volume Control
AWACS	Airborne Warning and Control System
BARRITT	Barrier controlled Injection and Transit Time (diode)

BASIC	Beginners All-purpose Symbolic Instruction Code
BBD	Bucket-Brigade Device
BCD	Binary-Coded Decimal
BCD	Bipolar-CMOS-DMOS
BGA	Ball Grid Array
BIFET	Bipolar Field Effect Transistor
BITE	Built-In Test Equipment
BIT	Built-In-Test (Bit-unit of binary data)
BIOS	Basic Input/Output System
BJT	Bipolar Junction Transistor
BORAM	Block-Oriented Random-Access Memory
b/s	Bits per Second
BTAB	Bumped Tape Automated Bonding
BW	Band Width
BYTE	8 bits

CAAP	Coprocessor Architecture and Protocols
CAD	Computer-Aided Design
CAL	Computer-Assisted Learning
CAM	Content-Addressable Memory
CAM	Computer-Aided Manufacturing
CATT	Controlled Avalanche Transit Time (diode)
CAT	Computer Aided Testing
CB	Citizens Band (radio)
CCD	Charge-Coupled Device
CCIR	Int. Radio Consultative Committee
CCITT	Int. Telegraph & Telephone Consultative Committee
CCTT	Cold Cathode Trigger Tube
CCTV	Closed Circuit Television
CD	Compact Disc
CDI	Collector Diffusion Isolation
CDMA	Code-Division Multiple Access
CEEFAX	Teletext (BBC)
CERDIP	Ceramic Dual-In-Line Package
CFC	Chloro-Fluoro-Carbon
CGI	Common Gateway Interface
CIRC	Cross-Interleaved Reed-Soloman Code (for CDs)
CISC	Complex Instruction Set
CML	Current-Mode Logic
C-MOS	Complementary-Metal-Oxide Semiconductor
CMRR	Common-Mode Rejection Ratio
COBOL	Common Business Oriented Language
CPU	Central Processing Unit
CRO	Cathode Ray Oscilloscope
CROM	Control Read-Only Memory
CRT	Cathode-Ray Tube
CRC	Cyclic Redundancy Check
CVD	Chemical-Vapour Deposition
CVT	Constant-Voltage Transformer

d–a	Digital to Analogue
DAC	Digital-to-Analogue Converter
DAP	Distributed Array Processor

IANA	Internet Assigned Numbers Authority
IARU	International Amateur Radio Union
IBE	Ion Beam Etching
IC	Integrated Circuit
ICE	In-Circuit Emulator
IDS	Input-Data Strobe
IEC	Infused Emitter Coupling
IEC	International Electrotechnical Commission
IEEE	Institute of Electrical and Electronics Engineers
IGFET	Insulated Gate Field Effect Transistor
I^2L	Integrated Injection Logic
ILB	Inner Lead Bonding
IMPATT	Impact Avalanche Transit Time (diode)
InP	Indium Phosphide
I/O	Input/Output
IP	Internet Protocol
IR	Infra Red
ISDN	Integrated Services Digital Network
ISO	International Standards Organisation
ITU	International Telecommunications Union
JEDEC	Joint Electron Device Engineering Council
J-FET	Junction Field-Effect Transistor
JTI	Junction Isolation
KBIT	Kilobit
LAPUT	Light-Activated Programmable Unijunction Transistor
LAN	Local Area Network
LARAM	Line Addressable Random Access Memory
LASCR	Light-Activated Silicon Controlled Rectifier
LASER	Light Amplification by Stimulated Emission of Radiation
LCA	Life Cycle Analysis
LCC	Leadless Chip Carrier
LCCC	Leadless Ceramic Chip Carrier
LCD	Liquid-Crystal Display
LED	Light-Emitting Diode
LEED	Low Energy Electron Diffraction
LIC	Linear Integrated Circuit
LIFO	Last In, First Out
LF	Low Freqency
LLCC	Leadless Chip Carrier
LNA	Low-Noise Amplifier
LOCMOS	Locally Oxidised C-MOS
LORAN	Long Range Air Navigation
LPCVD	Low Pressure Chemical Vapour Deposition
LPTTL	Low-power Transistor–Transistor Logic
LRU	Least Recently Used
LSB	Least Significant Bit
LSI	Large-Scale Integration
MAC	Multiply-Accumulate

MASER	Microwave Amplification by Stimulated Emission of Radiation
MBE	Molecular Beam Epitaxy
MCM	Multi-Chip Module
MCM-C	Multi-Chip Module—Ceramic Dielectric
MCM-D	Multi-Chip Module—Organic Dielectric
MCM-L	Multi-Chip Module—Laminate
MCT	Mobile Communications Terminal
MDS	Microprocessor-Development System
MESFET	Metalised Semiconductor Field-Effect Transistor
MF	Medium Frequency
MFLOPS	Millions of Floating-Point Instructions Per Second
MHL	Microprocessor Host Loader
MIMD	Multiple Instruction Multiple Data
MIME	Multipurpose Internet Mail Extensions
MIPS	Millions of Instructions Per Second
MIS	Metal Insulator Silicon
MISD	Multiple Instruction Single Data
MLA	Microprocessor Language Assembler
MLB	Multilayer Board
MLE	Microprocessor Language Editor
MLS	Microwave Landing Scheme
MMIC	Monolithic Microwave Intergrated Circuit
MMU	Memory Management Unit
MNCS	Multipoint Network Control System
MNOS	Metal-Nitride-Oxide Semiconductor
MODEM	Modulator/Demodulator
MODHIC	Modular Hybrid Integrated Circuit
MOS	Metal-Oxide Semiconductor
MOSFET	Metal-Oxide-Semiconductor Field-Effect Transistor
MOST	Metal Oxide Semiconductor Transistor
μP	Microprocessor
MPEG	Moving Pictures Expert Group
MPU	Microprocessor Unit
MSB	Most Significant Bit
MSI	Medium-Scale Integration
MSIN	Multi-Stage Interconnection Network
MST	Multiservice Terminal
MTBF	Mean Time Between Failures
MTD	Mass Tape Duplicator/verifier
MTTF	Mean Time To Failure
MTTR	Mean Time To Repair
MUSA	Multiple Unit Steerable Antenna
MUX	Multiplexer
NAND	Inverted AND gate
NASA	National Aeronautics & Space Administration
NDRO	Nondestructive Readout
NF	Noise Figure
n-MOS	n-Channel Metal-Oxide Semiconductor
NOR	Inverted OR gate
NRZ	Non-Return to Zero
NRZI	Non-Return to Zero Inverted
NTSC	National Television System Committee

OAM	Operation And Maintenance
OBL	One Block Lookahead
OCR	Optical Character Recognition
ODS	Output Data Strobe
OEM	Original-Equipment Manufacturer
OLB	Outer Lead Bonding (in TAB)
OPAL	Operational Performance-Analysis Language
OP/AMP	Operational Amplifier
ORACLE	Teletext
OTA	Operational Transconductance Amplifier
PABX	Private Automatic Branch Exchange
PACE	Plasma Assisted Chemical Etching
PAL	Phase Alternation Line (TV)
PAM	Pulse-Amplitude Modulation
PAR	Precision Approach Radar
PASCAL	Computer Language
PAR	Programme-Aid Routine
PBX	Private Branch Exchange
PC	Personal Computer
PC	Printed Circuit
PCA	Principal Components Analysis
PCB	Printed-Circuit Board
PCC	Plastic Chip Carrier
PCMCIA	Personal Computer Memory Card
PCM	Pulse Code Modulatlon
PCN	Personal Communication Network
PDP	Plasma Display Panel
PGA	Pin Grid Array (Package)
PIA	Peripheral Interface Adapter
PIND	Particle Impact Noise Detection
PLA	Programmable Logic Array
PLC	Programmable Logic Controller
PLL	Phase-Locked Loop
PM	Phase Modulation
PMG	Permanent-Magnet Generator
p-MOS	p-channel Metal-Oxide Semiconductor
PPI	Plan-Position Indicator
PPI	Programmable Peripheral Interface
PPM	Parts Per Million
PQFP	Plastic Quad Flat Package
PRACL	Page-Replacement Algorithm and Control Logic
PRESTEL	Teletext (BBC)
PROM	Programmable Read-Only Memory
PSPDN	Packet-Switched Public Data Network
PSTN	Public-Switched Telephone Network
PTFE	Polytetrafluoroethylene (plastic)
PTH	Plated-Through Holes
PUT	Programmable Unijunction Transistor
QA	Quality Assurance
QAM	Quadrature Amplitude Modulation
QC	Quality Control

QFP	Quad Flat Pack
QMB	Quick Make and Break (switch)
RACE	Research and Technology Development in Advanced Communication Technologies in Europe
RADAR	Radio Detection And Ranging
RDF	Radio Direction Finding
RAS	Row Address Select
RALU	Register and Arithmetic/Logic Unit
RAM	Random-Access Memory
RDP	Reliable Data Protocol
RFI	Radio-Frequency Interference
RIGRET	Resistive-Insulated Gate FET
RIM	Read-In Mode
RISC	Reduced Instruction Set Computer
RMM	Read-Mostly Mode
ROM	Read-Only Memory
RTB	Reverse Translation Buffer
RTC	Realtime Clock
RTL	Resistor–Transistor Logic
R/W	Read/Write
SAM	Scanning Auger Microscopy
SAR	Synthetic Aperture Radar
SAW	Surface Acoustic Wave
SBS	Silicon Bilateral Switch
SC	Semiconductor
SCA	Sub-Channel Adaptor
SCAT	Scanning Acoustic Tomography
SCR	Silicon Controlled Rectifier
SDH	Synchronous Digital Hierarchy
SDLC	Synchronous Data-Link Control
SDMA	Space-Division Multiple Access
SECAM	Sequential Couleur a Memoire (TV System)
SEM	Scanning Electron Microscope
S/H	Sample and Hold
Si	Silicon
SIL	Single In Line
SIMD	Single Instruction Multiple Data
SIO	Serial Input/Output
$SiO_2$2	Silicon dioxide
SIP	Single In-line Package
SISD	Single Instruction Single Data
SMA	Surface Mount Assembly
SMD	Surface Mount Device
SMART	Stress Marginality and Accelerated Reliability
SMPS	Switched Mode Power Supply
SMT	Surface Mount Technology
SMTP	Simple Message Transfer Protocol
SOIC	Small Outline Integrated Circuit Package
SOS	Silicon-On-Sapphire
SPICE	Simulated Programme with Intregrated Circuit Emphasis
SSB	Single Sideband Broadcasting

SSI	Small Scale Integration
SSIN	Single Stage Interconnection Network
SSSC	Single-Sideband Suppressed Carrier
SSTC	Single Sideband Transmitted Carrier
SUS	Silicon Unilateral Switch
Sx	Simplex

TAB	Tape Automated Bonding
TBMT	Transmitter Buffer Empty
TDD	Time Division Duplex
TDM	Time-Division Multiplexing
TDMA	Time-Division Multiple Access
TEM	Transmission Electron Microscope
TFT	Thin Film Transistor
TLB	Translation Lookaside Buffer
TMR	Triple Modular Redundancy
TOS	Top Of Stack
TRAPATT	Trapped Plasma Avalanche Transit Time (diode)
TTL	Transistor–Transistor Logic
T^2L	Transistor–Transistor Logic
TTTN	Tandem Tie Trunk Network
TTY	Teletypewriter
TWT	Travelling-Wave Tube

UART	Universal Asynchronous Receiver/Transmitter
UJT	Unijunction Transistor
ULA	Uncommitted Logic Array
ULSI	Ultra Large Scale Integration
UMTS	Universal Mobile Telecommunication System
UNCOL	Universal Computer-Orientated Language
UNI	User–Network Interface
URCLK	Universal Receiver Clock
USART	Universal Synchronous/Asynchronous Receiver/Transmitter
USRT	Universal Synchronous Receiver/Transmitter
UTCLK	Universal Transmitter Clock
UTP	Ultra-Thin Package
UUT	Unit Under Test

VCO	Voltage-Controlled Oscillator
VCT	Voltage-to-Current Transactor
VDU	Visual Display Unit
VHF	Very High Frequency
VHSIC	Very High Speed Integrated Circuit
VIL	Vertical Injection Logic
VLF	Very Low Frequency
VLSI	Very Large-Scale Integration
VMA	Valid Memory Address
VMM	Virtual Machine Moniter
V-MOS	V groove MOS
VSOP	Very Small Outline Package
VSP	Video Signal Processing
VTR	Video-Tape Recorder

WARC	World Advisory Radio Conference
WATS	Wide-Area Telephone Service
WIMPS	Windows, Icons, Menus and Pulldowns
WORM	Write Once—Read Many Times
WS	Working Set
WWW	World Wide Web
XOR	Exclusive-OR gate
YIG	Yetrium–Iron–Garnet (magnetic properties)

Chapter 13

List of Books on Inventions

A selection of books on the history of electronics and general history of technology and science is given for interest.

13.1 History of Inventions

'A History of Invention' (London: J M Dent & Sons and New York: Roy Publishers) (1971)

'Dictionary of Inventions and Discoveries' by E F Carter (London: Fdk Muller) (1966)

'Discoveries and Inventions of the 20th Century' by J G Crowther (London: Routledge & Regan Paul Ltd) (1966)

'Inventions and Discoveries' by Valerie-Anne Giscard d'Estaing (New York: Facts-on-File) (1993)

'Science & Technology in History' by Ian Inkster, Distribution Ltd, UK

'The Sources of Invention' by J Jawkes, D Sawers and R Stillerman (London: MacMillan) (1958)

'The History of Invention' by Trevor I Williams (New York: Facts-on-File) (1987)

'The Progress of Invention in the Nineteenth Century' by E W Bryn (New York: Munn) (1920)

'The Economics of Invention and Innovation' by F S Johnson (Martin Robinson) (1975)

13.2 History of Technology

'An Encyclopaedia of the History of Technology' edited by Ian McNell (London: Routledge) (1989)

'Children of Prometheus: a History of Science and Technology' by James MacLachlan, Collegiate edn (Toronto, Ontario: Wall & Emerson) (1989)

'Encyclopaedia of Modern Technology' (London: Hutchinson) (1987)

'Fontana History of Technology' by Donald Cardwell (Harper-Collins)

'History of Technology' by G M Short (Hollizter Publishing Ltd)

'Technological Change : Methods and Themes in the History of Technology' edited by Robert Fox (Amsterdam: Harwood Academic) (1996)

'Timetable of Technology' Michael Joseph (London) (1983)

13.3 History of Science

'A Biographical Dictionary of Scientists' by T I Williams (Bath: Pitman Press) (1976)

'A Short History of Science to the Nineteenth Century' by C J Singer (Clarendon Press) (1941)

'A History of Science' by W Waltham (Cambridge: Cambridge University Press) (1929)

'British Scientists of the 19th Century' by J G Crowther (London: MacMillan)

'Masters of Science and Invention' by F L Darrow (New York: Harcourt Brace) (1923)

'Men of Science in America' by Jaffe Bernard (New York: Simon & Schuster) (1944)

'Science Since 1500' by H T Pledge (HMSO) (1946)

13.4 History of Telegraphy

'From Machine Shop to Industrial Laboratory : Telegraphy and the Changing Context of American Invention, 1830–1920' by Paul Israel (Baltimore: Johns Hopkins University Press) (1992)

'History of Wireless Telegraphy' by J J Fahie (Edinburgh & London: Wm Blackwood) (1899)

'History of Electric Telegraphy to the Year 1837' by J J Fahie (London: F N Skoon) (1884)

'The Telegraph : a History of Morse's Invention and its Predecessors in the United States' by Lewis Coe (Jefferson, NC: McFarland) (1993)

13.5 History of Telephony

'100 Years of Telephone Switching (1878–1978)' *Pt 1: Manual Electromechanical Switching (1878–1960s)* by Robert J Chapuis (Amsterdam: North-Holland Publishing) (1982)

'The Beginning of Telephony' by F L Rhodes (New York: Harper) (1929)

'The Telephone and its Several Inventors: a History' by Lewis Coe (Jefferson, NC: McFarland) (1995)

'Who Invented the Telephone?' by W Aitken (London: Blackie) (1939)

13.6 History of Electricity and Electronics

'A History of the World Semiconductory Industry' by P R Morris (Peter Peregrinus) (1990)

'Bibliographical History of Electricity and Magnetism' by Motteley (London: Griffin) (1922)

'Bibliography of the History of Electronics' by G Shiers (Metuchen, NJ: Scarecrow Press)

'Early Electrodynamics' by R A R Tricker (Oxford: Pergamon Press) (1965)

'Fifty Years of Electricity' by J A Fleming (Wireless Press) (1921)

'Electronics—a Bibliographical Guide' by C K Moore and K J Spencer (London: MacDonald) (1965)

'Electronic Inventions and Discoveries' by G W A Dummer (Bristol: Institute of Physics Publishing) (1997)

'My Life with the Printed Circuit' by Paul Eisler (Cranbury, NJ: Associated University Press) (1989)

'The Conquest of the Microchip' by H Queisser (Harvard University Press) (1985)

'The History of Science and Technology' by K J Rider (London: Library Association of London) (1967)

13.7   History of Radio and Communications

'A History of Broadcasting in the United States' by E Barnouw (Oxford: Oxford University Press) (1968)

'A History of the Marconi Company' by W J Baker (London: Methner & Co) (1970)

'Behind the Tube : History of Broadcasting Technology and Business' by Andrew F Inglis (Boston: Focal Press) (1990)

'Broadcasting Technology : the Major Landmarks' by D P Leggatt *The Journal of the Institution of Electronic and Radio Engineers* Vol 56, p 303–310 (1986)

'Communications Miracle : Telecommunications Pioneers from Morse to the Information Superhighway' by John Bray (New York: Plenum) (1995)

'History of International Broadcasting' by James Wood (London: Peregrinus in association with the Science Museum) (1994)

'Invention & Innovation in the Radio Industry' by W J MacLaurin (New York: MacMillan) (1949)

'Pioneers of Electrical Communications' by R Appleyard (London: MacMillan) (1930)

'Pioneers of Wireless' by E Hawks (London: Methuen & Co) (1927)

'The History of Broadcasting in the United Kingdom' by Asa Briggs (London: Oxford University Press) Vols. 1 to 6 (1961 to 1995)

'The Beginning of Satellite Communications' by J R Pierce (San Francisco: San Francisco Press) (1968)

13.8   History of Radar

'Beginnings of Radar' by S S Swords (London: Peregrinus) (1986)

'One Story of Radar' by A P Rowe (Cambridge: Cambridge University Press) (1948)

'Radar Days' by E G Bowen (Bristol: Adam Hilger) (1987)

'Three Steps to Victory' by R A Watson-Watt (London: Odhams Press Limited) (1957)

13.9   History of Television

'Digital Television' by C P Sandbank (John Wiley & Sons) (1990)

'Birth of the Box: the Story of Television' by Ian R Sinclair (Wilmslow: Sigma Press) (1995)

13.10   History of computers

'A History of Computing Technology' by Michael R Williams (Englewood Cliffs: Prentice-Hall) (1985)

'Computers and Computing: a Chronology of the People and Machines that Made Computer History' by Mark W Greenia (Sacramento: Lexikon Services Publications) (1992)

'Computer: a History of the Information Machine' by Martin Campbell-Kelly and William Aspray (New York: Basic Books) (1996)

'Computer Pioneers' by Laura Greene (New York: Watts) (1985)

'Engines of the Mind: a History of the Computer' by Joel Shurkin, 1st ed. (New York: Norton) (1984)

'Japan's Computer and Communications Industry: the Evolution of Industrial Giants and Global Competitiveness' by Martin Fransman (Oxford: Oxford University Press) (1995)

'Landmarks in Digital Computing' by Peggy A Kidwell and Paul E Ceruzzi (Washington: Smithsonian Institution Press) (1994)

'The Computer Comes of Age: the People, Hardware, and the Software' by R Moreau, translated by J Howlett (Cambridge, MA: MIT Press) (1984)

'The Making of the Micro: a History of the Computer' by Christopher Evans, foreword by Tom Stonier (London: Gollancz) (1981)

'The Computer from Pascal to von Neuman' by H H Goldstine (Princeton University Press) (1972)

'The Early History of Data Networks' by Gerard J Holzmann and Bjorn Pehrson (Los Alamitos, CA: IEEE Computer Society Press) (1995)

'Transforming Computer Technology: Information Processing for the Pentagon, 1962–1986' by Arthur L Norberg and Judy E O'Neill; with contributions by Kerry J Freedman (Baltimore, OH: Johns Hopkins University Press) (1986)

'Understanding Computers—Illustrated Chronology and Index' (Alexandria, VA: Time-Life Books)

Chapter 14

List of Books on Inventors

A selection of books on inventors primarily in the field of electronics is given for interest.

AMPERE

'Andre-Marie Ampere and his English Acquaintances' by J R and D L Gardiner *British Journal for the History of Science* Vol 2 (July 1965) p 235

'Andre-Marie Ampere' by James R Holmann (Oxford: Blackwell) (1995)

BAIRD

'Baird of Television—the Life Story of John Logie Baird' by R F Tiltman (London: Selley Service) (1933)

'John Baird: the Romance and Tragedy of the Pioneer of Television' by Sydney Moseley (London: Odhams Press) (1952)

BELL

'Alexander Graham Bell: the Man who Contracted Space' by Catherine D Mackenzie (Houghton Mifflin) (1928)

'Bell, Alexander Graham Bell and the Conquest of Solitude' by R V Bruce (Victor Gollanz) (1973)

'Genius at Work: Images of Alexander Graham Bell' by Dorothy Harley, Eber. (New York: Viking Press) (1982)

BERLINER

'Grevile Berliner, Maker of the Microphone' by F W Wile (Indianapolis: Bobbs-Merrill) (1926)

BRAUN

'Ferdinand Braun : Leben und Wirken oes Erfinders der Brauchen Roehre, Nobel-Preistraeger' by F Kurylo (Heinz Moos Verlag) (1965)

CROOKES

'The Life of Sir Williams Crookes' by E E Fourner D'Albe (London: Fisher Unwin London) (1923)

EDISON

'My Friend Edison' by H Ford (Ernest Benn Limited)

'Edison' by M Josephson (New York: McCraw-Hill) (1959)

'Edison: Inventing the Century' by Neil Baldwin (New York: Hyperion) (1995)

FAHIE

'The Life and Work of John Joseph Fahie' by E S Whitehead (Liverpool: University Press of Liverpool) (1939)

FARADAY

'Faraday' by R & R Clark (Brit. Elec & Allied Mnfrs. Assoc.) (1931)

'Michael Faraday—his Life and Work' by S P Thompson (Cassel) (1901)

'Faraday, Maxwell & Kelvin' by D J MacDonald (New York: Doubleday) (1964)

'Michael Faraday and the Modern World' by Brian Bowers (Saffron Walden: EPA) (1991)

FESSENDEN

'Fessenden—Builder of Tomorrow' by H M Fessenden (Coward-McCann) (1940)

FITZGERALD

'The Scientific Writings of the late George Francis Fitzgerald' edited by J Lamor (Dublin: Dublin University Press) (1902)

FLEMING

'Memories of a Scientific Life' by Alexander Fleming (Marshal)

HEAVISIDE

'Oliver Heaviside' by G Lee (London: Longmans Green) (1947)

'Oliver Heaviside: a Biography' by H J Josephs (1963)

HENRY

'Joseph Henry—His Life and Work' by T Coulson (Princeton University Press) (1950)

HERTZ

'Gesammelte Werke' by P E A Lenard Ambrosius Borth (Leipzig) (1895) (Papers in three volumes) English translations (London: MacMillan)

'The Creation of Scientific Effects: Heinrich Hertz and Electric Waves' by Jed Z Buchwald (Chicago: University of Chicago Press) (1994)

LODGE

'Oliver Lodge—Past Years, an Autobiography' by Scribner (New York) (1932)

'Oliver Lodge and the Liverpool Physical Society' by Peter Rowlands (Liverpool: Published for the Department of History, University of Liverpool by Liverpool University Press) (1990) - iv. 310p. : ill, ports; 21cm - (Liverpool historical studies; No. 4)

MARCONI

'Marconi, the Man and his Wireless' by O E Dumlap (MacMillan) (1937)

'Marconi, Master of Space' by B L Jacot & D M B Collier (Hutchinson) (1935)

'Guglielmo Marconi, 1874–1937' by Keith Geddes (London: HMSO) (1974)

'My Father, Marconi' by D P Marconi (New York: McGraw Hill) (1962)

'Marconi, Pioneer of Radio' by D Coe (Julian Messner) (1935)

MAXWELL

'The Life of James Clark Maxwell' by L Campbell & W Garnett (London: MacMillan) (1882)

'James Clark Maxwell FRS 1831–1879' by R L Smith-Rose (London: Longmans Green) (1948)

'James Clark Maxwell : a Biography' by Ivan Tolstoy (Edinburgh: Canongate) (1981)

MORSE

'The Life of Samuel F B Morse, Inventor of the Electro-Magnetic Recording Telegraph' by S A Prime (New York: D Appleton) (1875), (New York: Arno Press) (1974)

REIS

'Philipp Reis' *Deutche Bundespost Archiv. fur Deutche Postgeschicht* No. 1 (1963)

RONTGEN

'Wilhelm Conrad Rontgen and the Early History of the Rontgen Rays' by O Glasser (Springfield, IL: Charles C Thomas) (1934)

'Rontgen Rays Centennial: Exhibition on the Occasion of the Discovery of X-rays in Wurtzburg on 8 November 1895 (Wurtzburg University) (1995)

RUTHERFORD

'Rutherford—Being the Life and Letters of the Right Honourable Lord Rutherford' by A S Eve (Cambridge: Cambridge University Press) (1939)

'Lord Rutherford' by Norman Feather; foreword by Sir Harrie Massey (New ed) (London: Priory Press) (1973)

TESLA

'The Inventions, Researches and Writings of Nikola Tesla' by T C Martin (New York: The Electrical Engineer) (1894)

THOMSON J J

'The Life of Sir J J Thomson O M Sometime Master of Trinity College, Cambridge' by Lord Rayleigh (Cambridge: Cambridge University Press) (1942)

'J J Thomson and the Cavendish Laboratory of His Day' by G P Thomson (London: Nelson) (1964)

'Joseph John Thomson : an Unfinished Social and Intellectual Biography' by Paul Georg Spitzer (Ann Arbor, MI: University Microfilms International) (1995)

THOMSON W

'The Life of William Thomson' by S P Thompson (London: MacMillan) (1910)

VARIOUS

'Ten Founding Fathers of the Electrical Science' by B Dibner (Norwalk, CT: Bundy Library Publications) (1954) *(GILBERT, GEURICKE, FRANKLIN, VOLTA, AMPERE, OHM, GAUSS, FARADAY, HENRY and MAXWELL)*

Index